Cosmopolitan Commons

Cosmopolitan Commons

Sharing Resources and Risks across Borders

edited by Nil Disco and Eda Kranakis

The MIT Press
Cambridge, Massachusetts
London, England

MIT Press books may be purchased at special quantity discounts for business or sales promotional use. For information, please email special_sales@mitpress.mit.edu or write to Special Sales Department, The MIT Press, 55 Hayward Street, Cambridge, MA 02142.

Set in Stone Sans and Stone Serif by the MIT Press. Printed and bound in the United States of America.

Library of Congress Cataloging-in-Publication Data

Cosmopolitan commons : sharing resources and risks across borders / edited by Nil Disco and Eda Kranakis.
    pages ;   cm.—(Infrastructures series)
Includes bibliographical references and index.
ISBN 978-0-262-01902-6 (hardcover : alk. paper) — ISBN 978-0-262-51841-3 (pbk. : alk.paper)
1. Infrastructure (Economics)—Europe. 2. Europe—Economic integration. 3. Natural resources—International cooperation. I. Disco, Cornelis, editor of compilation. II. Kranakis, Eda, editor of compilation.
HC240.9.C3C66   2013
333—dc23
2012038063

10 9 8 7 6 5 4 3 2 1

Just as we were completing the manuscript of this book, we learned of the death of Elinor Ostrom. We began the project as newcomers to the field of commons research. At every turn, we encountered her fundamental contributions, which in turn have shaped our thinking. We see this work as building upon the intellectual tradition that she did so much to establish. We dedicate this book to her memory, and to Commoners present and future, in the hope that they will maintain a vibrancy of thought and action in this domain worthy of Elinor Ostrom's lifetime of effort, commitment, and accomplishment.

Contents

Acknowledgments

This book has a long history and would not have survived as a project without the support of many individuals and institutions. The editors want first of all to thank the contributing authors for their diligence, their patience, and their insights as we went about constructing a "common" approach to the phenomenon of "cosmopolitan commons." Karl-Erik Michelsen and Bruce Hevly made valuable contributions to this discussion in its early stages. Chris Kelty kindly agreed to help us hone our tentative insights at a formative meeting in Lisbon in May of 2009.

The authors and editors have had the privilege of being able to hold a series of workshops at a number of European venues. Particular thanks are due to the Department of Philosophy and History of Technology at the Royal Technological Institute in Stockholm and the Institute of Social Sciences at the University of Lisbon. We also owe much to the embedding of this book project in the Tensions of Europe network and particularly in the EUROCORES Project "Inventing Europe," coordinated by the Foundation for the History of Technology at Eindhoven University and funded by the European Science Foundation. As a "shadow project" of the "Inventing Europe" program, and thanks to the unflagging support of Johan Schot of the Foundation for the History of Technology and Rüdiger Klein of the ESF, we had access to facilities and financial support that allowed us to hold meetings and coordinate our writing. Eda Kranakis also gratefully acknowledges research support from the Social Sciences and Humanities Research Council of Canada.

We would like to thank the staff at the MIT Press for seeing the book through the production process: editor Marguerite B. Avery, acquisitions assistant Katie Persons, and manuscript editor Paul Bethge. Our way at the MIT Press was paved by the acceptance of our manuscript into the Infrastructures series, edited by Geoffrey Bowker and Paul Edwards. We are grateful for that early show of confidence. Finally, we want to thank Lissa Roberts, who critiqued an earlier version of the introductory chapter, and the two anonymous referees who read the original manuscript. Their incisive and helpful criticisms encouraged us to make many changes in organization and content that we hope have made for a better and more readable book.

1   Introduction

Nil Disco and Eda Kranakis

The idea for this book emerged from Transnational Infrastructures and the Rise of Contemporary Europe, a project in which several of the contributors participated.[1] As we grappled with the emergence of transnationally managed infrastructures and resource-spaces in Europe in the nineteenth and twentieth centuries, we were struck by the resemblance to commons. Since the Industrial Revolution, it appeared to us, European nations had created— or had been forced to share—resources and infrastructures that transcended national boundaries. To manage the ensuing challenges and risks, they had had to create and abide by common rules for the maintenance and the use of those resources and infrastructures. Rather than a village pasture, we had resource-spaces such as the Rhine River, European airspace, global weather-forecasting systems, and the global pool of crop genetic diversity. Instead of peasants living in a village, we had technology-rich commons created, used, or governed by communities of nation-states, international organizations, non-governmental organizations, multinational corporations, and other collective actors. It seemed to us that viewing these scaled-up resource-spaces as elements of a new generation of commons would not only enable us to view the history of European industrialization in new ways, but also could enrich Science and Technology Studies and History of Technology, fields in which we were professionally involved. We sensed that commons could provide a framework for integrating the study of technology in society and history with issues such as controlling risk, establishing moral orders, building social and ecological sustainability, and in general for organizing technology, society, and space at aggregated levels, including the transnational.

However, we immediately faced a big challenge in mobilizing the existing scholarship on commons for this purpose. We found that scholarship to have a weakly developed sense of history and a blind spot for technology, which limited its value for understanding commons native to industrial and post-industrial societies.[2] Moreover, commons scholarship embraced theoretical frames that were not always suited to historical analysis, especially when the focus was on contemporary history. Clearly there was work to be done before we could cash in on our intuitions.

To account for this new species of technological, transnational commons, and to historicize commons theory, we needed to counter—or bypass—two stubborn tendencies in that intellectual tradition. The first of these, linked to mathematical economics, was a tendency to analyze commons as goods (e.g., public goods, private goods, club goods) according to idealized schemata more appropriate for mathematical than historical analysis. The second tendency, linked closely to development economics and anthropology, was to see commons as essentially an archaic form of social organization that enjoyed its European heyday during the Middle Ages and has survived only in local settings in what are now called developing countries. We assumed, instead, that commons evolve along with the rest of history and society. And we observed that the commons of contemporary industrial society were embedded in spaces harboring a diversity of related "goods." Common spaces such as the Rhine and the North Sea have been simultaneously used for transport, fishing, energy production, and effluent discharge, each use evolving over time in heterogeneous ways.

These conceptual shifts challenged us to extend commons theory in new directions. This effort was imperative because the novel characteristics of the new commons called for closer scrutiny and explanation. The new commons—seamlessly integrated into modernity and displaying modernity's technological, informational, and networking proclivities—appeared to be almost incongruous with the traditional face-to-face community-based commons that still figure prominently in the scholarly literature. They involved not only large-scale collective actors but also more abstract, multifaceted terms of access and use. Of great interest to us, as historians of technology, was that the new commons—like industrial society itself—were decidedly shaped around technologies. Also of interest to us, as historians of contemporary Europe, was that they covered increasingly large spaces that transcended national borders, and they seemed to us an important motor of the "hidden integration of Europe."[3]

We settled on *cosmopolitan commons* as a moniker for this new generation of scaled-up, technology-rich commons. That term draws on the theory of cosmopolitanism that has emerged since the 1990s in work by David Held, Daniele Archibugi, Ulrich Beck, and others.[4] Cosmopolitanism seeks to model the dynamics of an increasingly interdependent world characterized by "overlapping communities of fate defined by the spatial reach of transnational networks, systems, allegiances, and problems."[5] It holds that such communities of fate have called forth innovative, non-coercive, distributive forms of collective negotiation and action.

The new commons are particularly amenable to the approach of neo-medievalism—an important variant of cosmopolitanism. Neo-medievalism, which emerged co-temporally with cosmopolitanism, also is concerned with globalization and integration processes, and draws on some of the same intellectual roots.[6] Like cosmopolitanism, neo-medievalism seeks to move beyond rationalist-statist conceptions of international relations. Neo-medievalism has been applied successfully to the study of European integration, which it reveals as a set of multi-level networking processes that operate not only through the nation-state but also through structures above and below it. While not ignoring high politics, neo-medievalism calls for new studies of European integration that link issues of high-level governance with broader and deeper currents of social, economic, and political transformation.[7]

We saw an immediate fit with the new commons we were trying to understand: they too were "communities of fate," rooted simultaneously in technology and in nature's spaces and resources. And, in line with the neo-medievalist program, we saw these commons as a new way to understand European integration, and ultimately globalization. More than that: technology has been invoked as a driver of cosmopolitanization,[8] but its roles in enabling cosmopolitan (or neo-medieval) forms of governance have not been adequately delineated. In this book we purport to do just that. We contend that cosmopolitan forms of governance (including the new modalities of moral economy explained further in chapter 2) are integral to the new generation of commons under our purview, that technology undergirds cosmopolitan scales of governance in several distinct ways, and that cosmopolitan commons, in turn, are building blocks of regional integration and globalization. Nevertheless, our case studies also reveal the precariousness of these societal constructions: it is clear that cosmopolitan commons are inherently threatened by subversion and by clashing interests. One aim of this book is to understand the tensions between cooperation and conflict that make these new commons more precarious and more prone to dissolution than we might want.

After a theoretical and historiographical chapter that lays out the principal elements needed to frame a theory of cosmopolitan commons, we present eight case studies that, through their temporal breadth and diversity, add connective tissue and empirical flesh to our theory. Each case study is a unique laboratory for studying the dynamics of cosmopolitan commons in the making, but the chosen cases also contribute in a more structured way to theorization and to the expansion of historical understanding. The case studies are divided into three parts. (See table 1.1.) Those in parts I and II explore a fundamental tension in cosmopolitan commons: the tension between their role in the valorization of resources and their role in the protection of humans and nature (in many cases *against* overzealous valorization). Each pole of this duality is an answer to the question of what exactly motivates autonomous actors to join together in a cosmopolitan commons. Sometimes the predominant goal is collective action to do something powerful that could not be done by the actors operating individually; sometimes the driving force is mutual aid to overcome risks and dangers that humans face in their changing environments, and increasingly as a result of their own actions. The three case studies that make up part I—one on European commercial air travel, one on European radio broadcasting, and one on Vuoksi hydropower—elaborate on the aim of valorization. The three studies that make up part II—one on weather forecasting, one on preservation of crop genetic diversity, and one on transboundary air pollution—address the protection of nature and of humans. Each of these six case studies, however, also demonstrates that the goals of valorization and protection, despite their seeming incommensurability, are always intimately bound up together and continuously interact. Perhaps we may think of these dual aims as the yin and the yang of cosmopolitan commons.

Part III consists of two "capstone studies" that contribute to a theory of cosmopolitan commons by tracing, in two important cases, multiple commons that have been organized in relation to large nature-made resource-spaces in Europe: the North Sea and the Rhine. These capstone studies not only build on and extend comprehension of the yin and yang of cosmopolitan commons developed in parts I and II; through their integrative approach, they also elaborate on the phenomena of historical evolution, temporal layering, and interactions among commons.

Valorizing Nature

The three studies in part I explore cooperative efforts to make more productive use of nature's spaces and resources. Spanning the twentieth century, they are ordered chronologically on the basis of the commons-building eras covered. In "The 'Good Miracle': Building a European Airspace Commons, 1919–1939," Eda Kranakis charts the emergence and growth of a transnational commons aimed at making international air transport possible. She examines the initial framing of this commons by means of an international treaty, the governance regime (and additional treaties) that emerged to

Table 1.1
Organization and profile of case studies.

Chapter	Problems	Regime elements
Valorizing Nature		
Kranakis *Airspace*	Regulate/improve international air transport	International Air Traffic Association (1919)
	Maintain safety, reliability	International Commission for Air Navigation (1919)
	Manage sovereignty incursions	
Wormbs *Etherspace*	Mitigate radio interference	International Broadcasting Union (1925)
		International Telecommunication Union (1932)
Kurjonen *Vuoksi hydropower*	Maintain/improve hydropower production	Frontier Watercourses Treaty (1964)
	Alleviate flooding	
Protecting Humans and Nature		
Edwards *Weather forecasting*	Improve forecast term and accuracy	International Meteorological Organization (1870s) WMO (1950)
		European Center for Medium Range Weather Forecasts (1973)
Saraiva *Crop biodiversity*	Conserve crop biodiversity	Seed Banks/Svalbard Global Vault (2008)
		Global Crop Diversity Trust International Seed Treaty (2001)
Kaijser *Acid rain*	Conserve air quality	Long Range Transboundary Air Pollution Convention (1979)
Temporal Layering and Interlinking of Cosmopolitan Commons in Nature's Spaces		
With Andersen *North Sea*	Maintain/improve reliability of shipping and manage risks	Insurance companies (Norske Veritas) (1864)
	Conserve fish stocks	North East Atlantic Fisheries Convention (1959)
	Manage oil appropriation and cost recovery	Geneva Sector Agreement (1958)
Disco *Rhine River*	Regulate Rhine transportation	Central Commission for the Navigation of the Rhine (1815)
	Sustain salmon fisheries	Berlin Salmon Convention (1885)
	Conserve water quality	International Commission for the Protection of the Rhine (1950)

regulate this new commons, an emergent moral economy, and the many risks and challenges (physical, technical, political, and organizational) that had to be managed to make this commons work. Kranakis looks at the crisis of confidence that emerged in the prologue to World War II as the war-making potential of airspace loomed ever larger, the collapse of the airspace commons with the onset of war, and the enduring model that this commons established, which shaped the global airspace commons of the post-World War II era.

In "Negotiating the Radio Spectrum: The Incessant Labor of Maintaining Space for European Broadcasting," Nina Wormbs investigates the European radio broadcasting commons from its origin in 1925 to the 1950s. The radio-frequency spectrum is a unique kind of resource-space. Its natural potential exists everywhere, uniformly. However, the resource-space that humans *experience* as they seek to receive transmitted signals has a complex, changing spatiality. There is often no telling where radio waves will be strong and where they will be weak. The propagation and the intelligibility of radio signals depend on their frequency, on the characteristics of the transmitter that emitted them and the receiver receiving them, and on a host of other factors, including the time of day, the weather, geography, and the Earth's variable conductivity. This posed enormous practical challenges for optimizing reception of signals, something that quickly proved impossible without a coherent structure of radio governance across Europe's fragmented political space. Wormbs' analysis focuses on international efforts to maximize collective exploitation of radio broadcasting by eliminating interference among transmitters operating at identical or proximate frequencies. In a kind of "prisoner's dilemma" reminiscent of classical commons problems, the resource-space degenerates if its use is left to the discretion of the individual actors. Wormbs shows that this precarious resource can be salvaged only by international cooperation in a spirit of trust and give-and-take backed up by stringent norms and by mutual, transparent monitoring—that is, by the forging of a cosmopolitan etherspace commons.

In the final chapter in part I, "Conflict and Cooperation: Negotiating a Transnational Hydropower Commons on the Karelian Isthmus," Kristiina Korjonen-Kuusipuro analyzes the making of a cosmopolitan hydropower commons on the Vuoksi, a river straddling the shifting border between Finland and the USSR/Russia. She details how those two states, despite hostility and deep ideological differences, eventually worked out mutually agreed, beneficial ways of sharing the resource. The new commons had the aim of eliminating obstacles to the maximum utilization of the Vuoksi's hydroelectric potential. This did not mean that each country got the most it could possibly get under ideal circumstances; rather, it meant that the overall production did not suffer from mutually antagonistic water-level strategies or simple lack of coordination. The commons, in short, replaced counterproductive, emotion-laden bickering with a more economically rational regime that evolved into a wider system of trust and cooperation, and that found creative ways to balance the sometimes divergent needs of upstream Finland and downstream USSR/Russia.

Protecting Humans and Nature

The cosmopolitan commons that are explored in part II congealed around perceived threats to the integrity of nature, humans, or natural resources. Many traditional historical commons had rules that prevented degradation or depletion of valued resources; however, where markets, states, and private and state property came to rule, technological change and the exploitation of nature often proceeded without regard for long-term effects on resources and ecosystems. Negative environmental externalities proliferated, and it became evident to increasing numbers of perspicacious individuals that resources and human environments were in peril as a result of what we might call the "tragedy of the market." This was true of flows of air or water subject to pollution or other forms of degradation, which, in the European context, also increasingly crossed borders. Thus, in Europe, efforts and activities shifted to establishing effective solutions at the international level. Technology also offered more and more ways of grappling with these new risks, no less than with traditional risks such as bad weather.

In "Predicting the Weather: An Information Commons for Europe and the World," Paul Edwards looks at the atmosphere as a source of information about future weather. This information—if properly analyzed and distributed—can protect people and their economic activities and assets. Being able to predict what the weather will be like at some point in the future has become an increasingly valuable resource. The more precise, reliable, and long term the forecast, the greater its value. Edwards argues that, *ceteris paribus*, the reliability and the futurity of forecasts increase as more data from more locations, and more centralized processing power, are incorporated into the systems for distributing and processing meteorological data. Thus, weather forecasting is fundamentally tied to the reach and the capabilities of technological infrastructures. At least in Europe, this meant organization across national borders. Accomplishing such organization while respecting the sovereignty claims of national weather services has required the creation of a cosmopolitan weather commons with transnational rules and institutions. The main features of this commons are standardization of data formats and the pooling of data, sharing and continual upgrading of data-processing and data-transmission facilities, and development of shared meteorological models and processing power. Edwards traces the growth of this commons from the nineteenth century (notably with the founding of the International Meteorological Organization in 1873) to the late twentieth century, devoting particular attention to the European Centre for Medium Range Weather Forecasts, which began operation in the mid 1970s. He also examines the recently increasing tension between efforts to privatize this resource in various ways and continuing commitments to keep weather forecasting in the public domain as a public good provided freely to all and paid for by taxpayers.

Tiago Saraiva's chapter, "Breeding Europe: Crop Diversity, Gene Banks, and Commoners," is a story of three different commons based on the global resource of plant genetic diversity. Ready access to genetic diversity gives breeders and farmers more options in hybridizing existing plant varieties to improve yields, to enhance desired characteristics such as drought tolerance, or to increase resistance against specific diseases or pests. The traditional form of local farmers' commons, which was based on exchanges of the seeds of their "landrace" crops, had been in place for hundreds if not thousands of years by the twentieth century, when it succumbed to what Saraiva calls the "breeders' commons." This was a type of commons based on ready access by breeders to global crop genetic diversity stored *ex situ* (i.e., out of the fields) in large national seed banks. It was a commons because breeders freely exchanged hybrids in a kind of common pool. Saraiva traces the origins of the seed banks to Soviet and Nazi projects to improve agriculture by using collected genetic diversity to create and subsequently impose standardized "industrial" hybrids on the peasantry. The post-World War II breeders' commons did much the same, and in so doing they considerably diminished the pool of genetic diversity—their golden goose—by eliminating the variety formerly borne by landraces in farmers' fields. In turn, the breeders' commons succumbed to commercialization, proprietary varieties, and the capital demands of genetic manipulation as a new route to hybridization. Genetic variety has all but disappeared from the actual fields, and leads an artificial life in ever more remote and well-protected seed vaults. Saraiva nonetheless points to initiatives to restore traditional landraces and to build a new commons in which gene banks, breeders, and farmers cooperate as equal partners in conserving and building crop genetic diversity, arguing that *in situ* breeding is ultimately more robust and sensitive to the risks and opportunities of specific *terroirs*.

Arne Kaijser's contribution, "Under a Common Acid Sky: Negotiating Transboundary Air Pollution in Europe," examines a commons that has flourished since the 1970s. Industrialization brought with it an increasing burden of noxious atmospheric effluents. Initially perceived as a local problem that could be dealt with by building smokestacks to disperse and dilute pollutants, air pollution became international as smokestacks became ever taller, thus releasing effluents into higher layers that carried them across borders according to patterns of prevailing winds. In the late 1960s, waste products of industrial and municipal combustion were discovered to be the cause of acid rain, which harmed ecosystems, especially in the Nordic countries. Whereas in the case of a river the community of potential commoners is given by the river's unambiguous course, the relevant community in Kaijser's study was defined by initially unknown statistical flows in the atmosphere and by the differential vulnerability of national ecosystems to acid rain. The challenge of precisely tracking the paths of airborne pollutants made it difficult to organize an international commons. Adding to the difficulty, nations were net perpetrators or victims in only a stochastic sense. The ideological divisions of the Cold War, moreover, militated against an easy political solution. Worse, polluters often saw more costs than benefits in spending money to protect regions hundreds of kilometers and several borders away, particularly if the regions in question were more prosperous. Kaijser tells the remarkable story of how the victims nevertheless found ways to build a moral economy that encouraged more and more participation by a growing number of states and that has produced real improvements and a stronger and more widely diffused ethos of environmental protection.

Temporal Layering, Diversification, and Interlinking of Cosmopolitan Commons in Nature's Spaces

The two chapters in part III trace cosmopolitan commons making over large time spans within physically extensive, nature-made resource-spaces. Through their analysis of temporal layering and interaction among commons within specific resource-spaces, these capstone studies provide insights not only into the nature and historical evolution of cosmopolitan commons but also into the convoluted nature of sustainability challenges. They show that solutions (for both nature and human societies) must take into account the net consequences that ensue from humanity's simultaneous and ever more diverse uses of nature's spaces.

In "Changing Technology, Changing Commons: Freight, Fish and Oil in the North Sea," Håkon With Andersen considers three temporally overlapping commons within a large natural resource-space: one commons aimed at valorizing nature, one aimed at conserving it, and one that represents a failed commons—in fact an enclosure of what was in principle a shared resource. By the simple expedient of distinguishing among the surface of the sea (as a medium of shipping), the water column (as a fish habitat), and the sub-seabed (as a site for fossil-fuel deposits), With Andersen makes it clear that what looks like a single geographical space harbors a complex of contiguous resource-spaces and potential commons. The specifics of how the various resources have been exploited, the risks involved, and the ways they have been commonized are also instructive. We see that shipping, for example, was carried on by private parties, and that the North Sea transport commons emerged out of the quest to improve safety and reliability. Achieving this dual goal required governments to cooperate in the provision of maritime infrastructure, but it also required shippers and insurance companies to develop ways to distribute and mitigate the inevitable risks. Fishing in the water column initially seemed to be governed only by the challenge of appropriating as much of this "endless" natural bounty as was technologically possible. Finding and catching the moving shoals of fish was the overriding goal. At first, cooperation was limited to rules of decorum among fishers as they competed on the fishing grounds. However, as fishing technologies improved, it became apparent that this "public good" (in the sense used by economists) was rapidly becoming a precarious "common-pool resource," and the various governments began to realize that they would have to cooperate in limiting catches lest the resource disappear entirely. With Andersen's third topic is the seabed. In the 1960s the North Sea's sub-seabed was found to contain enormous reserves of oil and natural gas. On land such reserves would have been under the purview of one or another government, which could have granted concessions for exploration, drilling, and extraction, thus providing some guarantee that the enormous costs of exploration could be recovered. At sea, however, the discovery of oil led to great confusion, because the mineral resources were now located under many meters of international waters and were, by implication, a shared resource. In view of the immense commercial interest in this resource and the private risks entailed in its appropriation, a commons was out of the question; rather, the resource was simply enclosed by dividing the floor of the North Sea into national territorial segments, thus spatializing the fossil-fuel resources in accord with redefined national spaces.

Nil Disco's chapter, "'One Touch of Nature Makes the Whole World Kin': Ships, Fish, Phenol, and the Rhine, 1815–2000," also addresses three commons: a transport commons, a salmon-fishing commons, and a commons aimed at controlling the growing levels of pollutants in the river. Disco's account, spanning nearly two centuries and a heterogeneous range of resources and risks, aims to show how different cosmopolitan commons in the same resource-space become intertwined in a functional sense but also in a historical sense, and how these commons are also embedded in the larger history of the Rhine and Europe. It also shows how the idea of the Rhine as a system of flows that incorporate distant locations into "communities of fate" almost predestines it to become a corridor of modernity and a locus for cosmopolitan commons. The transportation commons, dating from 1815, hastened the transformation of the Rhine into an increasingly polluted industrial and urban conduit. The perverse effects of that transformation first became apparent in reduced catches of salmon. On the only partly correct assumption that overfishing was the culprit, riparian states along the Rhine banded together to create a salmon commons to regulate fishing practices, but to no avail. Pollution reared its ugly head again after World War II, and Dutch waterworks and surface-water management were overwhelmed by the rapidly increasing levels of pollution produced by the German *Wirtschaftswunder*. Efforts in the 1970s to combat toxic effluents by means of detailed treaty provisions were hampered by national interests, and it took the major environmental disaster of the 1986 Sandoz incident to create space for new initiatives. Salmon, long gone from the Rhine, became a symbol of a new ecological élan. The result was a moral economy that prized voluntary measures by national states to contribute to a riparian society that would make of the Rhine a habitat fit for salmon once again.

This book has been a shared effort by all the contributing authors—an effort in which the collective goal of conceptualizing and analyzing the new commons we were exploring was negotiated and refined over several years and, in turn, channeled the authors' individual approaches and efforts. In effect, we took a "commons" approach to the project. Through a nearly continuous process of dialog, comparison, debate, and reflection, we developed a larger theoretical comprehension of the phenomena with which we were grappling. Chapter 2, "Toward a Theory of Cosmopolitan Commons," presents the results of this theory-building effort. It situates the case studies within a more comprehensive framework, drawing attention to the patterns and dynamics that link them together, but also to the distinguishing characteristics that set each study apart and that, together, contribute to a deeper understanding of the evolution and variety of cosmopolitan commons.

Notes

1. The project's website, at http://www.tie-project.nl/, provides details on the history of the project and its research outcomes.

2. For simplicity's sake, we will refer simply to "industrial" societies in what follows, recognizing that this covers a broad spectrum of concrete historical forms and that it may no longer be a fully adequate characterization of contemporary society.

3. Thomas J. Misa and Johan Schot, "Inventing Europe: Technology and the Hidden Integration of Europe," *History and Technology* 21, no. 1 (March 2005): 1–19.

4. The extensive and growing literature on cosmopolitan theory comprises several distinct variants. One branch explores cosmopolitanism as a system of political ethics appropriate for a globalizing world. Other strands grow out of empirical and theoretical analyses of globalization, mobility, risk, and transnational networks. Ulrich Beck, for example, emphasizes processes of cosmopolitanization that have been underway for some time. Useful works on cosmopolitanism include the following: David Held, *Democracy and the Global Order: From the Modern State to Cosmopolitan Governance* (Stanford University Press, 1995); Daniele Archibugi and David Held, eds., *Cosmopolitan Democracy* (Polity, 1995); Pheng Chea and Bruce Robbins, eds., *Cosmopolitics: Thinking and Feeling Beyond the Nation* (University of Minnesota Press, 1998); Ulrich Beck, *World Risk Society* (Polity, 1999); Ulrich Beck, "The Cosmopolitan Perspective: Sociology in the Second Age of Modernity," *British Journal of Sociology* 51, no. 1 (2000): 79–105; Ulrich Beck, *Cosmopolitan Vision* (Polity, 2006); Ulrich Beck and Edgar Grande, *Cosmopolitan Europe* (Polity, 2007); Steven Vertovec and Robin Cohen, eds., *Conceiving Cosmopolitanism: Theory, Context and Practice* (Oxford University Press, 2002). For a more comprehensive bibliography, see Ulrich Beck and Natan Sznaider, "A Literature on Cosmopolitanism: An Overview," *British Journal of Sociology* 57, no. 1 (2006): 153–164.

5. Anthony McGrew, "Models of Transnational Democracy," in David Held and Anthony McGrew, eds., *The Global Transformations Reader* (Polity, 2000), p. 509. In the same volume, see also David Held, "Cosmopolitanism: Taming Globalization," pp. 514–529. The term "overlapping communities of fate" was originally formulated by Held.

6. Notably, both cosmopolitanism and neo-medievalism draw on Hedley Bull, *The Anarchical Society: A Study of Order in World Politics* (Macmillan, 1997).

7. Jörg Friedrichs, "The Meaning of New Medievalism," *European Journal of International Relations* 7, no. 4 (2001): 475–502. See also Jan Zielonka, *Europe as Empire* (Oxford University Press, 2006); James Anderson, "The Shifting Stage of Politics: New Medieval and Postmodern Territorialities?" *Environment and Planning D* 14, no. 2 (1996): 133–154.

8. See, e.g., Beck, *World Risk* Society, passim; Beck, *Cosmopolitan Vision*, p. 6; Held, "Cosmopolitanism: Taming Globalization," p. 525.

2 Toward a Theory of Cosmopolitan Commons

Nil Disco and Eda Kranakis

In the era of globalization, issues of the maintenance, production, and distribution of the common . . . in both ecological and socioeconomic frameworks become increasingly central. . . . Contemporary forms of capitalist production and accumulation in fact, despite their continuing drive to privatize resources and wealth, paradoxically make possible and even require expansions of the common.
—Michael Hardt and Antonio Negri, *Commonwealth* (Harvard University Press, 2009), pp. viii–ix

Cosmopolitan commons have proliferated and increased in size and scope since the onset of industrialization. If they are not yet the rule in the transnational and global ordering of resources, society, and space, neither are they any longer the exception. Their increasing salience makes it necessary to reframe commons theory to take account of historical evolution and the increased scale, scope, technization, and bureaucratization of commons regimes. In this chapter we review the existing foundations of commons theory and consider how to reframe and extend it to accommodate this new generation of commons. We discuss the need to historicize commons theory, briefly sketching out the broader historical context from which cosmopolitan commons emerged. We also argue that greater weight should be given to technology in the analysis of the origins, the functioning, and the regimes of access and governance of commons. After a brief review of relevant STS theories, we show how they can be applied and adapted to analyze the roles of technologies in cosmopolitan commons. Another aspect of cosmopolitan commons is their spatiality, and here we draw on a developed tradition of spatial theory to provide deeper insight into the multiform, overlapping spaces of cosmopolitan commons. Finally, we argue that the governance regimes of cosmopolitan commons are new kinds of moral economies that direct individual actions toward collective goals and help to achieve greater social and environmental sustainability. Based on the evidence of the case studies, we posit that cosmopolitan commons and their moral economies contribute to social and environmental resilience, but that they remain ultimately precarious, subject to dissolution by any number of human, technological, or environmental forces.

The Evolution of Commons Theory: From Traditional to Cosmopolitan Commons

Commons scholarship got off to a dramatic, not to say dramaturgical, start in response to an article by the biologist Garrett Hardin called "The Tragedy of the Commons," published in 1964.[1] This became one of the most widely cited papers of the twentieth century. It continues to shape thinking in numerous disciplines because it addresses a fundamental, continuing problem: how to create a sustainable society.[2] Hardin's account, redolent of the Middle Ages but set in no specific time or place, is a reflection on the fate of a pasture openly accessible to all villagers:

Picture a pasture open to all. It is to be expected that each herdsman will try to keep as many cattle as possible on the commons. . . . As a rational being, each herdsman seeks to maximize his gain. . . .[3]

Hardin concludes that such a setup would inevitably result in the ruin of the common pasture, the canonic example of the infamous "tragedy of the commons." Three apparently self-evident assumptions produced this conclusion: first, the grass on the pasture was a subtractable resource, i.e., consumption by any animal diminished the total supply and left less for the rest; second, the villagers were guided only by the aim of maximizing their short-term profit and so would seek to graze as many cattle as they could; third, the villagers lacked perspicacity about the perverse game structure in which they were ensnared, did not communicate with one another, and so could not reflexively modify their behaviors to collectively escape the tragedy.

Critics were quick to point out that this idea of commons as pristine natural spaces—unfettered by rules, common sense, or shared moral codes—was both mythical and misleading.[4] Hardin was in fact not portraying a historical commons, but rather something akin to a wasteland or wilderness beyond the pale of society. Historical examples of real commons show that they were resources governed by community rules. The etymology of *commons* reveals this link: the Latin root, *communis*, refers not only to something held in common by a group, but also to a user community bound by responsibilities as well as rights.[5] The collectively negotiated aspect of historical commons is also reflected in the meaning of the words *commonable*,

commonage, piscage, and *warrenage.* For example, the *Oxford English Diction-
ary* cites a use of *commonable* in 1640: "If any tenant doth surcharge the
common . . . by putting of cattel there not commonable." In Hardin's original model, the community and its rules were unrealistically absent—the
only actors were nature, cattle, and a collection of solitary, rational, profit-
maximizing human beings.[6]

Commons theorizing since Hardin has worked to redefine cooperative communities as the *sine qua non* of commons. Elinor Ostrom, Bonnie
McCay, and Charlotte Hess have taken the lead here. They have challenged
Hardin's assumption that commons or common property implies a regime
of unrestricted open access. Ostrom proposed the term "common-pool
resources" as a means of dissociating the governance of shared resources
from any assumption about what kinds of property or what rights of access
or use might be in place in any specific case. "Although *open access* is a distinct 'commons' problem," McCay writes,

> it is not definitive of common property. . . . In common property systems, there can
> be restrictions on who is a proper "commoner" and what people do. . . . Common
> property is a cultural artifact, socially constructed and contested. . . . The revisionist
> view recognizes that the natural environment may be dealt with in many different
> ways, with many different consequences. Common-pool resources may be under a
> variety of management regimes. . . .[7]

In line with this reasoning, McCay proposes a "social/community paradigm" for understanding commons that views people as "social beings trying to come to some collective agreement about common problems."

This social constructivist turn was a great leap forward in commons theory, and one that encouraged greater historical sensibility, but the field as
a whole still encompasses conceptual approaches that do not always mesh
with historians' aims. On the theoretical side, commons scholars have
drawn extensively from mathematical economics. On the empirical side,
they have drawn extensively on anthropological approaches. And in quite
different ways, the imprint of each of these domains on the corpus of commons scholarship has not always facilitated rapprochement with historical
understanding, particularly with respect to the nature-technology-society
complexes of the contemporary era.

Economic analysts have cultivated an approach that treats commons as
goods and that applies idealized schemata to specify essential properties of
distinct classes of these goods. The origin of this form of theorizing was a
series of papers published in the 1950s by Paul Samuelson, intended as a
contribution to the theory of public expenditure.[8] Classical economics is
built around analysis of market competition, but economists increasingly

saw a need to model the dynamics of production, exchange, and consumption of public-sector goods, for example in the context of nationalization of industries or welfare expenditures. Samuelson proposed "an extreme polar case" intended to aid theoretical reasoning. Against "a *private* consumption good like bread, whose total can be parcelled out among two or more persons, with one man having a loaf less if another gets a loaf more" he opposed "a public consumption good, like an outdoor circus, . . . which is provided for each person to enjoy or not, according to his tastes."[9] He observed that, unlike bread, one person's consumption of an open-air circus did not diminish the stock available to other consumers. Samuelson emphasized that his exaggerated model did not realistically depict all possible types of public and private goods, noting that "the careful empiricist will recognize that many— though not all—of the realistic cases of government activity can be fruitfully analyzed as some kind of a blend of these two extreme polar cases."[10]

Samuelson's dichotomy became a bedrock of commons categorization, embodied in the proposition that common-pool resources might be classified as either rivalrous and subtractable (like bread) or non-rivalrous and non-subtractable (like circuses).[11] Drawing inspiration from Samuelson's model, economists developed further insights and categorizations that also entered the theoretical corpus of commons scholarship.[12] In this way, resources (still theorized as goods) came also to be categorized as either excludable or non-excludable. The latter referred to goods for which it was "not possible, or at any rate economically feasible to exclude non-purchasers."[13] The precise terminology of excludable versus non-excludable seems to have been introduced by the economist James Buchanan in a paper that built on Samuelson's work by developing a theory of club goods to serve as a "missing link" between private and public goods. Club goods were available only to members of restricted groups. The concept of excludability was essential to Buchanan's theory because the formation of a "club" could not occur if there was no way to exclude non-members from enjoying the good.[14]

An example of excludability cited by Samuelson illustrates why schematic models are often not adequate containers for the historically grounded theoretical aims of this book. Referring to the consumption of television broadcasts, Samuelson observed that systems for scrambling and unscrambling television broadcasts made it possible to transform what had seemingly been a "perfect example of my public good" into a system in which "it is technically possible to limit the consumptions of a particular broadcast to any specified group of individuals."[15] An attentive historian of technology would immediately be drawn to the many complexities this example suggests. First, the example demonstrates that excludability is not an inalienable property of a good—owing to scramblers, television broadcasting *became excludable* to viewers. The implication for historian-theorists is that they ought to pay attention to how new excludability systems emerge and evolve. Second, Samuelson's example hints at the potential multidimensionality of systems of exclusion associated with goods. In the case of television exclusion of viewers may occur not only through scrambling, but also through parental control systems or through the prices of television sets. This implies that a historian-theorist should pay attention to the variety of such excludability systems, and to how that variety has changed and why. Third, Samuelson's example focuses on the case of a single broadcaster relative to a body of potential viewers, but historically rooted theorizing often approaches phenomena from multiple points of view. Television excludability might also be analyzed from the perspective of multiple broadcasters trying to use the same frequency in the same region (analogous to the case of radio), or from the perspective of multiple writers, producers, and advertisers trying to get their content on the air. Ultimately, a historian-theorist might want to explore how several systems of exclusion evolved in relation to one another—for example, how the use of parental control systems evolved relative to the relaxation of public censorship rules for television programming, or to the increasing number of television sets per family (with children increasingly having television sets in their rooms), or to the rise of pay-per-view systems. Samuelson and Buchanan were primarily interested in economic efficiency and optimization, but understanding the historical evolution of commons and theorizing the roles played by technology in their development and governance necessarily incorporates broader aims and conclusions rooted in *empirical* analysis that take account of the ways that messy reality departs from analytical models.

A further issue raised by Samuelson's example concerns the narrowness of defining television broadcasting—and shared resources in general—as goods. Our objective in this book is to analyze the emergence and the functioning of cosmopolitan commons, viewing these commons as large, collectively governed resources and resource-spaces amenable to multiple forms of exploitation. These commons are not, in general, analytically reducible to goods. The authors of the chapters in this volume have drawn insights from the goods approach, but each of them must also grapple with historical complexities that do not fit neatly within its framework of analysis. As an example of the difficulties that simplified models and dichotomies of types of economic goods pose for historians of commons, consider chapter 4, in which Nina Wormbs considers the many complexities—technological, political, and natural/physical—that attended efforts to create a system of exclusion that could be successfully applied to the radio-frequency spectrum in Europe.

The case of airspace offers a further example of the limitations of the goods approach for historians. Attempting to frame analysis of airspace within simplified schemata of public, private, or club goods, excludable or non-excludable goods, and subtractable or non-subtractable goods does not offer an adequate template for historical understanding of airspace as a commons. Airspace subtractability, for example, was partly dependent on type of use, unlike either bread or circuses. If a certain area of airspace was used for warfare, that same area could not be used over the same time period for commercial transport, because the risks would be too high. Even if warfare occurred only intermittently, commercial use still could not occur in the absence of stable peace. Nor was airspace, even when restricted to commercial aviation in peaceful periods, quite like either bread or circuses. Once a plane entered a certain portion of airspace, that portion was, for the moment, consumed, and could not be simultaneously used by another plane without a collision occurring. But as soon as a plane left that portion of airspace, it could be consumed by another plane, and the process could go on forever without any danger of overconsumption of the resource as long as planes consumed the airspace sequentially. Moreover, the faster and closer together the planes could fly without colliding, the more frequently each section of airspace could be re-consumed, up to a theoretical upper limit. In fact, air traffic control systems have evolved over the years with precisely this aim: to permit smaller time and space gaps between planes sequentially passing through portions of airspace. Yet such a line of analysis arguably contributes only in a limited fashion to holistic historical understanding of the emergence and evolution of airspace as a transnationally governed commons. It directs us primarily toward the study of air traffic control (or, increasingly, to the study of airport slot allocation), which are small parts of the larger story and which were of limited significance in the interwar period, when this commons first arose.

Alongside the sometimes incommensurable aims of historical and economic analysis of commons, the influence of anthropological approaches and development economics within commons scholarship has also produced, inadvertently, a deficit of historical sensibility that has diminished its value for analyzing the cosmopolitan commons of the modern and contemporary eras. The empirical side of commons scholarship somehow missed the huge technological and political transformations that modernity brought. The study of local, communal, face-to-face, natural-resource-based commons became the leading empirical paradigm of the discipline.[16] By the mid 1990s, however, the revisionist camp had begun to sense the limitations of this approach and to acknowledge an erosion of its monopoly on defining commons. As Hess and Ostrom noted,

> Most of the CPR [common-pool resource] examples discussed so far have been natural resource systems and human-made resources such as irrigation systems. In the past five years, more and more scholars have broken away from the erroneous idea that commons were antiquated institutions mainly prevalent in developing countries managed by indigenous peoples. Interdisciplinary researchers are finding great benefit in applying CPR analysis to a number of new and/or unrecognized common-pool resources.[17]

In 2001 Ostrom's revisionist school was still referring to these newly identified types of commons as "non-traditional common-pool resources," thus effectively treating them as deviant. By 2008, however, Charlotte Hess was unfurling the flag of the "new commons" in a more unabashed way, in part as a response to the popular ideal of the "information commons" based on the Internet and mobile communications.[18]

Our aim in this book is to contribute to the process of theorizing the "new commons" by calling attention to a generation of technology-rich, transnational commons that emerged in close association with the rise of nation-states and industrial capitalism, and that have become important components of globalization. These new commons are linked to the proliferation of international treaties. They are characterized by upscaling to transnational regional levels, in some cases to continental and global levels. Studying them, moreover, reveals an overlap among spaces defined at multiple levels: globally expansive resource-spaces such as biodiversity, the atmosphere, and the global weather forecasting system, regional spaces defined (for example) by natural structures such as the Rhine River or the North Sea, and spaces defined by national sovereignty. The global scale has the drawback (and perhaps the strength) that it is literally everywhere but also nowhere. At the European level, however, we can observe these shared, expansive resources *situated* in well-defined geomorphological, infrastructural, and political spaces. Taking up the challenge of theorizing cosmopolitan commons at the European level, while analyzing resource-spaces that occupy spaces both above and below this level, therefore offers a way to conceptually grasp the interlinked, multi-tiered spaces of cosmopolitan commons.

Meeting this challenge requires a major overhaul of revisionist commons theory, however. The way forward is suggested by Hess' and McCay's conception of a commons. Hess, struggling against the flamboyant heterogeneity of the "new commons," offers the following compact definition:

A commons is a resource shared by a group where the resource is vulnerable to enclosure, overuse and social dilemmas. Unlike a public good, it requires management and protection in order to sustain it.[19]

Though we have big qualms (as should Hess) about casually equating commons with resources, this definition has the virtue of stating plainly that a commons entails both a *shared, precarious resource* and a *regime* for "management and protection in order to sustain it."[20] This definition is echoed in McCay's assertion that the revisionist perspective distinguishes "between the features of the resource and those of the way people choose to relate to the resource and each other."[21] We readily admit that this dual structure of "shared, precarious resources under protective or facilitating regimes" describes traditional as well as cosmopolitan commons. The differences lie in the nature, the scale, and the diversity of the resources that are valorized, in the scale and complexity of the governance regimes, in the associated technologies, and in the way all these factors have evolved in interaction.

In order to theorize this new generation of cosmopolitan commons, we must focus not just on resources but also, more broadly, on resource-spaces. That term highlights the fact that resources exist within various kinds of spaces, which may have unique features that shape access to and use of the resources. Airspace is a very different kind of space from a river, for example: its terrestrial layout is different, its flow characteristics are different, and the physical and resource valorization challenges it poses are different. The term *resource-space* also helps to convey the idea that a particular terrestrial space, such as a river, may contain a multiplicity of resources that come to be governed within distinct commons regimes, so that multiple commons exist within the same geomorphological space. The North Sea and the Rhine are two examples explored in this book. The term *resource-space* also helps us to recognize that commons may be created to govern not just resources themselves but also the spaces in which they are located—that is, *resource-spaces*. Airspace is one example analyzed in this book. Finally, the notion of a resource-space accommodates the fact that commons may be organized at levels not strictly defined either by the geographic extent of a resource or by the geographic profile of a nature-made space, but within some other space defined by a combination of technology, politics, resource properties, and associated geomorphologies. Polluted air exists within a resource-space whose geographical extent depends on many factors, including the height of smokestacks. The resource-space of hydropower production on the Vuoksi River comprises only a portion of the river, but to manage this hydropower resource effectively the entire river has to be governed under a unified regulatory regime. The resource-space of weather forecasting is a particularly interesting example. Its profile and its extent are measured by the capabilities, the reach, and the speed of the technological systems that serve to gather, transmit, process, and diffuse weather information. In fact, Paul Edwards suggests that there is a need to conceptualize a resource-space/time of weather forecasting, because the speed at which weather data are transmitted is every bit as important for creating valuable forecasts as the geographical extent over which the data are collected.

Historicizing Commons

Among the shortcomings of the revisionist theory of commons is its failure to historicize commons—to link the past and the present in a meaningful way. Recently commons scholarship has begun to explore new kinds of collectively governed resources that are clearly creatures of modernity, yet neither these nor more traditional commons have been made part of history—neither of the history of technology, nor of the history of nation-state formation, nor of histories of regional integration like that of Europe. Our project of cosmopolitan commons thus aims to re-embed commons more firmly in history, beginning with the history of Europe.

The increasing technization of commons reveals their innate historicity. Norbert Elias defined technization as the unplanned, accretionary process by which humans, through learning, have continually expanded their technological capacities, and have, moreover, used these capacities in ways that foster ever more far-reaching interdependencies, or forms of integration.[22] Technization encapsulates not only the increasing variety of technologies affecting human utilization and governance of resource-spaces but also the increasing use of technology to structure new commons or to reconfigure and upscale existing commons. Applying the concept of technization to commons implies the understanding that commons are temporally layered as well as spatially expanding.

We hold that cosmopolitan commons are endemic to industrial society, but we find that their importance in international cooperation, integration, and globalization has not been adequately recognized or documented. On one side, Science and Technology Studies (STS) and History of Technology have charted the technological systems of industrial society, but they have not analyzed these systems as commons or as elements of commons regimes. On the other side, commons research has tended to look away from industrial modernity. Yet the search for social, economic, and environmental sustainability would profit from recognition of the historical process of consolidation of commons at an increasing scale.

A case-study approach provides a good way to unravel the dynamics of the emergence and governance of cosmopolitan commons; case studies offer footholds for theorizing. Yet we must also position these case studies, however sketchily, within their broader historical context. In line with our understanding of commons as "resources under regimes," it is evident that forms of collective organization and action for purposes of resource valorization and protection emerged in ancient times, sometimes on a large geographical scale. The imperial water-management systems of the Egyptian, Han Chinese, and Roman empires were regulated by governance regimes spanning these empires and show how much could be achieved with a little technology and a lot of organization. Some of the ancient systems were not surpassed in extent, in organizational complexity, or in potential economic value until the nineteenth century. The medieval commons of Europe were typically organized on a smaller scale, often at the level of a manor or a village, but the expansion and strengthening of monarchies in early modern Europe tended to increase the scale on which commons like forests were organized.

The definitive shift to a new era of commons making came with capitalism, nation-states, and industrialization. The "new commons" were linked to the emergence and growth of an economic and political context unlike anything seen in the ancient world: the rise of capitalism and nation-states. Capitalism and nation-states were deeply implicated in the destruction of older commons. Without putting too fine a point on the matter, it is evident that the modern nation-state, as the upholder of capitalist institutions, has been no friend to the idea of shared or common property. Its "all-absorbing authority," as the Russian anarchist Peter Kropotkin put it, has suppressed traditional commons in favor of private property regimes and various forms of state or public property ruled not by a community of users but by state bureaucrats presumptively acting in the general interest. As usurper of the community and a guarantor of free markets, the modern state seemed to usher in a new dark age of the commons. Kropotkin saw this as a systematic assault on the civilizing principle of mutual aid:

[T]he States, both on the Continent and in these islands [Great Britain], systematically weeded out all institutions in which the mutual-aid tendency had formerly found its expression. The village communities were bereft of their folkmotes, their courts and independent administration; their lands were confiscated. The guilds were spoliated of their possessions and liberties, and placed under the control, the fancy, and the bribery of the State's official. The cities were divested of their sovereignty, and the very springs of their inner life—the folkmote, the elected justices and administration, the sovereign parish and the sovereign guild—were annihilated; the State's functionary took possession of every link of what formerly was an organic whole.[23]

Yet as traditional commons declined, in part through state-imposed "enclosure," capitalists were pioneering new collective organizational forms—joint stock companies and, later, corporations—that made possible resource valorization at increasing scales and the harnessing of more complex and costly technologies, and that required new systems of collective regulation.

Equally important, large-scale capitalist undertakings depended heavily on ever more costly and extensive infrastructural systems (roads, canals, postal systems, railways, telegraph systems, and so on), the creation and governance of which became a primary mission of nation-states.[24] This was the hidden furniture of industrial capitalist society, heavily subsidized by taxpayers. It allowed many capitalist enterprises to emerge and expand. Resource valorization thereby came to intersect the public and private sectors in new ways. On the one hand, capitalist undertakings often depended on direct and indirect financial support from governments and taxpayers; on the other hand, governments developed vast new competences and regulatory powers to manage both capitalist enterprises and the complex infrastructures on which the latter depended.

Ideologically, these new forms of public-private collective action came to be justified by a discourse emphasizing "the common good." An early example can be seen in the ideology of the "cameral sciences" (*Kameralwissenschaften*) in eighteenth-century Prussia. Although many projects carried out under the cameralist banner proved to be motivated as much by private gain as by concern for the public weal,[25] what was significant about its ideology was the early recognition that a growing culture of public "improvements" demanded renewed, complementary forms of collective governance. As Albion Small (the godfather of American sociology) noted in his classic study of cameralism, "The salient fact about the cameralist civic theory was its fundamental assumption of the paramount value of collective interests, or in other words the subordination of the interests of the individual to the interests of the community."[26] The Progressive Era at the outset of the twentieth century in both the United States and Europe can be understood in a similar light.[27]

Industrial capitalism, like hurricanes and earthquakes, had little truck with national boundaries, but borders were the *sine qua non* of nation-states, and it was within this space of tension between imposition of borders and transcending or permeating through borders that cosmopolitan commons became full-fledged. That is, cosmopolitan commons flourished in the space between (on one side) nation-states' widening protection of borders and national sovereignty and (on the other side) the more expansive undertakings that technization permitted, and which capitalists and others desired. Within the boundaries of nation-states, large-scale infrastructures, related services, and large-scale schemes for resource valorization or protection could be carried out either as forms of state initiative or as state-regulated (often subsidized) private enterprises, which might or might not be classified as commons. However, as soon as it became a question of carrying out projects over spaces that fell under the authority of multiple, independent nation-states, the commons form became essential. Commons require unitary governance regimes, and transnational commons had to be negotiated among participating nation-states, which accordingly entered into a new role as commoners.

As long as resource-spaces (and attendant risks) remained embedded within the geographical confines of a nation-state, the tendency was to exploit them as private property or to institute bureaucratic state ownership. We agree with Brett Frischmann's characterization of commons as both "the antithesis of private property and an alternative to government ownership or control." Proprietorship—whether private or public—entails a *non-consensual authority* to define who has access to a resource-space and how the resources it contains may be used.[28] Our view is that commons in the strict sense exist only when the users of a shareable resource develop consensual rules on how in fact to manage and share it—that is, when they develop a moral-economy regime (as explained in a subsequent section). And that is precisely the condition that existed in resource-spaces that transcended yet challenged national sovereignty. One of the earliest examples of this new kind of commons was the transnational Rhine River commons created by an international treaty in 1815 (further consolidated in 1831 and 1868). That treaty valorized the river as an integrated, uniformly regulated transport corridor. Only then could it become an increasingly important avenue for moving the raw materials of industrialization.

Technization figured prominently in defining new areas of tension between national sovereignty and expansive, international enterprise. It opened up entirely new resource-spaces—including airspace and the radio-frequency spectrum—that had not previously been subjected to any form of human ownership or governance, but which seemed to threaten national sovereignty. Radio broadcasts or air pollution could not be stopped at borders by traditional or technological means, and airplanes could easily fly above customs posts. In such cases, nation-state governments, working in collaboration with national and international organizations and companies, interposed themselves as authorities to regulate access to these new resource-spaces and to lay down the terms of their use. And they extended this authority to control new risks and perverse logics of subtractability that emerged at a transnational scale, often as by-products of industrialization. These dynamics first emerged clearly in Europe, where a combination of rapid industrialization, many contiguous modest-sized yet independent nation-states, and a number of border-spanning natural features launched the new cosmopolitan commons into a decisively transnational phase earlier than elsewhere. Our book's focus on Europe thus makes it possible to explore transnational commons making at its most densely configured points of origin.

The rise of cosmopolitan commons tended to place individuals farther away from commons making and management, and indeed often burdened them with new restrictions inspired by experts' assessment of the common good rather than based on their own sense of efficiency or justice. In general, cosmopolitan commons evolved complex governance structures, with governments, intergovernmental organizations, private organizations, and public non-governmental organizations mutually shaping and participating in governance regimes, often with no direct involvement by individual citizens. In principle, national governments represented the interests of their citizens in negotiating the rules of cosmopolitan commons; in practice, however, states acted as enforcers of cosmopolitan commons regimes at the national level. Achieving unitary governance regimes for transnational commons meant imposing new regulations on the citizens of participating nation-states, sometimes against their will. Cosmopolitan commons are thus by no means only egalitarian Edens innocent of power differentials, unfair exclusion, and economic inequalities.

No matter where the new commons making occurred, the process was far from innocent. Our hypothesis is that cosmopolitan commons, because they were bound up with the formation of states and state power, with large, well-financed capitalist enterprise, and with strong technological networks, had a tendency to destroy or profoundly alter older commons. Most often, as commons expanded in scope and extent, local commons as well as smaller-scale privately managed resources were reorganized and incorporated into larger national and transnational systems. The emergence of internationally integrated river, weather forecasting, and etherspace commons are cases in point. Yet we would also expect to find some older commons still surviving, and to find new local commons established using tools generally associated with cosmopolitan commons.[29] We would also expect to see a dialogue between older and newer forms of commons and their associated ideologies, or between commons functioning on different geographic scales, with the assumption that such a dialogue influences both sides. Some obvious examples include the "appropriate technology" movement, the perennial "small is beautiful" concept, and the reinvigoration of interest in pre-capitalist property systems and traditional resource-management systems of indigenous peoples.[30]

Theorizing Technology

The authors represented in this book are historians and theorists of science and technology and newcomers to the study of commons. Though this clearly burdens us with having to catch up with a dense tradition of commons scholarship, we feel that this is more than made up for by our understanding of the ways technologies and societies mutually shape one another. We can thus mobilize a cognizance about the roles of technologies in commons that has been lacking in the field up till now. This technology deficit is not simply our own professional prejudice; it has been recognized by commons scholars themselves. In 2002 the editors of a landmark book-length "review of knowledge about commons" sponsored by the U.S. National Research Council highlighted technology as one of the "key understudied issues" in commons scholarship; alas, the volume itself says very little about what a technologically informed view of commons might offer.[31]

Although we think that commons theory can profit from the insights of Science and Technology Studies, we certainly do not see this as a one-way street. This book is a sustained argument for the need to adapt the theory of commons to industrial societies and hence to incorporate technology in the making and governing of commons; however, it is equally an argument for confronting existing theories of technology and society with their relative indifference to issues of cooperation, equity, reciprocity, and sustainability, which are at the heart of commons scholarship. By reinvigorating theories of technology and society in this way, we can help re-tie the always slipping knot between technologies' stories and the moral order. The case studies in this volume provide a range of examples of how the commons framework can contribute to the aims of STS. Our immediate focus here, however, is on how STS informs a theory of commons.

Commons and STS Methodology

STS celebrates the premise that technologies are socially (or, better, societally) constructed. Social constructivism claims that societies produce "ready-to-wear" technologies—that is, technologies designed specifically to fit into, reinforce, and reproduce social orders. Mature versions of social constructivism by no means deny that technologies *also* influence social action in ways that may create new (perhaps unforeseen) social orders. Specifically, technologies embody "scripts" that shape social action simply by virtue of their configurations and the way they work.

Social constructivism comes in many flavors, some of which are more relevant for our purposes here than others. The most prominent varieties of constructivism in STS are social construction of technology (SCOT), large technical systems (LTS) theory, actor-network theory (ANT), the consumption junction, script theory, gender and technology, users and design, and sociotechnical regime theory.[32] The last has already been put to work in this chapter; elements of the others pervade this chapter and the case studies. LTS theory and actor-network theory are singularly useful for understanding cosmopolitan commons, and for that reason they justify some further explanatory comments.

LTS theory, developed by Thomas Hughes and others since the mid 1980s, explains how heterogeneous technological artifacts are linked by human agency into systems of artifacts and humans that invade ever-greater expanses of space and time. Hughes' explanatory matrix relies heavily on notions of capitalist entrepreneurship and on military metaphors associated with advancing across terrain. Though Hughes shows how human system builders initially create the systems as they pursue entrepreneurial interests, he adds the twist that expanding systems eventually acquire their own developmental momentum simply by accumulating sufficient amounts of technological mass, capital, and humans to become, in effect, like powerful vested interests.[33] Their subsequent expansion has progressively less to do with entrepreneurial initiative and more to do with the desire to maintain and extend the existing infrastructure and economic investment as going concerns. Several of the commons discussed in this book (such as airspace, genetic diversity of crops, and weather forecasting) exhibit features of large technological systems.[34]

Actor-network theory proposes an ontology of the social marked by the ceaseless struggles of actors to achieve and maintain robust orders in the face of the innate precariousness of negotiated ties.[35] "Society" is built, according to ANT, through the formation of actor networks that become more powerful as more and more actors are "enrolled"—voluntarily or otherwise—into them, and as the latter's performances are reconfigured, or "translated," to make the network more robust and less subject to erosion and deconstruction from various causes. This basic stance meshes with our conviction that cosmopolitan commons are always precarious achievements that are able to defy the perennial seductions of short-term self-interest only thanks to incessant technical, organizational, and moral labor.

ANT treats nature and technologies as integral elements of actor networks, putting them on an equal footing with human beings, laws, customs, and morality. It thus programmatically concedes to the "other"—nature and artifacts—an equal place with humans in the constitution of society. In his classic account of the scallops of St. Brieuc Bay, Michel Callon portrayed nature as a volatile and unreliable actor that could be enrolled in networks only conditionally, and then only by dint of hard "translating" work. Scientific researchers attempted to merge the reproductive potentials of the scallops of St. Brieuc Bay into an actor network that also included fishermen and the scientific community.[36] Callon describes an aquaculture program intended to restock the bay with scallops. At first the program seemed destined to succeed, but the scallops displayed a will of their own. Expressed more broadly, nature showed its ontological priority and autonomy when the scallop larvae refused to behave as human researchers predicted and desired. We find that, in the same way, unruly nature incessantly threatens to undermine the precarious networks of cosmopolitan commons, despite persistent efforts by network builders to bolster the latter's resilience and robustness.

ANT is also notable for its conceptualization of space. The geographer Jonathan Murdoch has observed that ANT portrays space as emerging within human-constructed actor networks. He notes the affinity of this position with theories of spatial-temporal relativism (discussed in the next section), encompassing such phenomena as time-space compression through novel transportation and communications technologies. "Networks," Murdoch observes, "are fundamental to [ANT] . . . as the means by which the world is both built and stratified. [ANT] thus sees space as constructed within networks. And not only spaces; times are also forged within network configurations. . . ."[37] Murdoch's observation implies, therefore, that humans construct specific time-space topologies within and through actor networks. This perspective resonates with several of the case studies in this volume, most notably Paul Edwards' study of the growth of a transnational information commons for weather forecasting. In Edwards' example, the valuable resource is accurate weather reports, and the relevant resource-space in which it is found is human-built, consisting of the infrastructures used to collect, transmit, process, and diffuse weather data.

Nevertheless, the studies in this volume—extending ANT's view—also challenge us to recognize the ways in which nature's spaces and environments contest, structure, and inform commons making—that is, how nature's spaces function as obdurate realities with which human actors must continually grapple in the constitution and the expansion of cosmopolitan commons. The issue is the "reluctance of the cosmos,"[38] or, in ANT terms, the resistance to and costs of translating natural actors into dependable elements of new cosmopolitan commons. Humans must reach compromises in this regard. The terrestrial and waterbound commons they build are inevitably shaped around the natural features of the terrain. Even airspace and etherspace—as evidenced by the case studies of Kranakis and

Wormbs—require certain kinds of actor networks adapted to their unique characteristics as resource-spaces with specific environmental characteristics. The kinds of technologies employed, whether for resource valorization or surveillance, are shaped in large part by pre-given geophysical traits, topologies, and natural rifts in the landscape. A North Sea transport network cannot be the same as a land-based network; it will have to contain very different kinds of elements. The nuts and bolts of the network change visibly the moment you reach the water; less visibly, so do the institutions. The same holds when we look at a large river such as the Rhine. Behind the human-made actor networks focused on the Rhine and its locus of resources, the Rhine itself figures as a nature-made rift or fold in space-time that has proved to be only partly mutable by human agency. The river, with its specific discharge regime, gradients, shoreline topography, and aquatic species, became the focus of commons regimes that had to be structured in relation to it—of the Rhine, by the Rhine, and for the Rhine.[39]

In its project to place nature and artifacts on an equal footing with humans in the constitution of social order, ANT has also tried to redefine morality. Bruno Latour launched this line of inquiry by investigating artifacts as "moral placeholders."[40] Referring to the way artifacts can be designed to compel humans (and even other artifacts and nature) to behave along proscribed lines, he argued as follows:

In spite of the constant weeping of moralists, no human is as relentlessly moral as a machine, especially if it is (she is, he is, they are) as "user friendly" as my Macintosh computer. We have been able to delegate to nonhumans not only force as we have known it for centuries, but also values, duties, and ethics. It is because of this morality that we, humans, behave so ethically, no matter how weak and wicked we feel we are.[41]

A decade later, Latour backed off from this perhaps immoderate position by allowing that, although artifacts could incorporate morality, it was probably stretching things a bit to confer an active morality on them (understood to encompass a sense of duty).[42] Instead, Latour now sought to grant to both technology and morality "the same ontological dignity." He proposed to do this in part by divorcing the idea of technology from the simple notion of "means" and the idea of morality from the simple notion of "ends." On this reading, technology implies action by means of detours, and the outcomes of these actions are always contingent and unpredictable. In other words, technology is never *simply* a means. In the same vein, morality is not just about the quality of the ends; it is also about the quality of the means. It is about recovering suppressed agency and responsibility—or, in ANT terms, about recovering creative and autonomous mediators where there appeared to be only passive and predictable intermediaries. Hence, morality, like technology, "detours" headlong action:

> Wherever we want to go fast by establishing tracks so that a goal can race along them whistling like a high-speed train, morality dislocates the tracks and recalls to existence all the lost sidings. The goal-oriented train soon comes to a stop, burdened, powerless. As is often said, morality is less preoccupied with values than with preventing too ready an access to ends.[43]

As will become obvious in the case studies, ANT's particular perspective on technology and morality echoes the way we understand technologies interacting with cosmopolitan commons, and particularly Latour's notion of morality as a process of slowing down or even derailing "high-speed trains." Cosmopolitan commons may incorporate technologies that "slow down" or redirect resource exploitation to abate risk and achieve socially and environmentally sustainable patterns of use. Cosmopolitan commons are material actor networks, but are also—inevitably—moral ones, and from this the fundamental yin-and-yang duality of cosmopolitan commons springs: their simultaneous roles in valorizing nature and in protecting nature and humans from unwanted consequences of that valorization.

Technologies in Cosmopolitan Commons

Commons theory has tended to view property rights as the foundation of access regimes, and hence as the basis for sustainable governance of common risks and resources, but we hold that technologies are just as important. Technology is innately protean, so it should come as no surprise that its sundry artifactual expressions play diverse roles in creating, maintaining, and governing cosmopolitan commons—roles that property rights alone cannot play. For example, property rights depend in practice on artifacts (chains, compasses, theodolites) that can physically delineate property rights in a topographical resource-space. More recently, Earth-observation satellites help chart and monitor the health of commons or detect illegal users (pirates). Even more radically, technologies have opened up new resource-spaces to human exploitation, including airspace, ether-space, and the infrastructural space that became the locus for improved weather forecasts.

Technology's roles in commons derive from the latter's dual structure: *resources under regimes.* Resources sooner or later come down to nature and its spaces. Technologies are, of course, the strategies we humans use to bend nature to our purposes, to transform natural potentials into resources. So, insofar as cosmopolitan commons are about the exploitation of resources, technologies are always an essential element. But, under commons, technologies and their development are also encapsulated in moral-economy regimes, the basic feature of which is that the interests of the parts (actors, users, relevant social groups) take a back seat to the interest of the whole, so as to achieve sustainable, collaborative management of shared resources and risks. This duality means that technologies are (on the one hand) incorporated into Latour's "high-speed train" and (on the other hand) incorporated into arrangements that seek to channel the train, to slow it down, and perhaps to make it also serve the interests of those who merely live beside the tracks, so to speak.

However, as our stories show, much depends on grasping the nuances. A useful concept here is that of "technicity," which among its myriad meanings has one that suits our purpose:

Technicity refers to the extent to which technologies mediate, supplement, and augment collective life; the extent to which technologies are fundamental to the constitution and grounding of human endeavor; and the unfolding or evolutive power of technologies to make things happen in *conjunction* with people.[44]

The main point is that "technicities" are highly specific in that the unfoldings and the evolutive power of technologies are sensitive to nuances of design and to the contexts in which they are implemented. This will become evident in the various chapters. Here we will have to strike a pragmatic balance and limit ourselves to describing three basic types of roles that technologies play in cosmopolitan commons, to wit: valorization, access control, and regulation of use.

Technologies as Agents of Resource Valorization

Technologies are, first of all, used to build the "high-speed train" and keep it on the rails—in other words, they are used to create, find, define, extract and otherwise process resources to produce economic and/or social benefits. This role of technology is not specific to commons, and the multifarious technicities of resource valorization have been well documented in the field of STS. But although many resources are valorized outside the context of cosmopolitan commons, there certainly are ways of valorizing that are typical for commons, and these require appropriate technologies. Fishing nets are a simple example. Nets are technologies of choice for efficiently landing large catches of fish, as in commercial trawling. If the mesh size of the nets is too small, many juveniles may be caught along with mature fish. Though this increases the size of the present catch and thus optimizes valorization relative to costs, it is poor strategy for the long term. Caught juveniles cannot reproduce, and hence the future fish population will be smaller, reducing overall catches for everybody. Moreover, waiting till juveniles mature before they are landed means more weight per fish caught. This is, of course, a typical "tragedy of the commons" dilemma. It can be avoided only by reflexive investigation, communication, and the institution of a moral economy, one of whose priorities must be to avoid catching juvenile fish and hence to eschew the use of small-mesh fishing nets. In a similar fashion, early radio broadcasters optimized their valorization of etherspace by using high-power transmitters tuned (more or less) to frequencies that seemed opportune. These machines left much to be desired with regard to bandwidth and frequency stability. This economically rational approach soon produced chaos as transmitter signals increasingly interfered with one another, causing poor reception. The only remedy was for broadcasters to agree on (and to follow) technical standards for transmissions and to submit to collective management of frequency allocation. In other words, the emergent moral economy of broadcasting stimulated the adoption of new technologies to improve radio transmission and reception.

Technologies as Agents of Access Control

A typical way to ensure a robust and enduring commons is to limit access to it. And in the case of cosmopolitan commons, legal and procedural methods of controlling access are ever more tightly interwoven with technological systems. Access control not only defines the community of commoners (who is in, who is out, and under what conditions); it also prevents too many commoners from using the resource simultaneously. Systems that regulate access to resource-spaces necessarily embody tradeoffs between individual and collective preferences. Partly for this reason (but also because of incessant technological change), cosmopolitan commons are subject to periodic re-negotiation of access terms, but few solutions ever meet every commoner's individual preferences. Particularly in the case of cosmopolitan commons, end users of resource-spaces and their products may find themselves removed from major decision-making processes concerning access, their only (indirect) contribution being through purchasing decisions. Commercial airline passengers, for example, have never had a direct seat at the bargaining table of the airspace commons; the commoners negotiating airspace access were, rather, airlines and governments, who nevertheless saw themselves as acting in the "best interests of" end users and as aiming to continually expand access for end users.

One example of both technization and increasingly high-level regulation of commons access is the global seed-bank system described by Tiago Saraiva in this volume, which artificially conserves the world's plant genetic diversity under "sustainable" and ostensibly carefully controlled storage conditions. In view of the mission of these storehouses, it is clear that access to them must be limited. In fact, ordinary farmers and consumers have been excluded; typically only formally recognized plant breeders are allowed draw on the stored diversity. Beyond seed banks, technologies of genetic conservation and technologies of exclusion have achieved their apotheosis in the new arctic fortress of the Svalbard Global Seed Vault. This backup facility, designed to withstand terrorist attacks and even a nuclear war, denies access even to breeders. The membership of this inner-circle commons is limited to authorized national and regional seed banks that can submit requests for seeds in order to augment or restore their own local diversity. Saraiva suggests that these methods of restricting access to ensure sustainability may not ultimately prove successful, precisely because they leave farmers out of the equation and do not preserve sustainability in farmers' fields.

Technologies for Regulation of Use

Once actors are granted access to a shared resource, the survival of a commons depends on how the resource is actually used. In the classic case, if Hardin's villagers had agreed to a system of sustainable limits, and had developed techniques to ensure that each villager's use was known to the group and remained within the allocated limits, they could have prevented the demise of their collective pasture. Building a sustainable commons thus rests on a pragmatic respect for rules and prohibitions, with trust and cooperation tending to follow in the footsteps of transparency and communication. A robust commons must also be a *symmetrical* panopticon, allowing *mutual, open surveillance*. Ideally, each commoner should be able to ascertain precisely what the other commoners are up to, in an open fashion. With the advent of cosmopolitan commons, the modalities of surveillance and policing have become ever more technologically sophisticated, employing increasingly complex systems of artifacts as "moral placeholders." The latter range from a transnational system of strategically located measuring stations for toxic pollutants in the Rhine to mundane artifacts like water-level indicators along the Vuoksi River. Additional examples include the technologies of seed storage and retrieval in seed banks, the management of transmitter frequencies and bandwidths, and the use of Earth-observation satellites to measure *Waldsterben* (the death of forests) and the dispersion patterns of toxic particles in the atmosphere. Organizational surveillance systems also continue to play important roles in cosmopolitan commons; one example is the code system that uniquely and visibly identifies every aircraft.

In the past few decades, technologies of surveillance and measurement have been tremendously augmented by the digital information revolution, which has made possible not only real-time monitoring of many natural processes and infrastructures but also their simultaneous publication and dissemination to interested parties and even to the general public. For example, now anyone with access to the Internet can monitor water levels on Lake Saaima and the Vuoksi River from hour to hour, monitor the precise locations of all the world's cargo vessels or commercial aircraft in real time, or monitor global weather patterns from moment to moment. All these monitoring systems, from the technologies of sensors to those of calculation and dissemination, appear to be making even the largest cosmopolitan commons more transparent and durable.

Once established, cosmopolitan commons provide specific ecotopes for the design and the implementation of further surveillance and control technologies. Emblematic examples include "instruments of accountability" such as airborne pollutant detectors and metal detectors in airports. Such technologies are inherently "moral" in the sense that they help constrain the behavior of individual actors in the interest of maintaining a collective good—even if that morality is materialized as magnets, wires, dials, images, and so on and the constraints are "only" the inconvenience of enduring an airport security check or the fear of public shaming (as when a company is reported to be responsible for a toxic spill).[45]

The Spaces of Cosmopolitan Commons

If commons are creatures of (historical) time, they are also creatures of space, and not just of abstract geographical space, or homogeneous, Euclidean space, but also of the spaces defined by the reach of infrastructure systems, by technological capabilities (e.g., the cruising altitudes of aircraft or the ability to send telegraphic messages through undersea cables), or by physical phenomena pertinent to commons management (e.g., patterns of Earth's conductivity that affect the propagation of radio waves). The spatiality of commons is, in the first instance, rooted in the extent and the configuration of the resources around which they are structured. By their very nature, resources occupy certain definite—though historically relative—spaces on, in, or above the Earth. The extent of these spaces at any given time depends on natural forces and geomorphologies, as well as on technological capabilities. Etherspace may, in principle, be uniform and omnipresent, but radio transmitters have definite geographical locations, and they propagate differently over particular reaches of geographic space, depending on weather, time of day, transmitter power, Earth's conductivity, and other things. As Nina Wormbs' case study shows, the complex spatiality of radio wave propagation must be appropriately managed in geographical (and political) space to make etherspace a viable commons. The North Sea is fixed in geographical space, yet North Sea fish have a mobile and non-homogeneous spatiality, which must be taken into account to manage this shared resource sustainably.

In general, we may think of resource-spaces in either an absolute sense or a relative sense. In an absolute sense, the entire Rhine may be considered as a nature-given resource-space with unique features. But in a relative sense, many commons have been built around smaller resource-spaces defined by a combination of resource geography, resource characteristics (e.g., average rainfall on a meadow), human organization, and technological capacities. Snowmobiles, for example, enable a more geographically extensive hunting commons than dog sleds, and outboard motors enable larger fishing and lobstering resource-spaces for local communities than sails and paddles. In general, however, the resource-spaces of traditional commons in Europe have either been contained within the geographical perimeter of some local community or contiguous to it and easily accessible by available technological means.[46]

The complex spatiality of cosmopolitan commons is evident in the case of European airspace. In chapter 3, Eda Kranakis argues that dramatic improvements in airplanes during World War I set the technological stage for continental-scale commercial air transport, which depended on radio communication. Not only did air transport have to be technologically embedded in the turbulent European atmosphere, however; it also had to be politically harmonized with Europe's patchwork quilt of nation-states. Figure 2.1 shows the level of European exploitation of airspace in 1935 represented as a network of point-to-point routes between large population centers.[47] This network covers nearly all of geographical Europe, though hardly in a uniform way. Unlike railways and roads, the lines can be straight, because there is little interaction with terrestrial topography; there are already even a few transalpine routes. It is also striking that, with the exception of the Soviet Union, the routes appear to ignore national frontiers altogether; they seem to be determined solely by the locations of big cities and their airports. In its "flat" geographical configuration, the resource-space of European airspace is thus largely indifferent to the underlying natural and political terrain. Clearly the network of airline routes creates a new kind of space defined by its layout of network connections (which, for example, connected London more closely to Paris than to Manchester). Nonetheless, as Kranakis goes on to show, this new space came into existence only by dint of elaborate political negotiations and the willing mutual suspension of national sovereignties in the context of an airspace commons.

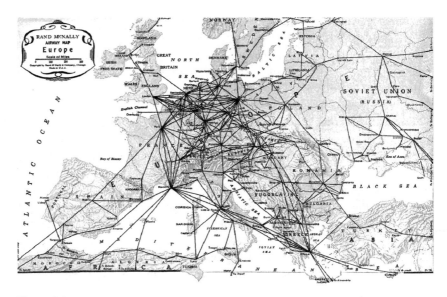

Figure 2.1
European airspace, 1935. Copyright: Rand-McNally Publishing Company.

In chapter 9, Håkon With Andersen invites us to think about how several resources can coexist within a large, nature-given resource-space. In fact, each specific resource (fish, undersea minerals, water as transport medium) engenders a distinct version of the North Sea (although these necessarily interact). Not only is there a North Sea of fish; there is a North Sea of herring, and even a North Sea of herring of different ages. Each of these specific resources defines the space and the extent of the North Sea differently. Figure 2.2, for example, shows the geographical distribution of landings of North Sea herring for thirteen seasons. Herring move about the North Sea in highly peripatetic schools, and their location can never be predicted precisely. The resource-space described in figure 2.2 is thus the geographical distribution of the performance of trawlers looking for and landing mobile schools of herring between 1991 and 2004. No doubt there is some relationship between the size of the catches at any given location and the number of fish present there, but clearly much also depends on the acuity of detection systems and the efficiency of fishing gear. The mobility of the fish, coupled with the extraterritoriality of the North Sea's water (the sea floor is another matter) creates unique challenges for maintaining a fishing commons, not the least of which is monitoring the hundreds of fishing vessels criss-crossing the sea in pursuit of the fish. This problem of *spaces of unidirectional flows* (as we shall see in other case studies on the Rhine and the Vuoksi and on medium-range air pollution) creates generic types of challenges for commons making. In the case of the North Sea, the flow problem is mitigated somewhat by the reversibility of the flows (that is, the fact that the North Sea is, in a way, a wet equivalent of a village pasture: it is a centrally located, easily accessible, and generally overseeable natural space). Thanks to modern communication and detection technologies, the North Sea has become something of a symmetrical panopticon, and it is not impossible for the Fisheries Inspectorate of a given country to know what foreign fishers are up to.

Crop biodiversity exhibits a quite different and vastly more complex spatiality than either airspace or North Sea fish. The absolute, nature-given resource-space of crop biodiversity is bounded only by the extent of the Earth's surface. However, as Tiago Saraiva argues in his chapter, it is possible to define multiple and interpenetrating spaces of crop biodiversity, both natural and artificial. The three that figure in his account are the natural Vavilov centers (regions of the world that harbor the greatest stocks of genetic diversity for distinct crop types), networks of farmers' fields, and the networks of seed banks. Figure 2.3 shows the location of the Vavilov centers. They began to be caught up in regional and national artificial spaces of crop genetic diversity as expeditions were dispatched from seed banks (the first of which were in Europe) to collect the seeds of many plant varieties for use in programs of crop hybridization. These national and regional seed banks have recently achieved a global spatiality with the creation of the (backup) Global Seed Vault on Norway's Svalbard Archipelago. Concurrently, there are also local and regional networks of genetic diversity exchange. These networks, which are based on the spontaneous diversity appearing in farmers' fields, are a grassroots challenge to the politically buttressed monopolies of the breeders and the seed banks. The situation is even more complex than this because, as we have noted, not all actors (farmers, breeders, governments, non-governmental organizations) are able to access these various resource-spaces in the same way.

The examples of airspace, the North Sea, and crop biodiversity, on which we could easily elaborate using examples from other chapters in this book, illustrate how technology-mediated resource-spaces, with their distinct spatialities or flows, penetrate and overlie conventional geographic space. This parallels Manuel Castells' diagnosis that the classical "space of places" has in recent times become penetrated and transformed by a "space of flows."[48] Castells' analytical arrows are aimed primarily at flows of data through the Internet as the informational backbone of a radically new phase of

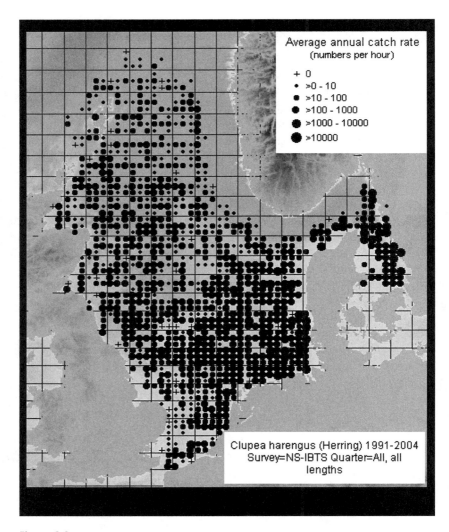

Figure 2.2
Distribution of landings of North Sea herring (all ages), 1991–2004. Source: International Council for the Exploration of the Sea.

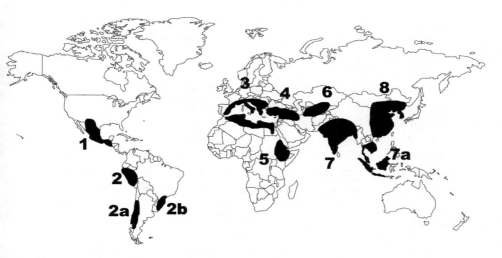

Figure 2.3
Vavilov centers of biodiversity.

modernity in which "time sharing" among geographically distant locations has become the norm (and time is thus "annihilated"). That may or may not be the case. Here we are concerned with the ontological underpinnings: the idea that we exist in multiple spaces and the idea that our realities are defined by the interaction of those spaces through time.

The idea of existing in multiple, mutable spaces has been given a radical turn by Roland Wenzlhuemer in a provocative essay on the concept of space in global history. Wenzlhuemer argues that spaces are not absolute and given, but that—in a neo-Kantian vein—they may literally be produced by human agency and purposes, including those of analysts and theorists like ourselves. He advocates "thinking of space as a theoretically infinite number of spaces defined by our research questions and scientific interest" and asserts that "the nature of relations between the individual objects defines the nature of space—and there are as many possible spaces as there are potential sorts of relationships." "From this perspective," he continues, "geographic space is but one variety of space in which objects are arranged according to their geographic distance."[49] Wenzlhuemer argues that our proclivity to absolutize and reify "geographic space" (often conceptualized in practice as homogeneous, Euclidean space) blinds us to the actual diversity and complexity of spatiality. Geographic space is the abstract space defined by directions and distances among objects, be they buildings, cities, people, or natural features. It is expressed in terms of standard distances and directions and is indexed to the physical features of the Earth by conventions of north and south and latitude and longitude. This is the spatial framework of surveying and mapmaking, to which, among other things, the tessellation of the Earth's surface into national territories is referenced. But Wenzlhuemer argues that spaces also can be defined in many other ways. His examples include "communication space" and "transport space"—that is, the length of time it takes, at any historical juncture, to communicate a message or to transport people and objects between two points. But time is only one criterion for defining space. We could just as easily substitute costs or the "carbon footprints" of different modes of communication or transportation. More to the point, we can substitute spaces defined by the extent and the accessibility of resources or infrastructures, as in the examples of airspace, North Sea herring, and global crop genetic diversity.

Wenzlhuemer and Castells agree that intersections or overlays of spaces produce novel social and political dynamics. Wenzlhuemer's analysis, in particular, invites us to consider the myriad ways that resource-spaces can intersect with, consolidate, or violate other kinds of spaces and the specific types of opportunities, constraints, and externalities that result. Our particular interest is to establish why and how "space-making" may induce actors to engage in commons-making. Though resource-spaces intersect with a variety of other spaces, the most salient for our purposes are the spaces defined by the extent of national sovereignties and, often, by nature-given resource-spaces like enclosed seas, mountain ranges or rivers. As we have emphasized, cosmopolitan commons have achieved their apotheosis in response to the tension between the spaces of national sovereignty, the spaces of rivers and other terrestrial structures, the (desired) spaces of resource valorization or technological infrastructures, and the (undesired) spaces of risk.

There are two basic forms of transnational cosmopolitan resources: either the resource is beyond the pale of regular national territoriality and can be claimed by all (like fish in the North Sea or air in the atmosphere) or the resource-space overlies multiple national territories at the same time and is hence subject to multiple sovereignties (like European airspace or salmon in the Rhine).

In principle, extra-territorial resources are easier to commonize, because they are not burdened by prior sovereignty, being located in various kinds of no-man's land. Often they start out as plentiful public goods—divine bounty simply out there for the taking—and become scarce common-pool resources and candidates for commonizing only when increasing demand and new technologies push rates of extraction to the point that the subtractability and the precariousness of the resource become obvious. This is

exactly what happened to North Sea herring stocks in the 1970s, culminat-
ing in a complete fishing ban. Once there is consensus among users that
their long-term interests are well served by agreements to regulate exploitation of the shared resource, the basic condition for a commons is met.

However, though the resource-space is cosmopolitan and transnational,
the nation-state still remains the pivot of cosmopolitan commons. Historically, national governments have garnered legitimacy as actors capable of
entering into higher-level international agreements that require revision
of national law, pooling of sovereignty in defined contexts, and/or negotiated limits on the exercise of national sovereignty in specified situations.
Accordingly, cosmopolitan commons are typically rooted in international
treaties. These treaties commit states to abide by mutually agreed-upon
rules about access to and use of the shared resources (or resource-spaces),
and to coerce their citizens to comply by rewriting domestic laws to suit
those rules.

Many of these remarks also apply to situations in which cosmopolitan
resources are overlays on existing nationalized resources. Here the cosmopolitan and the national or private aspects of the resource are conflated;
the nationally situated (and perhaps privatized) resource is at the same time
also an element of a cosmopolitan resource. Take the case of the Rhine.
Any given stretch of the river "belongs" to the nation-state through which
it flows. According to principles of national sovereignty, that state may
regulate access to and use of its portion of the river as it sees fit (or as its
citizens see fit). But of course the Rhine flows through several nations, and
the resources it harbors (fish, fresh water, transport route, hydropower, and
so on) create numerous interdependencies within the riparian nations. For
example, any upstream nation that spoils the quality of the water in its
stretch of the river also spoils it in foreign stretches further downstream.
The impact of a cosmopolitan commons here is that local practices (such
as waste discharges of factories or municipalities) are regulated by national
law, which is now shaped and constrained by international agreements
based on an appreciation of the fact that the water in the river does not
simply flow through a concatenation of national stretches but rather is one
cosmopolitan resource shared by actors in all the riparian nations. A similar argument applies to the case of a radio transmitter run by a broadcasting company. At some point in the development of transmitter power and
receiver sensitivity, the freedom of a local broadcasting company to choose
a frequency and a power level begins to produce perverse effects in the
form of interference across national borders. The only way to deal with this
form of precariousness is to subject decisions about the local transmitter to
cosmopolitan agreements on frequency allocation. Given the transnational

range of transmitters, it is only at this supranational level that efficient frequency allocations can be specified on the basis of the spatial distribution of transmitters.

The Moral-Economy Regimes of Cosmopolitan Commons

Our argument in this book is that the governance regimes of cosmopolitan commons are new kinds of moral economies, and that this way of governing shared resources differs substantially from the abstract and indifferent rule of markets and also from the legal-bureaucratic governance of the nation-state (based on law backed up by police power).[50] Moral economies are normative systems intended to align action by directing social reproduction toward group-oriented aims and values such as reciprocity, cooperation, transparency, mutual aid, and equitable sharing of resources, risks, and responsibilities. An essential starting point for understanding moral economies in a historical sense is Karl Polanyi's book *The Great Transformation*. Polanyi did not use the term "moral economy," but he observed that economies were *embedded* within social systems oriented toward redistribution, reciprocity, and social protection.[51] There was a certain contradiction in Polanyi's thought, however. He argued that the rise of market society led to *disembedded* economies in which market relations *superseded* the social relations that had regulated embedded economies. Yet he also argued that disembedded market economies were torrentially destructive of both nature and human society, and that they invariably and necessarily called forth new social regulation and protection mechanisms—that is, the return of embedded economies.[52]

A certain confusion in treating moral economy systems as either inherent in all economies or restricted to pre-industrial, pre-market societies continues to exist. E. P. Thompson championed the disembedded view when he depicted the "moral economy of the crowd" as a traditional system of community morals fighting against an emerging market system that systematically de-moralized human interactions by turning them into abstract market transactions and profit-and-loss calculations.[53] More recently, Andrew Sayer has observed that "the term moral economy has usually been applied to societies in which there are few or no markets."[54] Neo-liberal preoccupation with the role of markets in industrial society has reinforced the disembedded view by touting the powers of social regulation inherent in the free play of markets. Implicit in such views is the notion that market mechanisms impartially neutralize and mediate among conflicting interests, producing social balance, a kind of market justice, and market-based collective security.[55] Yet some authors have taken up the other strand of Polanyi's thinking, arguing that all economies are always embedded. Mark Granovetter's "Economic Action and Social Structure: The Problem of Embeddedness" is a classic example.[56]

More recently, with renewed evidence of the inability of markets alone to regulate society effectively, scholars have begun to re-examine the normative foundations of markets, and have found that market economies did not simply replace moral economies. The relationship was more complex. Though some market economies did replace some moral economies (as E. P. Thompson showed), these scholars argue that market economies also depended on moral economies, indeed, that they were grafted onto them.[57] They argue that market societies cannot function without the foundations of trust, fair play, cooperative spirit, and individual subservience to codes of law and normative values that moral economy systems provided.[58] This new view of the dependence of market economies on moral economies helps overcome (at least to some degree) the otherwise contradictory historical understandings of moral economies. It suggests that moral economies evolve. The urbanization, industrialization, and myriad technological changes that brought the rise of market societies also brought a transformation of the scale and forms of moral economy. Thus, as Thompson's "moral economy of the crowd" was declining, others were arising or adapting.

There is evidence that pre-market societies and market societies put moral economies to rather different uses, however. In modern capitalist states, moral economies seem to be the underpinnings of (on the one hand) markets and cooperative capitalist ventures and (on the other hand) various forms of public property, rather than serving as the animating spirit of user-built governance regimes for widely accessible and hence shared resources. Our thesis is that history is being turned around yet again at the transnational level: cosmopolitan commons are in some ways a renaissance of traditional communal ways of managing shared resources, though on a different scale of size and complexity.

There is a further reason for the persistence and evolution of moral economy systems within industrialized societies, and it also underscores their importance for cosmopolitan commons: the continued need to band together to overcome hazards of nature and life. Writing in the early twentieth century, the American philosopher Ralph Barton Perry described moral economies as "the massing of interests against a reluctant cosmos":

The moral drama opens . . . when interest meets interest; when the path of one unit of life is crossed by that of another. Every interest is compelled to recognize other interests, on the one hand as parts of its environment, and on the other hand as partners in the general enterprise of life. Thus there is evolved the moral idea, or principle of action, according to which interest allies itself with interest in order to be free-handed and powerful against the common hereditary enemy, the heavy inertia and the incessant wear of the cosmos. Through morality a plurality of interests becomes an economy, or community of interests.[59]

Generalizing these ideas, we observe that moral economies emerge when pursuit of mutual interests offers more important benefits than pursuit of self-interests, or when self-interests need a larger, cooperative framework in order to succeed. The principle of mutual interest is readily visible in the modern capitalist nation-state in the forms of joint-stock companies, corporations, cartels, and trade associations. But it is perhaps even more essential for the sustenance of border-spanning resources (such as water in the Rhine) and for the management of border-spanning risks (such as trans-border air pollution). In transnational cases like these, mutual interest assumes the form of cosmopolitan commons.

Cosmopolitan commons can be understood in this vein as a communal way of managing shared resources, risks, and cooperative undertakings, though on a vastly different scale, involving high-level institutional actors and associated with very different mechanisms of surveillance and control than would be found in pre-industrial moral economies and commons. What is important here is that, rather than simply the market or the state, it is the actors themselves—no longer villagers but now national governments, international firms, and non-governmental organizations—that invent the rules and engage in the necessary scrutiny and sanctioning. States may exercise certain powers of surveillance, oversight, and enforcement of the standards and rules of cosmopolitan commons within their territories (as in the case of airspace violations), but states are not at liberty to set those standards and rules; the standards and rules of cosmopolitan commons have to be negotiated with other cosmopolitan actors. In practice, of course, inequalities of power exist and may shape negotiations and their outcomes. Yet the case studies presented in this volume offer multiple examples of how the moral economies of cosmopolitan commons can help to reduce the impact of power differentials.

Kristiina Kurjonen-Kuusipuro's case study of hydropower production on the Vuoksi River illustrates both why cosmopolitan commons need moral economies and how these moral economies can help to balance power differentials. It shows how a governance regime changed from a national legal-bureaucratic mode to a cosmopolitan moral-economy mode in the curious situation where a national resource-space suddenly became bi-national because of shifting borders. Efficient hydropower production on the Vuoksi required close coordination of water levels and production among the power plants, which were originally built and run by Finnish private and public interests when the Vuoksi was still in its entirety a Finnish river. The (national) regime to manage this resource could draw on a dense matrix of national law, a culture of trust and order, and common cultural and economic expectations. Its stipulations could also be enforced by threat of legal sanctions backed up by the power of the state. After the 1939–1940 Winter War, the hydropower-rich section of the river was cut in two by a new Soviet-Finnish border, thereby cosmopolitanizing the resource in one blow. There was no longer a unified rule of law, common cultural perceptions, nor even a basic trust on which to base a regime, so that, for both sides, sub-optimal hydropower production was the default outcome. The only way out was for the two parties to negotiate agreements to govern mutual behavior in the expectation that increased levels of valorization of the resource would more than compensate the constraints on behavior. This new moral-economy regime—negotiated after years of unilateral actions had failed to achieve the needs of either side—was codified in a Convention on Transboundary Water Courses signed by Finland and the Soviet Union in 1964, and further agreements have since been built upon that convention.

As the Vuoksi case shows, moral economies emerge as underpinnings of cosmopolitan commons because the latter are ultimately *voluntary*: states and NGOs are not *obliged* to cooperate, and many examples discussed in the case studies show that cooperative agreements (typically in the form of international treaties) have stalled until mutually agreeable solutions have been devised that balance the needs and priorities of the powerful and the weak. Yet case studies like Kaijser's analysis of transboundary air pollution in Europe also show that voluntary cooperation through moral suasion can overcome divergent perspectives and interests, and can encourage wider adherence to a common governance regime as well as a strengthening of the regime over time.

Yet cosmopolitan commons are not utopias. They are subject to intransigence, rule breaking, and inertia (phenomena evident in all the case studies), and they may be susceptible to capture by vested interests. Moreover, cosmopolitan commons may embody tensions between individual rights and community norms, or tensions related to the rules of community membership, and such tensions may be exacerbated by power inequalities. It also is worth emphasizing that the moral economies of cosmopolitan commons cannot be taken as synonymous with morally *good* regimes. A useful way to think about moral economies is to view them as systems that seek to benefit and safeguard *communities*, so the issue of how these communities and their interests are defined and delineated becomes crucial. Moral economies judge individual and group actions relative to community needs rather than in relation to a universal system of individual rights, and normative principles may be defined or applied in ways that give advantages to certain groups. In fact, moral economy systems may neglect manifest needs of individuals and groups if they do not conform to the system of rules and values in force, or if these individuals or groups do not have seats at the table where commons are negotiated. Finally, community needs depend on, and evolve in relation to, many variables that may periodically provoke new tensions or instability. Both internal tensions and evolving, destabilizing circumstances contribute to the fragility of commons.

Despite their inevitable flaws and shortcomings, however, the fact that commons are socially negotiated and regulated resource-spaces makes it evident that, if appropriately designed and governed, they are essential building blocks of environmental and societal sustainability. In contrast, Hardin's original conceptualization of commons—as Edenic, unregulated natural spaces prone to tragic destruction by the actions of greedy individuals—implied that they were a fundamental *cause* of sustainability crises, which could only be solved through privatization or hegemonic state control.

Notes

1. Garrett Hardin, "The Tragedy of the Commons," *Science*, n.s. 162, no. 3859 (1968): 1243–1248.

2. Since its 1968 publication in *Science*, Hardin's short text has, according to the Web of Science database, been cited in more than 4,200 scholarly articles that range across the disciplinary spectrum. (For comparison, Watson and Crick's pathbreaking 1953 article in *Nature* on the structure of DNA has been cited in approximately 3,900 scholarly articles.) Behind the citations, researchers have analyzed, challenged, tested, extended, and transformed Hardin's theory of tragedy-prone commons.

3. Hardin, "Tragedy of the Commons," p. 1244.

4. See, especially, S. V. Ciriacy-Wantrup and Richard C. Bishop, "'Common Property' as a Concept in Natural Resources Policy," *Natural Resources Journal* 15, no. 4 (1975): 713–727; Susan Jane Buck Cox, "No Tragedy on the Commons," *Environmental Ethics* 7, no. 1 (1985): 49–61. Hardin himself modified his position in a 1977 volume he co-edited: G. Hardin and J. Baden, eds., *Managing the Commons* (Freeman). In response to sustained criticism, Hardin stated that his earlier arguments about the inevitable tragedy of the commons had been based on the counterfactual

ideal of an anarchistic "unregulated" commons. Nevertheless, as in the classic game of "telephone," in which repeated stories are elaborated and distorted, some later researchers not only reproduced Hardin's original (thought-experiment) model but portrayed it as an accurate *historical* model of English commons, even falsely attributing the enclosure of English commons to infertility due to overgrazing—i.e., the playing out of a Hardinesque tragedy. Cox cites examples on pp. 52–53 of "No Tragedy on the Commons."

5. The term *communis* carries the implication of a group being "bound under obligation," from the root *munis* (whence, for example, "municipality"), its antonym, *immunis*, meaning "free of obligation."

6. There is good evidence that commons-management systems were well developed in hunting and gathering societies. See, e.g., Fikret Berkes, "Indigenous Knowledge and Resource Management Systems in the Canadian Subarctic," in *Linking Social and Ecological Systems*, ed. F. Berkes, C. Folke, and J. Colding (Cambridge University Press, 2000). See also FMA Heritage Resource Consultants, Traditional Ecological Knowledge and Land-Use Report, Deer Creek Energy Limited, Joslyn North Mine Project, January 2006 (http://www.total-ep-canada.com). The latter report focuses on traditional, cooperative land use systems by aboriginal peoples in the region of Fort McKay, Alberta.

7. Bonnie J. McCay, "Common and Private Concerns," in *Rights to Nature: Ecological, Economic, Cultural, and Political Principles of Institutions for the Environment*, ed. S. Hanna, C. Folke, and K.-G. Mäler (Island, 1996), p. 113.

8. Paul A. Samuelson, "The Pure Theory of Public Expenditure," *Review of Economics and Statistics* 36, no. 4 (1954): 387–389; "Diagrammatic Exposition of a Theory of Public Expenditure," *Review of Economics and Statistics* 37, no. 4 (1955): 350–356; "Aspects of Public Expenditure Theories," *Review of Economics and Statistics* 40, no. 4 (1958): 332–338.

9. Samuelson, "Diagrammatic Exposition of a Theory of Public Expenditure," p. 350.

10. Ibid.

11. The pairs "rivalrous/non-rivalrous" and "subtractable/non-subtractable" have the same meaning, both pairs referring to the difference identified by Samuelson between consumption of bread and consumption of circuses. A Google Scholar search shows that usage of the first pair of terms emerged within the economics community in the early 1970s, and that usage of the second pair of terms emerged later, within the commons theory community.

12. A particularly influential book that drew on Samuelson's ideas and helped bring the economics approach into commons theory is Mancur Olson, *The Logic of Collective Action: Public Goods and the Theory of Groups* (Harvard University Press, 1965).

13. Mancur Olson and Richard Zeckhauser, "Collective Goods, Comparative Advantage, and Alliance Efficiency," in *Issues in Defense Economics*, ed. R. McKean (National Bureau of Economic Research, 1967), pp. 25–26.

14. James M. Buchanan, "An Economic Theory of Clubs," *Economica*, n.s. 32, no. 5 (1965): 1–14.

15. Samuelson, "Aspects of Public Expenditure Theories," p. 335. Samuelson did not use the terminology "excludable/non-excludable" in this paper, but he offered examples that embodied this duality.

16. See, e.g., the online Digital Library of the Commons (http://dlc.dlib.indiana .edu/dlc/). Its extensive bibliography reveals an overwhelming interest in commons on the local scale, often in pre-industrial settings. That said, it should be noted that contemporary reflection on commons sees the emergence and interrelations of commons at different scales as a central conceptual and empirical problem. See Elinor Ostrom et al., eds., *The Drama of the Commons* (National Academy Press, 2002); Nives Dolsak and Elinor Ostrom, eds., *The Commons in the New Millennium: Challenges and Adaptation* (MIT Press, 2003). However, this concern seems to sprout from new challenges of globalization and remains rather presentist. In other words, a *historical* approach to commons, including the gradual upscaling of commons and entwinement with the emergence of new polities and collectives is still wanting.

17. Charlotte Hess and Elinor Ostrom, "Artifacts, Facilities, and Content: Information as a Common-pool Resource," paper presented at Conference on the Public Domain, Duke Law School, 2001 (available at *law.duke.edu*), p. 63. Note the reference to "antiquated institutions," suggesting that the prevalence of commons in developing countries is due to the existence of traditional social arrangements. Obliquely this suggests that commons would also have been prevalent in Europe in the pre-modern period. The fact that this prejudice is described as "erroneous" suggests a belief that commons still persist in Europe or in other developed countries, though we are left quite in the dark as to what they might look like—in particular, as to whether they would also be *modern* commons. An issue of the *International Journal of the Commons* dedicated to "The Commons in Europe: From Past to Future" (volume 2, no. 2, 2008) is rather disenchanting in this regard. Though the focus is trans-historical, the commons chosen all seem to be throwbacks to an earlier era of communal living and to replicate the traditional commons which are still the object of the vast bulk of commons studies—somewhat to the editors' dismay, we might add. It is, furthermore, interesting to note that, for some unknown reason, the sentence referring to the "erroneous idea" was omitted from the published version of the conference paper cited above. See Charlotte Hess and Elinor Ostrom, "Ideas, Artifacts, and Facilities: Information as a Common-Pool Resource," *Law and Contemporary Problems* 66, no. 1/2 (2003): 111–145, at 127–128.

18. Charlotte Hess, "Mapping the New Commons," paper presented at Twelfth Biennial Conference of the International Association for the Study of Commons, Cheltenham, England, 2008 (available at http://ssrn.com and at http://dx.doi.org).

19. Hess, "Mapping the New Commons," p. 37.

20. We borrow the concept of "regime" from the literature on technological transitions and niche management. Richard Nelson and Sidney Winter used the term in 1982 to describe the stable patterns of rules and knowledge that guided engineers in design work; see their book *An Evolutionary Theory of Economic Change* (Harvard University Press). In the 1990s, transition theorists broadened the concept to include a wider variety of social groups and rules, including those in the "selection environment" that made decisions on the adoption of technology. For the seminal formulation, see Arie Rip and René Kemp, "Technological Change," in *Human Choice and Climate Change*, ed. S. Rayner and E. Malone (Battelle, 1998); also see René Kemp, Johan Schot, and Remco Hoogma, "Regime Shifts to Sustainability through Processes of Niche Formation," *Technology Analysis and Strategic Management* 10, no. 2 (1998): 175–95; René Kemp, Arie Rip, and Johan Schot, "Constructing Transition Paths through the Management of Niches," in *Path Creation as a Process of Mindful Deviation*, ed. R. Garud and P. Karnøe (Erlbaum, 2001). Recently, Schot and Geels have once again broadened the concept by emphasizing the importance of moral and ethical rules and conventions. They describe a socio-technical regime as involving a "community of social groups and the alignment of activities," and they assert that "regimes not only refer to cognitive routines and belief systems, but also to regulative rules and normative roles." See Johan Schot and Frank Geels, "Strategic Niche Management and Sustainable Innovation Journeys: Theory, Findings, Research Agenda and Policy," *Technology Analysis and Strategic Management* 20, no. 5 (2008): 537–554. This conception of "socio-technical regime" resonates with our notion of "moral-economy regimes," though we emphasize that stability and robustness come not so much from simply following set rules as from the mutual negotiation, accounting, and sanctioning of behaviors to maintain equity and moral equilibrium.

21. Bonnie J. McCay, "Common and Private Concerns," p. 112.

22. Norbert Elias, "Technization and Civilization," *Theory, Culture, and Society* 12 (1995): 7–42. The playing out of technization in ways that reconfigure networks and geography is evident in Jonathan Murdoch, "The Spaces of Actor-Network Theory," *Geoforum* 29, no. 4: 357–374.

23. Peter Kropotkin, *Mutual Aid: A Factor of Evolution*, revised edition (Heinemann, 1904), p. 226.

24. Karl Marx and Friedrich Engels' famous quip in the *Communist Manifesto* that the "executive of the modern state is but a committee for managing the common affairs of the entire bourgeoisie" takes on new meaning in this light. See Ralph Miliband, "Marx and the State," in *Democracy and the State*, ed. G. Duncan (Cambridge University Press, 1989).

25. See Andre Wakefield, *The Disordered Police State: German Cameralism and Science and Practice* (University of Chicago Press, 2009). We thank Lissa Roberts for calling our attention to this study.

26. Albion Small, *The Cameralists, The Pioneers of German Social Polity* (University of Chicago Press, 1909). See also David Blackbourn's splendid chapter on the reclamation of the Oderbruch marshes under Frederick II in *The Conquest of Nature: Water, Landscape and the Making of Modern Germany* (Norton, 2006).

27. Daniel T. Rogers, *Atlantic Crossings: Social Politics in a Progressive Age* (Harvard University Press, 1998); Mikael Hård and Marcus Stippak, "Progressive Dreams: The German City in Britain and the United States," in *Urban Machinery: Inside Modern European Cities*, ed. M. Hård and T. Misa (MIT Press, 2008).

28. Brett M. Frischmann, "An Economic Theory of Infrastructure and Commons Management," *Minnesota Law Review* 89 (2005): 917–1030, at 921.

29. De Moor and Bravo, the editors of the special issue of *International Journal of the Commons* cited above, take a first step toward analyzing the historical layering of commons.

30. On indigenous knowledge, see David Turnbull, *Masons, Tricksters and Cartographers: Comparative Studies in the Sociology of Scientific and Indigenous Knowledge* (Harwood Academic, 2000). The locus classicus of the "appropriate" or "intermediate" technology movement is, of course, E. F. Schumacher, *Small Is Beautiful: A Study of Economics As If People Mattered* (Blond and Briggs, 1973).

31. Ostrom et al., eds., *The Drama of the Commons*, pp. 469–479.

32. Early, now classic texts include Wiebe E. Bijker, Thomas P. Hughes, and Trevor J. Pinch, eds., *The Social Construction of Technological Systems: New Directions in the Sociology and History of Technology* (MIT Press, 1987); Wiebe E. Bijker and John Law, eds., *Shaping Technology/Building Society: Studies in Sociotechnical Change* (MIT Press, 1992); Thomas Hughes, *Networks of Power: Electrification in Western Society, 1870–1930* (Johns Hopkins University Press, 1983); Bruno Latour, *Science in Action: How to Follow Scientists and Engineers through Society* (Harvard University Press, 1987); John Law, ed., *Power, Action and Belief: A New Sociology of Knowledge?* (Routledge & Kegan Paul, 1986).

33. The term "vested interested" is used here in the sense that abandonment of the systems would entail enough financial loss and social disruption to warrant strong societal efforts to avoid that outcome.

34. After *Networks of Power*, Hughes also put LTS theory to work analyzing big government projects. See especially *American Genesis: A Century of Invention and Technological Enthusiasm, 1870–1970* (University of Chicago Press, 1989) and *Rescuing Prometheus* (Pantheon, 1998).

35. The literature on actor-network theory is vast and contentious. However, the approach as a general metaphysics for sociology was well summarized by Bruno Latour in his 2005 Clarendon Lectures. See Bruno Latour, *Reassembling the Social: An Introduction to Actor-Network Theory* (Oxford University Press, 2005). Also see Latour, *The Poli-*

tics of Nature: How to Bring the Sciences into Democracy (Harvard University Press, 2004). Latour's project for recasting sociology as the science of *reassembling* the social (by tracing the constitution of heterogeneous actor networks), resonates at many points with our project of delineating the emergence of cosmopolitan commons.

36. See especially Michel Callon, "Some Elements of a Sociology of Translation: Domestication of the Scallops and Fisherman of St. Brieuc Bay," in *Power, Action and Belief: A New Sociology of Knowledge*, ed. J. Law (Routledge & Kegan Paul, 1986).

37. Jonathan Murdoch, "The Spaces of Actor-Network Theory," *Geoforum* 29, no. 4: 359.

38. Ralph Barton Perry, *The Moral Economy* (Scribner, 1909), p. 13.

39. This is not to say that the Rhine's "nature" hasn't been seriously tampered with by generations of users and engineers. In some respects, notably the morphology of its bedding and its ecology, the present-day river bears little resemblance to what it was even in the eighteenth century. See Marc Cioc, *The Rhine: An Eco-Biography, 1815–2000* (University of Washington Press, 2002); Horst Johannes Tümmers, *Der Rhein—ein europäischer Fluss und seine Geschichte* (Beck, 1994); Cornelis Disco, "Taming the Rhine: Economic Connection and Urban Competition," in *Urban Machinery: Inside Modern European Cities*, ed. M. Hård and T. Misa (MIT Press, 2008).

40. Bruno Latour, "Where Are the Missing Masses? The Sociology of a Few Mundane Artifacts," in *Shaping Technology/Building Society: Studies in Sociotechnical Change*, ed. W. Bijker and J. Law (MIT Press, 1992).

41. Ibid., p. 232.

42. Bruno Latour and Couze Venn, "Morality and Technology: The End of the Means," *Theory, Culture and Society* 19, no. 5–6 (2002): 247–260.

43. Ibid., p. 257.

44. Martin Dodge and Rob Kitchin, "Code and the Transduction of Space," *Annals of the Association of American Geographers* 95, no. 1 (2005): 162–180, at 169. See also Adrian Mackenzie, *Transductions: Bodies and Machines at Speed* (Continuum, 2002); Adrian Mackenzie, "The Meshing of Impersonal and Personal Forces in Technological Action," *Culture, Theory and Critique* 47, no. 2 (2006): 197–212.

45. Lewis Mumford bore witness to the "morality of technology" in a number of impressive books. More recently, the power of technology to impose, maintain, or privilege certain normative orders, whether deemed to be "good" or "bad," "just" or "unjust," has been explored by Langdon Winner. See, notably, "Do Artifacts Have Politics?" *Daedalus* 109, no. 1 (1980): 121–136.

46. Pertti J. Pelto, *The Snowmobile Revolution: Technology and Social Change in the Arctic* (Cummings, 1973).

47. Of course, the intensity of utilization of airspace is not simply a matter of the number and configuration of routes, but also of the traffic levels on those routes.

48. Manuel Castells. *The Informational City: Information Technology, Economic Restructuring and the Urban-Regional Process* (Blackwell, 1989).

49. Roland Wenzlhuemer, "Globalization, Communication, and the Concept of Space in Global History," *Historical Social Research* 35, no. 1 (2010): 19–47, at 27.

50. This holds *a fortiori* for authoritarian state forms such as the state socialisms of both fascist and communist persuasions. On the latter, see Paul R. Josephson, *Resources under Regimes: Technology, Environment and the State* (Harvard University Press, 2005).

51. Karl Polanyi, *The Great Transformation: The Political and Economic Origins of Our Time* (1944).

52. Kurtulus Gemici, "Karl Polanyi and the Antinomies of Embeddedness," *Socioeconomic Review* 6 (2008): 5–33.

53. E. P. Thompson, "The Moral Economy of the English Crowd in the Eighteenth Century," *Past and Present* 50 (1971): 76–136.

54. Andrew Sayer, "Moral Economy and Political Economy," *Studies in Political Economy* 61 (spring 2000): 79–104, at 79.

55. On the comparative functioning of embedded economies (moral economies) and liberal market economies, see William James Booth, "On the Idea of the Moral Economy," *American Political Science Review* 88, no. 3 (1994): 653–667.

56. Mark Granovetter, "Economic Action and Social Structure: The Problem of Embeddedness," *American Journal of Sociology* 91, no. 3 (1985): 481–510.

57. See, e.g., Walter Powell, "Neither Market nor Hierarchy: Network Forms of Organization," *Research in Organizational Behavior* 12 (1990): 295–336; Jean-Philippe Plateau, "Behind the Market Stage Where Real Economies Exist—Part I: The Role of Public and Private Order Institutions," *Journal of Development Studies* 30, no. 3 (1994): 533–577; "Behind the Market Stage Where Real Economies Exist—Part II: The Role of Moral Norms," *Journal of Development Studies* 30, no. 3 (1994): 753–817. Another approach that seeks to bring moral economy back into contemporary economic analysis focuses on social capital, a "term encompassing the norms and networks facilitating collective action for mutual benefit." See Michael Woolcock, "Social Capital and Economic Development: Toward a Theoretical Synthesis and Policy Framework," *Theory and Society* 27, no. 2 (1998): 151–208, at 155. The concept of moral economy has been extended to analyze a range of value-laden human activities not usually treated as "economic," one example being science. See, e.g., Lorraine Daston, "The Moral Economy of Science," *Osiris* 10 (1995): 3–24.

58. The global credit crisis that erupted in full force in 2008–2009 has underscored the continued need for trust, fair play, and individual subservience to higher codes

of law and ethics. Financial critics have repeatedly pointed to the breakdown of trust, transparency, and adherence to law and ethics as causes and by-products of the financial crisis, which in turn must be restored to rebuild the health of the financial sector.

59. Perry, *The Moral Economy*, pp. 13, 14. Perry observed that life "has been attended with discord and mutual destruction, but this is its failure. The first grumbling truce between savage enemies, the first collective enterprise, the first peaceful community . . . these were the first victories of [moral economy]. They were . . . victories in that they organized life into more comprehensive unities, making it a more formidable thing, and securing a more abundant satisfaction." (pp. 14–15) Interestingly, contemporary analysts of moral economy have overlooked Perry's observations, which recognize the continuing need for moral economies in industrial market societies.

I Valorizing Nature

3 The "Good Miracle": Building a European Airspace Commons, 1919–1939

Eda Kranakis

In pulling us free from the Earth . . . aviation materially opens the pathways of the skies to us. . . . And this sudden admission of mankind to the bird's domain entails so many consequences, for peace and for war, that a shared concern has pushed all States, at the same time, to regulate the miracle, to try to make it, as far as possible, a good miracle.

—O. Jallu, "L'espace aérienne," *Revue générale de droit aérienne* 6 (1937): 348–360, at 348

Today there are humans in airspace every hour of every day. Two billion of us spend time there every year, along with millions of tons of mail and freight.[1] We have become so accustomed to our use of this resource-space that we tend to overlook its challenges. We take it for granted that we can fly wherever and whenever, our checked baggage meeting us at the end of the trip, or that we can routinely and safely send and receive items by air to and from just about anywhere. In the early decades of human colonization of airspace, none of these things could be taken for granted.

Sustained use of airspace as a transport medium posed an array of environmental, technological, political, and organizational challenges, particularly in the case of international air travel. To become as attractive as other forms of transport, international air travel had to be made safe, seamless, reliable, efficient, comfortable, and widely accessible, and had to be accepted as such by users. Yet few of us appreciate what went into achieving these goals. Most of us don't know the inner workings of this system, or what has kept it running smoothly despite ceaseless growth and change. The specialist literatures on air transport do not analyze airspace and aviation as an integrated, organic system. The picture that emerges is, rather, of a plethora of arcane regulatory structures (such as airworthiness certification) and complicated apparatus (e.g., radiogoniometry systems). What we need is an evolutionary, interactive view of such elements, framed within an understanding of airspace as a geophysical space, as a space of national sovereignty with strategic importance, and, increasingly, as a collaboratively managed, transnational resource-space governed by a moral economy—a cosmopolitan commons.

This commons did not exist *a priori*; it had to be built through political negotiation, infrastructure development, and organizational effort on many fronts. The aim of this chapter is to explain its emergence, its growth, the nature of its moral economy, its tensions, its precariousness, and its resilience. At the core of the building of this commons was a deep, underlying conflict of interest: the clash between the security aims of nation-states on one side, and the commercial and logistical aims of airlines and the desires of airline travelers on the other. For both military and commercial reasons, nation-states were wary of opening their sovereign skies to foreign aircraft and passengers, and they were not eager to adapt their border-control regimes to benefit international air travel. States also used airspace sovereignty as a bargaining chip in international diplomatic negotiations, which impeded the development of air routes. Airlines were more keen to build an integrated transnational network, and lobbied their governments to help make that possible; at the same time, however, airlines wanted to avoid free-market competition (which they felt to be destructive), and each airline wanted to enjoy special economic advantages in its home country.

Beyond states and airlines, the development of international air transport was such a complex, multi-dimensional undertaking that numerous groups and organizations—related to telecommunications, weather, aircraft, insurance, international liability, tourism, airports, and so on—became involved; and they sought ways to collaborate, divide tasks, and overcome inevitable tensions and frictions that attended the system's growth and expansion. Achieving seamless international air travel across borders and between airlines involved gauge problems (analogous to conflicting railway-gauge standards) and network-coordination problems (such as luggage-allowance rules); it also required common technical support systems (e.g., for pilots to receive weather reports in foreign territories over which they flew).

In tracing these developments, I focus on Europe from 1919 to 1939, because it was in interwar Europe that the world's densest network of international air routes first emerged and grew, that the first transnational airspace commons was established, and that the enormous effort and unrelenting cooperation required to achieve such a result can be seen most clearly. Further, through analysis of the concomitant strains and frictions, we also see the precariousness of this commons, made dramatically visible by its collapse at the outset of World War II. Finally, analysis of interwar

Europe's cosmopolitan airspace commons sheds light on its importance as a model for the global airspace commons of the post-World War II era.

On a theoretical level, this chapter offers insight into the dynamics of "cosmopolitanization" and its relationship to European integration. Though many studies have posited a role for technology in this process, they have not recognized that technology-rich commons are themselves a form of cosmopolitanization. In addressing the question "What is cosmopolitanism?" Ulrich Beck and Edgar Grande found that it involved "attempts to conceive of new democratic forms of political rule beyond the nation-state."[2] The creation of Europe's airspace commons in the interwar period is an early and important example of this process. Airspace was important politically, strategically, economically, and socially. It linked high politics and international diplomacy to strategic concerns, international business and market issues, and matters of everyday life. Exploitation of airspace, moreover, led to the emergence of a transnational, "neo-medieval" governance web, characterized by overlapping, multi-level organizations and networks, both formal and informal.[3] It preceded the onset of formal European integration by decades, yet it involved many of the same elements, such as international treaties, supranational organizations, attempts to harmonize areas of European law, and continual travel to meetings around Europe by state officials and others trying to carry out and oversee the transnational governance process.

Airspace as a Geophysical and Ecological Space

Although we routinely think of airspace as bare space beyond human society, it is in fact a unique geophysical and ecological space with properties that affect its exploitation by humans. Airspace extends upward from the Earth's surface and contains the atmosphere, which becomes thinner the farther up you go. Three important consequences follow from this. First, airspace does not have a weight-resisting surface. Things cannot rest on top of airspace, supported by it in as they would be supported by land or water. For organisms or objects to spend time there, entirely off the ground, they must have some system of buoyancy and/or propulsion. Many creatures developed these systems through evolution, whereas humans developed them through technology. The second consequence pertains to the atmosphere's role in providing lift, permitting heavier-than-air objects to fly. Lift requires a flow of air relative to an airfoil; stopping aloft is thus a problem for aircraft that lack the ability to hover. A car on a road or a ship on the sea encountering mechanical trouble can safely stop and sit still, with engine shut down, waiting for repairs or assistance, but the properties of the atmosphere and airspace do not allow this for most fixed-wing aircraft. The atmosphere's role in flight translates not only into a need for appropriately designed aircraft, but also into a need for infrastructures to allow planes to transition safely between land (or water) and sky—to move in and out of flight mode. Such "gateway technologies" also help monitor and control access to airspace.[4] Third, the fact that the atmosphere becomes thinner the higher up you go, and intensely cold, also creates unique vulnerabilities. Some birds have evolved lungs that extract oxygen efficiently at high altitudes, which allows them to fly as high as 32,000 feet during migration, but conditions at that altitude would soon kill an unprotected human, and aircraft intended for higher altitudes must therefore be designed to replicate certain terrestrial conditions.[5]

Airspace also contains liquid and gaseous water—enough to cover the Earth's surface with an inch of water.[6] Unlike the water in a river, however, airspace water is not confined to a fixed channel or to a single direction of flow. Atmospheric water agglomerates sporadically in clouds that block visibility. Moreover, it is carried by air currents whose occasionally tempestuous movements are difficult or impossible to predict. The water also cycles back and forth between the Earth's surface and the sky. However, at 40,000–60,000 feet, the atmosphere is too cold and thin to hold significant volumes of water, and above that limit (known as the tropopause), clouds generally do not form and storms do not occur.[7] But the very fact that atmospheric water is both trapped and unpredictably mobile—endlessly rising, condensing, falling, and moving with air currents and changes in temperature and pressure—represents, along with tempestuous winds, turbulence, and other phenomena, a great risk for users of airspace.[8]

Airspace is an important resource-space for many creatures, and recognizing this clarifies its utility for humans. Geese and monarch butterflies use airspace to reach a greater range of terrestrial resources than would otherwise be possible. Eagles use it to locate and attack prey that live in pastures, rivers, or oceans. Without access to airspace, these creatures could not valorize the terrestrial resources they need to survive. Human use of airspace is similar: we use it to travel, to gain access to resources, and to carry out life and survival activities more efficiently or effectively than would otherwise be possible. Air transport was seen from the outset as a way to establish new markets, enhance revenues and market access, increase business efficiency, and enhance valorization of other resources. In the very first week of commercial air transport between Britain and Paris, "an urgent consignment of millinery" was sent by air from its Paris manufacturer to a high-end London shop. Dutch flower and bulb sellers followed suit. Perishable delicacies such as grouse and Devonshire cream began to be sent by air to markets farther afield. Sending newspapers and films by air allowed topical materials to be distributed with less delay, which enhanced their economic value.[9] And all these activities brought revenue to airlines.

Yet airspace also offered a new way to make war, and European aviation got its first major impetus through war rather than commerce. Moreover, the continued possibility of using airspace for hostile purposes profoundly affected the way this new resource-space was governed, most notably by cementing formal adherence to the principle of national sovereignty over airspace and by fostering reliance on open, transparent surveillance systems to ensure that users of airspace would behave properly.

Founding an Airspace Commons: The 1919 Paris Treaty and ICAN

World War I highlighted the risks associated with mastery of airspace but also showed the growing promise of air travel, and the 1919 Paris Peace Conference established an Aeronautical Commission to address these issues. It drafted a Convention Relating to the Regulation of Aerial Navigation (often referred to as the 1919 Paris Treaty) that came into force in 1922, having gained formal approval from 15 states, most of them European.[10]

The goal of the Paris treaty was to promote international air travel while minimizing the attendant risks. The treaty proclaimed airspace a domain of national sovereignty, yet included provisions for each nation's airspace to be accessible to other member states under specified conditions.[11] The preamble asserted that airspace stakeholders would be best served by a unitary system of governance: "the establishment of regulations of universal application will be to the interest of all." It called for rules and principles "to prevent controversy" and it signaled the desire to "encourage the peaceful intercourse of nations" through "aerial communications." The remainder of the treaty established a moral-economy foundation for European airspace governance by promulgating rules to be obeyed by all and "applied without distinction of nationality." It specified collective safety measures, and set down a principle of mutual aid: "Aircraft of the contracting States shall be entitled to the same measures of assistance for landing, particularly in case of distress, as national aircraft." And to ensure that airspace users obeyed the common rules, it established systems of transparent surveillance of all by all. These included mandatory crew credentials that had to be produced upon request; obligatory passenger and cargo manifests; rules for establishment and maintenance of aircraft logbooks, registrations, airworthiness credentials; and aircraft identification markings.[12]

Technological characteristics of interwar aircraft reinforced the need for a moral-economy system of airspace governance. In the 1920s and the 1930s, airplanes flew at low altitudes (generally below 5,000 feet), where turbulence and weather were constant concerns, and had comparatively short ranges (on the order of 200–500 miles). Weather or mechanical problems often forced pilots to land, sometimes in foreign territory.[13] Interwar aircraft were vulnerable in so many ways that sustained international commercial aviation was inconceivable without active support from all countries along a route.

Paradoxically, the vulnerability of aircraft allowed the moral economy of interwar European airspace to depend largely on an honor system at the international level, backed by the threat of penalties applied by individual states. The Paris Treaty provided for dispute resolution when states differed in their interpretation of rules or the treaty's intent, but in the case of infractions the treaty stipulated only that member states could punish offenders in their own airspace. The risks and technical limitations of interwar air transport made such an approach workable.[14]

The moral economy established by the Paris Treaty nevertheless had two failings. First, it excluded the Great War's vanquished states—Germany, Austria, Hungary, the Ottoman Empire, and Bulgaria. Those states were not allowed to participate in the treaty (i.e., they were not accorded the status of "commoners" and were thereby denied access to the airspace sharing regime). They were also temporarily denied sovereignty over their own airspace (generally until January 1923), and their aircraft were confiscated. Moreover, the treaty forbade any member state from opening its airspace to aircraft from any non-member state. In short, the original Paris Treaty established a system of total exclusion of non-member state aircraft from the totality of member states' airspace. Its other failing was that its voting rules gave the "great powers," which dominated the treaty-making process, more votes than other states.[15]

As a consequence of these two provisions, Denmark, Finland, Spain, the Netherlands, Sweden, Norway, and Switzerland, which had remained neutral in World War I, banded together and refused to sign the treaty until the rules were altered. Their opposition to the airspace exclusion rule stemmed in part from the geographic and economic importance of Germany with respect to air transport. But the rule was also an affront to the principle of neutrality, since it forced every potential signatory to choose sides: if

Denmark established air connections with Germany, for example, then it would not be allowed to establish air connections with Britain. With regard to voting, the neutral countries demanded equality.[16]

The neutral states won their case, and the changes were initiated as soon as the original treaty was ratified. Besides altering the voting procedure, the revised treaty allowed states outside the system, including Germany, to participate indirectly via bilateral agreements with Treaty member states. Side agreements between member states and non-member states permitting use of each other's airspace were henceforth allowed, as long as they conformed with the technical provisions of the Paris Treaty and its appendixes pertaining to aircraft markings, crew credentials, flight rules, and other matters.[17] In this way, Germany, Austria, Hungary, and other non-member states were allowed to enter the commons "through the back door." In later years, negotiations were initiated to revise the Paris Treaty further to enable Germany, Austria, and Hungary to become signatories, yet this matter was still not fully resolved at the outbreak of World War II.[18]

Providing a back-door route into the commons had practical importance for the building of a European air transport system. For example, during the period when Germany's sovereignty over its airspace was still curtailed, France developed an air route to Prague that crossed 200 miles of German airspace to avoid dangerous mountain terrain. But as soon as Germany regained control of its airspace, in 1923, it began enforcement actions against the French airline that flew the route. On multiple occasions, when French aircraft flying between Paris and Prague had to land in Germany (because of weather or mechanical problems), the aircraft were confiscated and their pilots arrested. The problem was resolved only when the Allied powers reached a new agreement with Germany that lifted restrictions on that country's development of commercial aviation. Obtaining that agreement was Germany's real aim, and once that had been achieved Germany and France soon reached a general agreement on commercial air transport.[19]

Beyond establishing a moral-economy foundation (however flawed) for the governance of European airspace, the Paris Treaty set down the basis for a European airspace commons in three ways. First, it specified who could have access to member states' airspace and under what conditions (e.g., aircraft properly marked and registered). Second, it set out member states' rights and responsibilities regarding safety and commercial issues. On the commercial side, each state had the right to monopolize economic use of its own airspace. Thus, an airline from one state did not *a priori* have the right to carry air passengers, freight, or mail originating and traveling within another state—that is, in general *cabotage* was not allowed. In this sense the treaty established property rights within the commons. Finally, the Paris Treaty created a new institution—the International Commission for Air Navigation (ICAN)—to regulate use of the group's airspace. Its primary mandate was to oversee common rules and standards set down in lengthy technical appendixes to the treaty. The Paris Treaty gave ICAN supranational authority to amend and update the appendixes, using a system of qualified majority voting. In the words of its Secretary-General, Albert Roper, ICAN functioned "as a sort of international Parliament having the power at any time to adapt the technical regulations to the requirements of aerial traffic."[20] It also had a mandate to resolve disputes between member states concerning the rules and their application and interpretation.

A treaty is a structure for action rather than action itself. It was one thing for a group of states to establish a basic framework for joint management of airspace; it was quite another to extend this framework more widely over Europe, let alone to iron out the many risks and continuing political impediments to the establishment of a practical, functioning system of international commercial air transport. Achieving these larger aims was a more complex process, involving a range of actors, additional agreements, and cooperation on many fronts. A review of the larger dynamics of cooperation, conflict, and change in the first twenty years of European commercial aviation—focusing on expansion and use of the commons, its governance system, and collective risk-management practices—will substantiate and clarify this point.

Beyond ICAN: The Governance Web of Europe's Airspace Commons

Mastering airspace and building an international air transport system were such expansive missions that ICAN became a node in a larger web of organizations and networks. Already by 1931, ICAN Secretary General Albert Roper counted nineteen separate organizations involved, some with multiple committees devoted to matters of civil aviation. He noted that the large number of these bodies might "seem astonishing" but that "it would be difficult to eliminate any of them,"[21] although he admitted that the expanding matrix was hard to keep track of: "International aeronautical organizations: conferences, congresses, commissions, committees, have become so numerous and hold such frequent meetings . . . that it has become very difficult if not impossible to know . . . which aeronautical commission met in London or Copenhagen last week and what will be the program of the committee that must meet in Geneva or Paris next month."[22] Throughout the 1920s and the 1930s, Roper represented ICAN at more than one international meeting a month, ranging from Stockholm down to Rome and from Lisbon eastward to Warsaw and Bucharest.[23] Other organizations had analogous meeting schedules. To cope with this complexity, Roper founded *La revue aéronautique internationale* to report on the meetings of organizations shaping aviation governance. By 1939 it was covering the work of at least 24 organizations.[24]

An overview of European aviation governance, in combination with a closer look at several important organizations in this and following sections, illustrates how the system functioned and how aviation's moral economy spread beyond ICAN. The visual summary in figure 3.1 displays the major organizations. Table 3.1 briefly summarizes their major functions. In figure 3.1, organizations concerned solely with aviation matters are shown in hexagonal boxes placed near the center of the figure. Organizations that devoted some attention to aviation issues (among other activities) are shown in oval boxes placed farther afield. The boxes are shaded to indicate whether the entities were governmental organizations (dark gray)—i.e., organizations mandated to draft or oversee international treaties; official organizations (light gray), in which government experts were represented; or non-governmental organizations (white).[25] The boxes are also arrayed to indicate whether each organization predominantly focused on technical and risk issues (upper half of the diagram) or commercial issues (lower half). Three organizations represented by hexagonal boxes located in a row along the mid-line of the diagram worked roughly at the intersection between risk and commerce.

The airspace governance web depended on a growing series of international treaties regulating air transport, airmail, aviation telecommunications, public health issues ("sanitary aviation"), private international air law, and so on. Sometimes existing organizations framed new treaties. Governments also arranged special plenipotentiary conferences to draft and approve new treaties. These, in turn, gave rise to subsidiary organizations, networks, and conferences. ICAN itself convened 15 special decision-making conferences between 1932 and 1939 that brought member state delegates together with representatives from other organizations and from non-ICAN states.[26]

Division of labor between organizations, another feature of the airspace governance web, tended to follow organizations' domains of authority and expertise; a good example may be seen in the complementary roles played by the CAI and by ICAN. The CAI was one of a trio of regional European organizations; for convenience, I have grouped them together under the formal name of the earliest and most influential: the International

Technical and Risk Issues

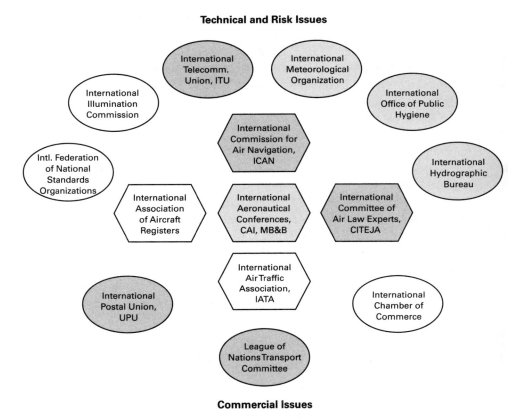

Commercial Issues

Figure 3.1
The governance web of Europe's airspace commons.

Aeronautical Conference, known by its French acronym, CAI. The CAI
was the model for the "Mediterranean Aeronautical Conference" and the
"Aeronautical Conference of Baltic and Balkan States." These aeronautical "conferences" were regular decision-making meetings of aviation ministry officials from participating states aimed at surveying and laying out
each international commercial air route, including establishment of the
necessary risk-control infrastructures, such as emergency landing fields,
weather forecasting arrangements, and coordination of radio communications across borders. The CAI brought together aviation ministry officials
from Northern European states. Initiated by Britain and France, the CAI
soon expanded to include representatives from Belgium, Czechoslovakia,
Denmark, Germany, Ireland, the Netherlands, Poland, and Switzerland.

Table 3.1
Principal functions of the primary organizations involved in European commercial aviation and airspace governance.

Organization	Status	Primary roles and activities
International Aeronautical Conferences, CAI; Medit. Aero. Conf.; Baltic and Balkan Aero. Conference	Official	Surveying and planning of international commercial air routes and their technical infrastructures; harmonization of customs regulations for international air travel
International Air Traffic Association, IATA	NGO	International schedules, fares, ticketing, reservation procedures, cost accounting rules, baggage and air cargo rules
International Association of Aircraft Registers	NGO	Development of aircraft registration and airworthiness certification systems
International Chamber of Commerce	NGO	Promotion of the business community's interests related to aviation
International Commission for Air Navigation, ICAN	Governmental	Development of Paris Treaty's technical appendixes; statistics; liaison and dispute resolution
Int'l Federation of National Standards Organizations	NGO	Coordination of international standards related to aviation
International Hydrographic Bureau	Official	Mapping; standardization of symbols and abbreviations for aviation maps
Int'l Meteorological Organization	Official	Development and diffusion of weather data and weather forecasts for aviation
International Office of Public Hygiene	Official	Regulations to control the spread of infectious diseases through air travel
International Postal Union	Governmental	Airmail arrangements
Int'l Technical Committee of Air Law Experts, CITEJA	Governmental	Preparation of international treaties to harmonize private air law
International Telecommunications Union	Governmental	Regulation of the use of radio communications for aviation
League of Nations, Communications and Transit Committee	Governmental	Promotion of freedom of transit, peaceful use of airspace; monitoring of national aviation legislation; dispute resolution; research; intermodal transport

The CAI organized practical trials of equipment and standards before ICAN attempted "to impose [them] on a large number of States."[27]

The CAI thus planned and charted details of specific air routes, whereas ICAN had a much wider mandate. The latter established the general technical regulations that *all* civil users of airspace had to follow on such matters as airworthiness certificates, aircraft logbooks, medical exams for pilots, emergency medical kits to be carried on commercial aircraft, lights and signals, and radio communication. ICAN also focused on meteorology and on the development of a unified system of aviation maps. Moreover, it produced international air traffic statistics and acted as a general clearinghouse for information on civil aviation matters, including national legislation.[28]

Equally important, ICAN strengthened the broader airspace governance web by its attentive cooperation with other relevant organizations and by playing the role of "honest broker," finding ways to resolve conflicts, to increase unity, and to expand ICAN's mission and reach. Perhaps most notably, ICAN sought changes to the Paris Treaty to encourage Germany and other non-member states to accede to it. In 1929 it organized a formal conference to examine "German proposals" formulated "with a view to facilitating the adhesion of all States to this international agreement, and ensuring the unity of air navigation regulations."[29] And new states did join, although Germany, Hungary, Turkey, and the USSR remained beyond the pale. When the Paris Treaty came into force in 1923, it had 15 ratifications. By 1938, 33 states had acceded to it, including 22 European states, five states that had entered as British dominions, three Asian states (Iraq, Japan, and Siam), and three South American states (Argentina, Peru, and Uruguay).

Another participant in European airspace governance—from the commercial side—was the International Air Traffic Association (IATA). Established in 1919, it brought together CEOs of European international airlines and acquired the prestige of an elite gentleman's club. "It is an awe-inspiring thing," one journalist quipped, "to find oneself a member and not unlike being elected to the Athenaeum. Most Englishmen express their emotion on joining I.A.T.A. by purchasing an Anthony Eden hat of portentous blackness."[30] The meetings themselves—"skillfully chosen as regards climate, cuisine, wines, and local sports"—moved around Europe. In this way, IATA created "an atmosphere of confident and friendly personal relations between the heads of the principal European systems."[31] IATA liaised with other aviation and transport organizations (including ICAN, CITEJA, the International Chamber of Commerce, and the International Railway Union), in part by inviting certain of these organizations to send representatives to IATA meetings.[32]

Despite its social club image, IATA worked tirelessly to extend international air travel and manage risk. IATA helped both ICAN and the CAI develop safety rules and standards. It established a Technical Committee and a Radiotelegraphic Committee, and their members developed a close camaraderie while tackling problems together:

Between members, the question of rivalry and competition bows down before the spirit of technical collaboration in the battle against the material obstacles which they each meet and the attempt to vanquish nature in a domain still full of unknown dangers which man has only just begun to penetrate. In this struggle of technique against the barriers of nature, we feel that we are all together, understanding perfectly that the progress we make will be all the more profitable to the individual if it can be used by all.[33]

IATA contributed to the establishment of ICAN technical standards and participated in establishing the regulatory system for radio communications for aviation. It also developed standards in areas beyond ICAN's mandate, including aircraft cockpit arrangements, throttle controls, indicator dials, rudder controls, and refueling connections.

Overlapping memberships and inter-organizational networking stretched the governance web of Europe's airspace commons further than it could reach through any particular organization. Ultimately not a single European state lay outside its reach. Germany, despite its isolation from ICAN, was integrated through membership in virtually every other organization on the chart. The USSR was also integrated into the system. It never sought membership in ICAN, but it participated in IATA and was indirectly linked to the CAI through an airline (Deruluft) that it operated with Germany. The USSR was also represented in CITEJA, and was a member of the International Postal Union, the International Telecommunications Union, the International Meteorological Organization, and the International Federation of National Standards Organizations. In practical terms, it was possible during the interwar period to fly commercially between Moscow and most major European capitals, a further indication of Soviet integration into the European airspace governance system.[34]

Cooperative Commerce: European Airline Pools

European airspace was potentially an "anticommons" because each state was "endowed with the right to exclude others."[35] Specifically, the Paris Treaty gave nations the right to exclude others from economic activity in their sovereign airspace. The problem was that national chunks of European airspace were small, whereas air travel only became commercially advantageous over long distances or difficult terrain (such as sea crossings), which in Europe meant over international routes.[36] The challenge for states was to find acceptable mechanisms to share and cooperate enough to allow international air transport to flourish, while still maintaining adequate strategic control of their airspace and access to it.

This anticommons tendency, rooted predominantly in the strategic perspectives and *realpolitik* actions of nation-states, inhibited development of international air travel in Europe. We have already noted the way Germany enforced its airspace sovereignty against a French airline in order to extract broader political concessions. Italy impeded Britain's civil aviation efforts for years: Britain needed to use Italian airspace to reach its colonies, but permission came only after a decade of negotiation, through which Italy managed to exact from Britain what amounted to subsidy payments for its airline in return for opening its airspace. Italy did not treat states equally, either: it awarded air route permissions to other states almost immediately. Turkey also played political games with its airspace, as did Franco's Spain, which began barring French, British, and Dutch airlines from Spanish airspace while according liberal privileges to German and Italian airlines.[37]

Despite the anticommons problem, commercial use of European airspace reached an impressive level of development in the interwar period. Beginning with a few airlines serving a few short, international routes that together carried fewer than 2,000 passengers during the first year of operation (1919–1920), the network expanded by 1929 to include dozens of airlines that together flew more than 45 million passenger-miles annually. By 1939, the passenger-miles flown by European commercial air carriers had increased a further 700 percent, to roughly 325 million passenger-miles annually. Airmail and air transport of freight also grew.[38] The entire European system was, moreover, increasingly coordinated to allow passengers and goods to travel more and more seamlessly across borders and between airlines. It was only through years of international cooperative effort and coordination across Europe's airspace commons that Air France could, by the mid 1930s, proudly claim its presence "in all skies" ("dans tous les ciels"—see figure 3.2).

In their efforts to overcome the anticommons problem, European airlines devised a pool system of shared or jointly operated routes. To explain, it must first be noted that European airlines in the interwar period did not engage in price competition. IATA established all the fares and freight rates for European routes. Pooling agreements were based on IATA rate structures, but went further. Airlines serving a shared route divided their revenues according to an agreed formula, such as a 50–50 split, or a *pro rata*

Figure 3.2
This 1935 Air France poster by artist Robert de Valerio reflects the interaction of nationalism and cosmopolitanism in Europe's airspace commons. Image copyright, Air France Museum. Reproduced with permission from the Air France Museum Collection.

regime that considered both distances flown and passengers and goods carried. Some pools involved multiple airlines and countries, such as the Paris-Brussels-Amsterdam-Hamburg-Copenhagen-Malmö route, which linked six countries, with stopovers in each one. And in at least one case, a pool involved pilots of one airline flying the planes of a partner airline. The pool structure thus involved variants adopted by negotiation on a case-by-case basis.[39] Pools also included detailed agreements on facilities and services for each airline in the other airline's country—maintenance and repair facilities, political and administrative representation, ticket sales, advertising, office space, access to telecommunications services, aircraft cabin cleaning, and baggage handling.[40]

Together with IATA, governments played a central role in the pool system (although ICAN did not). Pools became possible only *after* states had signed bilateral treaties and additional protocols for commercial use of each other's airspace. Pooling contracts between airlines had to be consistent with the underlying treaties and protocols and had to have government approval. In contrast, ICAN's sphere of activity focused on general risk and safety issues that *all* airlines and aircraft had to follow.

Airline pools suppressed competition, yet European governments supported them as a way to reduce costs. All interwar European airlines needed government subsidies to survive. States supported their airlines by providing the infrastructures they used, by subsidizing airmail and the aircraft industry, and through direct subsidies (which often amounted to more than half of airlines' revenues). States, moreover, organized their airlines as national monopolies to keep costs in check. Britain consolidated its multiple competing international carriers into a single "chosen instrument" in 1924, Germany did the same in 1926, and France in 1933. Most European states never had more than one international airline to begin with. Most states also became full or part owners of their "chosen instruments," a trend that intensified during the depression years of the 1930s. Pool agreements, which reduced overhead costs through shared facilities, helped control subsidy costs. Growing internationalization of airspace management was thus ironically linked with growing nationalization (and nationalism) of airlines, and in this respect airline pools represented a strange public-private and nationalist-internationalist amalgam.[41]

By the 1930s, the pool system had become the norm in Europe. "The vast majority of European international airlines," Henri Cornelius, chief operating officer of the Belgian airline Sabena, observed in 1937, "are operated as pools by two or several companies. These inter-company agreements are generalized in Europe to such an extent that, outside of a few exceptions, we may consider that the operation of air transport is based on collaboration between companies grouped into an international association, IATA."[42] The pool system received the unofficial blessing of the League of Nations through its Air Transport Cooperation Committee. A 1930 report sponsored by this committee "strongly recommended" the development of airline pools "to avoid superfluous competition, to augment the economic efficiency of international air services," and to develop a "spirit of understanding between the diverse companies."[43]

But were airline pools, IATA rate setting, and government monopolies and subsidies ultimately beneficial for European society? Could an airspace commons designed in this fashion really be considered a success? After all, IATA acted as a price-fixing cartel, pools suppressed competition, and interwar airline travel remained so expensive that most Europeans could not afford to fly. Yet without these government-mandated arrangements there would have been no European air transport system at all. Cartels are widely disliked, but they nevertheless constitute "moral economies" in the sense that they seek to constrain individual behavior to achieve collective goals. As was noted in chapter 2, a "moral economy" cannot be taken as synonymous with a morally *good* regime. Cartels and pools evoke popular disapproval because they are believed to amass extra profits at consumers' expense, but European airlines ran at a net loss, so in this case pools and IATA rate setting served to lessen the total level of government subsidies required to keep the system running rather than to extract excess profits. And technical experts of the period saw airline pools as a way to make the whole air transport system more cost-effective. Whether this pool and cartel system was the best airspace commons that could have emerged seems a moot point; in any event, our task here has been to describe the commons that did emerge, with all of its warts.

The Quest for Seamless European Air Travel

Although acting as a cartel, IATA's member airlines knew their commercial opportunities would improve by taking air travelers' desires into account, notably by making European air travel smoother, more nearly seamless, and more efficient. IATA took the lead in either orchestrating or pressing for the necessary changes, a quest that encompassed timetables, passenger ticketing, luggage handling, cargo and mail handling, air-rail links, and border formalities. Developing convenient, efficient timetables not only required regular meetings, which became increasingly demanding as the number of air routes grew; it also required harmonization of calendars. European countries initiated daylight-saving time in 1920, but each chose its own changeover dates. IATA took the lead in campaigning for Europe-wide changeover dates, and by the end of the 1920s only one government had not yet joined the uniform system. Similarly, during the interwar period airlines had separate summer and winter timetables (geared to weather and traffic patterns), but each airline determined its own changeover dates. On this point, uniformity was not achieved until 1935.[44]

Seamless air travel required air-rail coordination to make airports easily accessible, fill in network gaps, and allow trans-shipment over railways where and when needed. Achieving the necessary coordination became another of IATA's many missions, yet it sometimes won only halfhearted cooperation from railways. IATA took the position that air and rail systems complemented one another, but the railways saw airlines as competitors, often with good reason. IATA nevertheless assiduously liaised with the International Railway Union and invited that organization to send a representative to IATA meetings.

Seamless travel also required a standardized ticketing system that could accommodate journeys involving multiple countries and airlines. Achieving agreement on the basic terms of this "simple" document—an international airline ticket—took years of negotiation and an international treaty. The content of an international airline ticket was a concern for national border-control agencies. It was, moreover, a legal contract, so its precise content had important legal liability implications, which offered particular challenges in the case of international air travel. For example, if a Swedish citizen bought a ticket to fly from London to Paris in a British commercial aircraft, which then crashed in France, how would the legal liability problems be handled? If each country regulated air carrier liability differently and in its own way, the resulting legal chaos might make international air travel impossible.

The International Technical Committee of Air Law Experts, known by its French acronym CITEJA (Comité international technique d'experts juridiques aériens), was mandated to draw up an international treaty to address these issues,[45] which would, in turn, establish the rules for standardized international air travel tickets. CITEJA worked on the treaty from 1926 to 1929 in close collaboration with governments and with input from IATA, ICAN, IPU, the League of Nations, the International Chamber of Commerce, airlines, shippers, and travel agents. The final document, the Convention for the Unification of Certain Rules Relating to International

Carriage by Air, was signed in 1929 in Warsaw.[46] Once the Warsaw treaty was concluded, IATA stepped in further to establish the ticket's practical design and layout and to achieve uniformity in all details.[47]

In the early years of European commercial aviation, international trips involving multiple airlines often could not be booked through one ticket office or travel agency, and to solve this problem, IATA had to establish uniform rules on agents and commissions. A one-stop international booking system also required a means to allocate and transfer payments from booking offices to multiple airlines, which in turn required development of uniform and transparent accounting systems among IATA airlines. Developing and fine-tuning unified accounting practices kept IATA busy for years. Equally important, booking offices needed a way to determine seat availability. Governments stepped in to help, and by the mid 1920s seat reservation messages were being sent as "service messages" through government telegraph systems. (The same privilege was similarly extended to weather services, as Paul Edwards notes in his chapter.) The telegraph messages had to be short, so IATA worked out a coding system geared to airlines' needs. The IATA Code, implemented between 1930 and 1933, became the European standard.[48]

Passengers' luggage also posed practical and legal liability challenges. The Warsaw Convention established the basic terms of a luggage ticket, which, like its passenger counterpart, represented a legal contract. IATA implemented the working details of this system as well. It established uniform rules for checked and carry-on baggage (maximum number, weight, and size of bags, allowable carry-on items, etc.), so that passengers would not experience problems in transferring between airlines. Pets had to find their place in this system too. IATA ruled that they could travel as freight or as carry-on baggage with passengers (held in small travel hampers), but they could not be classed as ticketed luggage. IATA also standardized the charges for excess baggage, both to achieve seamless travel, and to guarantee equity among IATA airlines. One factor that made excess baggage rules (and air-rail coordination) particularly important in this era was that luggage had not yet been redesigned for air transport. Hatboxes and travel trunks could not always fit into the planes and sometimes had to be sent separately, leaving travelers stranded for days as their luggage moved on railways or sat in customs offices far away.[49]

Border-control regimes hindered seamless travel in ways that seem unfamiliar today. Some countries required a commercial pilot to have a visa for every flight, and sometimes a pilot's visa was refused. Jean Bastaki, an executive of the French airline CIDNA, claimed that thirty of its pilots "could not get into Turkey" and that "fully half of its flight force could not enter Rumania."[50] Some states required air passengers to get transit visas to fly over a country even if no landing was scheduled. In instances of forced landings on foreign soil (due to weather or mechanical problems), air travelers were sometimes jailed; more often they were harassed and delayed by government authorities. W. A. Bristow, a consulting aeronautical engineer for the early British airline Instone, explained the drama that unfolded in the event of a forced landing:

First the pilot must find a constable, who on arrival in effect arrests all the passengers and conveys them to the nearest police station, perhaps several miles away. Here they are put through their paces by the Superintendent, who then retires to consult his numerous notices and regulations in order to see the next move. He then proceeds to impound all the passports and send for the nearest Customs officer, which may involve a hunt through all the hotels for miles around. After he has been found and has conducted his examination, passengers are free to proceed to the nearest main line and catch a train, if there is one left, but without their passports, which remain at the police station, from whence they are sent to some authority in London.[51]

Another problem was that customs inspectors often levied import duties on goods and on passengers' luggage in transit. In principle, every country along an international air route could impose import duties on airfreight, on passengers' luggage, and on the aircraft itself, its equipment, and its fuel.[52]

Finally, seamless travel had to be worked out not only for passengers and their luggage but also for mail and goods. For airmail, IATA had to cooperate with the International Postal Union to establish rules for air carriage of mail, with governments to regulate airmail surcharges and subsidies, and with the International Railway Union to work out air-rail coordination for carriage of mail. With respect to goods, the Warsaw Convention established the terms of an international waybill or consignment note, with IATA and airlines working out the practicalities of goods shipping. Beyond the bureaucratic aspect, it was necessary to learn how to handle new types of airfreight: livestock, circus animals, gold and silver bullion, jewelry, highly perishable commodities, and so on. Memoirs of the period are full of accounts of unusual cargos and the difficulties they posed.[53]

Seamless, efficient airmail transport also required coordinated preparations for international night flying. Airmail was an important source of subsidies, but airlines understood immediately that they would be able to compete against ground transport systems only if night flights were instituted so that mail could travel around the clock. Yet international night flights could not occur without a high level of international coordination to ensure appropriate lighting and navigation systems over international routes. In 1930 IATA drew up plans and a route map for a Europe-wide night airmail system, but only parts of the system were completed in the interwar years.

Managing Airspace Risks

The quest to build a robust international air transport network in interwar Europe met with a wide range of physical risks. Today, on average, the number of fatal accidents per million flights by scheduled commercial European air carriers is 0.36; in the early 1920s, it was more than a thousand times that.[54] In the first years of British commercial air transport, from 1919 through 1924, there were an average of 600 accidents with fatalities per million commercial flights.[55] The prevalence of accidents was such that each of the airlines of leading European states generally experienced at least one fatal crash per year in the interwar period. In 1921, French commercial airlines suffered six fatal accidents out of a total of 6,513 flights, with a further ten accidents resulting in injuries.[56] In the years 1927–1936, the German carrier Deutsche Luft Hansa averaged three crashes with fatalities per year; French commercial air carriers together had an average of five crashes with fatalities per year over the same period.[57] So familiar was the image of airplane crashes in the first decade of commercial aviation that the French inventor Albert Sauvant developed shock-proof fuselages intended to allow passengers and crew to survive crashes. Sauvant's invention gained publicity when he walked away unscathed after his experimental aircraft was pushed off cliffs in two trials.[58] Over the 1930s, the rate of occurrence of accidents with fatalities decreased, but as aircraft became larger the number of fatalities per accident grew.[59]

Several factors contributed to high accident rates. One was the short range of commercial aircraft. More than 80 percent of air accidents occur during takeoff or landing, so the fact that air travel in the early period included stops every 200–500 miles meant more takeoffs and landings relative to distances traveled than would be the case today. A trip from Paris to Istanbul today can be a direct flight; in the interwar period, it required seven intermediate stops. In addition, the aircraft of the 1920s and the 1930s were less reliable than present-day aircraft. Engines failed; propellers, rudders, and crankshafts broke in flight; planes lost stability unexpectedly or caught fire for unknown reasons. Newspaper accounts from the period are filled with reports of crashes due to such occurrences.[60]

Low cruising altitudes also hindered safety. Commercial aircraft of the 1920s flew at 2,000 feet or less, so they were often unable to fly above storms or fog. Storms can reach heights of 30,000 feet and more, and fog or mist can extend at least to 10,000 feet. International meteorological services were rudimentary and communication services between aircraft in flight and ground staff were limited, so pilots often encountered storms or fog unexpectedly. Attempts to fly under, around, or above clouds and storms carried further risks. Navigation became more difficult with diminished visibility, and pilots were more likely to get lost or to encounter unknown obstacles if they deviated from known, prepared routes. In 1925, a French commercial airliner en route from Paris to London, after a clear crossing of the English Channel, suddenly encountered dense fog to the south of London. The pilot tried to fly low and follow railroad tracks, but the visibility was too poor even for that. The pilot then tried unsuccessfully to rise above the fog, and finally had to make a forced landing. In doing so, he hit a tree that became visible only at the last moment. The pilot survived, but one passenger was killed.[61]

Forced landings were so frequent in the interwar period that governments organized emergency landing fields along air routes. Britain's policy was to establish them at ten-mile intervals. In France they were placed "at intervals of 50 to 100 km, according to the importance of the line," and each was equipped with "a lodging place, a workshop, and a gasoline cellar."[62] For international routes, airlines posted technical experts on the foreign segment of the route, who provided assistance in cases of forced landings. The British flight engineer Lloyd Ifould, stationed at Le Bourget airport (Paris) to oversee maintenance of British aircraft on the French section of the Paris-London route, recalled:

I can safely say that not only did I know all of the aerodromes between the French coast and Paris, but all the large fields en route that an airplane could put down in, and I did engine changes in most of them, on all types of aircraft. In fact so frequent were forced landings at that time, that the pilots and mechanics always flew with their eyes 'peeled' for the next suitable field, in case anything should go wrong. . . . Before leaving on a flight, each pilot was always handed a sealed envelope to be opened up only in case of forced landings. This envelope contained French currency to pay for petrol, hotel accommodations for passengers, or railway tickets if the voyage could not be continued.[63]

Responding to the diverse risks associated with aviation, states and airlines cooperated to develop safety and support systems along international routes. Such coordination began at the onset of commercial flying. The British Air Ministry's first report on the progress of civil aviation, published only a few months after the first London-Paris commercial flights began operating in 1919, reviewed the joint development of that route:

Wireless liaison has been established with the French for the London-Paris air route. By this means, machine reports and weather reports have been successfully exchanged, the number of such messages on this route averaging 30 a day. . . . On the navigational side, the whole of the London to Paris air route has been surveyed, an experimental strip map prepared for aerial purposes and "flying directions" compiled containing information as to aerodromes, landing grounds, wireless and meteorological data. . . . Proposals have been considered for the installation of an aerial lighthouse system on the London to Paris route. . . . Preparations have also been made for the lighting of the aerodromes on this route, in the event of firms setting up services which necessitate night flying.[64]

Risk-control systems developed on pioneering routes were subsequently extended to other routes and airlines. For example, the wireless communication system tested on the London-Paris route was used on routes linking London and Paris with Amsterdam and Brussels.[65] ICAN also framed general rules and standards to control risks. It proposed guidelines for ground markings that a pilot could correlate with markings on aeronautical maps to determine his location. The system was based on a universal numbering system for aeronautical maps. Landmarks on the ground were to display their appropriate map number in large lettering visible from the sky, and another symbol system would show roughly where on the map the landmark could be found. ICAN similarly proposed a visual system for airports to display weather information on large cards for aircraft flying overhead. ICAN's attention to detail in these matters can be seen in standards it developed for emergency kits to be kept aboard all commercial aircraft, which stipulated what items were to be included, how each item in the kit was to be used, and where on the plane the kit was to be stored.[66]

Non-ICAN European states were ultimately pulled into the organization's system of technical rules because agreements of ICAN member states with non-ICAN states had to respect ICAN rules. One such treaty between Britain and Germany followed the language of the 1919 Paris Treaty regarding airworthiness, registration of aircraft, and pilot certification, and stipulated that each state's aircraft were "entitled to the assistance of the meteorological services, the wireless services, the lighting services, and the day and night signaling services"[67] of the other state.

Conflict and Rule Evasion

Despite deep underlying tensions, the European airspace commons achieved an impressive level of cooperation, rule abidance, and unity of purpose in the interwar period. There was not a single hijacking or case of destructive sabotage of a commercial airplane in Europe in that period. The airlines cooperated closely, abided by IATA fares, and, with very few exceptions, respected airspace rules. States, despite their security concerns and nationalistic tendencies, worked together to enhance safety and promote the expansion of air travel. Sefton Brancker, who represented Britain at ICAN, viewed it as an effective organization:

I am proud to say I represent the British Empire on the International Commission for Aerial Navigation. . . . It has been a most happy family. We have many clashing interests, but personal interests have been sunk, and on every question we have arrived at a satisfactory solution, to which everybody has agreed; and the more I see of it the more I am convinced that flying is going to do more towards the peace of Europe than anything else, as it develops and goes on. There is an extraordinary *camaraderie* in this matter which doesn't exist in any other walk of life.[68]

Yet we know that commons do not always work well, and Europe's airspace commons did not work as well in the post-World War II period. Beginning from the late 1940s there were cases of hijacking, sabotage, violations of airspace rules, undercutting of ticket prices, and many other types of rule evasion. Why did such problems not emerge more strongly the interwar era? There are several possible explanations.

First, in the 1920s and the 1930s aviation was trendy, elite, and widely seen as a wonderful pioneering adventure; it invited a cooperative spirit. Many commons, in contrast, require governance regimes that emphasize limits, restraints, and maintenance of the status quo, or return to an earlier status quo. They convey a certain negativity: don't fish so much; don't pollute; don't drive your car, and so on. The airspace commons, however, seemed to be more about lifting restraints and transforming the status quo in a positive way. It may be that "positive" commons more easily inspire cooperation and rule abidance. One also wonders if aviation's avant-garde image and camaraderie, in combination with the fact that only the upper classes could afford to fly in interwar Europe, helped keep passenger aircraft from becoming sites of violent political action.

Second, technological conditions help account for the relative success of Europe's airspace commons in the interwar years. Aircraft characteristics and dependence on terrestrial infrastructures at frequent intervals helped keep everyone honest. It is foolhardy to violate someone's airspace if you cannot fly at night, always have to fly low (with the plane's markings easily visible from the ground), and must stop frequently. Interwar passenger planes were easy targets for control by governments. We saw, for example, that France was unable to sustain a commercial air route between Paris and Prague in the face of Germany's enforcement of its airspace sovereignty.

Third, the short range of interwar aircraft enhanced cooperation. Long routes, such as Paris-Brussels-Amsterdam-Hamburg-Copenhagen-Malmö, required multilateral agreement on landing rights and on the use of airspace, with states demanding reciprocity as a condition for according the necessary permissions. Thus, airspace sovereignty, in combination with short-range aircraft, produced multilateral cooperation. Yet these fundamentals changed in the post-World War II period. Development of longer-range, higher-flying aircraft made direct flights between distant countries possible, so the dynamics of airspace governance became more bilateral, weakening the commons.[69] Yet many echoes of the old system persisted in the post-World War II period. One example is the establishment of the cooperative Scandinavian airline SAS, a direct heir of the pool system.[70]

The dynamics of airspace politics in interwar Europe also fostered something like a social support system, with stronger states (in economic or technological terms) aiding weaker states. For example, the Versailles restrictions forbade Germany from engaging in international air transport or selling commercial aircraft to foreign entities. To evade those rules, German airlines and aircraft manufacturers *gave* aircraft to new airlines that the German companies helped establish in surrounding countries. In return, the German companies gained stock in the new airlines, and the latter established links to other countries, as well as to airports within Germany, where connections could be made to German carriers. Many of Europe's airlines were founded in this way, including those of Denmark, Switzerland, and the Baltic states. Germany's strategy, moreover, was a foundation of the airline pool system that subsequently became the norm throughout Continental Europe. The strategy worked because Germany's self-interest meshed with everyone's mutual interest.[71]

However, the two major tensions underlying the commons, rooted in nationalism and militarism, increasingly subverted its moral economy in the 1930s. Europe's airlines became national monopolies and nationalist symbols. A notable example was the 1933 formation of Air France, which merged five private airlines none of whose names had originally invoked nationalism. In addition to the name change, the French state became a major shareholder of the new airline. We also saw how nationalistic tensions hindered efforts to establish international air routes. The frustrations of dealing with diplomatic obstructions gave rise to frequent comments about the negative effect of nationalism on European civil aviation.[72]

The second tension in the commons—militarism—was manifest in the vast resources that European states devoted to military aviation and in the growing fear that airspace would be used to make war.[73] The specter of military use of airspace shaped the framework and details of the Paris Treaty. It was the origin of states' adhesion to the doctrine of national sovereignty over airspace in defiance of the maritime model of freedom of the seas that many had hoped would be applied to airspace. It also was at the heart of the initial exclusion of Germany and its World War I allies from the Paris Treaty (which marked an early rift in European airspace in some ways comparable to the Iron Curtain). For a while, the rise of commercial aviation seemed to point in a more concordant direction. Yet despite its achievements, its public visibility, and its status as a symbol of modernity, governments devoted only 10 or 15 percent of their aviation budgets to the commercial side; the rest went to the military side. Many saw commercial aviation as a handmaiden to military aviation. Commercial pilots could become military pilots, and what better way to familiarize them with another state's airspace than have them fly commercial planes through it regularly? Commercial aviation's infrastructure could be seen as military aviation infrastructure in disguise, biding time while being tested and improved until needed for military purposes.[74]

Potential military use of European airspace was a focal point of the Geneva Disarmament Conference, which took place from 1932 to 1934 under the sponsorship of the League of Nations. But nationalism, militarism, and security concerns discouraged effective solutions, leading a participant to refer to the talks as the "Disenchantment Conference." Insofar as the conference ended in abject failure, unable to finalize a treaty that a Preparatory Commission had spent more than six years drafting (1925–1932), that sardonic characterization was on the mark.[75] In the conference committee dealing with air power, a French proposal suggested that nationalist and militarist forces might be controlled within civil aviation by creating a single international airline and an international air police to patrol the airspace and keep it safe. These ideas went nowhere, and today they seem pure fantasy. Meanwhile, intellectual, popular, and administrative imaginations increasingly raised the specter of aerial bombardment of European cities.[76]

The sense that aviation nationalism and its military potential posed a danger to Europe was expressed visually by a Deutsche Luft Hansa poster from the early 1930s. (See figure 3.3.) This image is unusual because it conveys a certain aura of foreboding, whereas most airline posters elicit positive, uplifting, or even lighthearted emotions. German aviation was a powerful element of Europe's airspace commons, and German airspace occupied an important position within the European air travel system. The poster expresses this power. The view of a Deutsche Luft Hansa aircraft bathed in sunlight evokes nationalist pride. Yet the enormous, dark shadow it casts over the European continent is menacing. It hints at concerns about the threat of military use of airspace that were prominent at the time. Intentionally or not, the poster visually challenges precepts of cosmopolitanism. It suggests a nationalist will to power more than a system of cosmopolitan cooperation.

Fears about military use of airspace became reality with the onset of World War II, and war brought the collapse of the airspace governance web. Nearly an entire page of the last issue of the *Revue aéronautique internationale*, published in June of 1939, was taken up by a list of scheduled meetings of aviation organizations that were never to take place.[77] The beginning of the end had already become visible some years earlier. The aforementioned Lloyd Ifould read the signs at Le Bourget airport in Paris, where he was stationed. There he saw Luft Hansa "building air power under the guise of civil aviation."[78] In the years after Hitler's accession to power, Luft Hansa vastly increased its technical and administrative staff at Le Bourget, and scheduled more night flights, ostensibly to carry mail and cargo. Although the amounts carried on the flights were small, the planes had unusually large crews. The pilots also began landing and taking off in darkness, using only their wing lights for visibility rather than asking to have the airport floodlights turned on, as was the norm. It was reported that similar developments were also occurring at other European airports, including London's major airport, Croydon. Ifould recalled the reaction of friends of his who came to see Le Bourget at night:

On various occasions I had brought friends back to the aerodrome with me in the evening, French friends, who wished to see the airliners land and take off at night. . . . They would watch Imperial Airways and their own Air France in admiration, but as soon as the Luft Hansa machines were brought up for departure, their expressions would change. . . . A look of alarm and sadness would cross their faces—for although these people knew little about aeroplanes, they saw the meaning of it all. The little freight that was loaded into the Luft Hansa machines was all loaded from underneath (as bombs would be loaded) and into the fuselage. The freight compartments of these machines were all designed to take bomb racks, which could be installed in a matter of a few hours. My friends did not know that, of course, but they understood what they saw all right. Then they would notice that all the passengers getting on board would be men, and of very military appearance, and, finally, the machines taking off in complete darkness. For them, all this meant one thing—preparation for war. . . . They could not get it out of their minds. "What is the matter with our Government that is allows such things to go on?" they would say. . . . What could anyone do about it? Germany was not yet at war with England or France—there was nothing in International Law which could be used to prevent Luft Hansa. . . . I could see quite clearly that the situation was a delicate one, and that all anyone could do was to watch and wait. . . .[79]

Figure 3.3
A Deutsche Luft Hansa poster, by Jupp Wiertz, from the early 1930s. Reproduced with permission from a poster in the collection of the Smithsonian Institution's National Air and Space Museum.

Although Ifould tells us that Luft Hansa did not technically break any rules in its war planning activities, he saw their behavior as a form of rule evasion or illicit activity that subverted the commons and broke its foundation of trust.

European airspace ceased to function as an internationally governed commons during World War II, yet a rebuilding of its foundations was underway even before the war's end. In 1944, 52 states signed a new international treaty, the Chicago Convention, establishing a global analogue of ICAN, the International Civil Aviation Organization (ICAO). The Chicago Convention represented a global reworking of the Paris Treaty to accord with new political and aviation circumstances. The new treaty retained the fundamental approach and foundational structures of the Paris Treaty, however, including the principle of national sovereignty over airspace and its safety and surveillance systems, such as registration of aircraft, airworthiness, flight crew training and credentials, and aircraft logbooks. The Chicago Convention kept the Paris Treaty's structure of technical appendixes that set down and periodically updated rules and minimum requirements in several categories.[80] Furthermore, IATA was resurrected after the war, and for some decades thereafter it continued many practices that first emerged in the old IATA, including setting fares and standards for tickets and baggage. In Europe, the old aeronautical conferences (such as the CAI) were resurrected as a single European Civil Aviation Conference, which networked with the ICAO. European airline pools were also resurrected, including their prewar practice of revenue sharing. Furthermore, virtually the entire European airspace/aviation governance web of the interwar era continues to exist in a more global form today, albeit with the participation of some different organizations, such as the United Nations, which replaced the League of Nations. Europe's airspace commons was thus restructured as a global commons, and that commons continues to function today.

Conclusion: Reflections on the "Good Miracle"

For most of human history, upper airspace was a "wasteland" beyond human society and law. Technological development opened the possibility of exploiting the "bird's domain" as a new resource-space that permitted rapid travel without the encumbrances of terrestrial geography. In Europe, however, two circumstances mandated the creation of a cosmopolitan commons to achieve this end. First, the military potential of aviation pushed nation-states to "nationalize" the airspace above their territories, a decision ratified by international treaty in 1919. Only a negotiated system for pooling these nationalized resource-spaces under a moral economy backed by mutual surveillance could overcome states' reluctance to open up their skies to outsiders. That is, the use of airspace for commercial transport presupposed a pacified, shared cosmopolitan space, based on mutual consent. Second, within Europe a commercial air travel system made economic sense only on an international scale. Unlike most of the commons analyzed in this book, which emerged more gradually in response to some perceived growing resource degradation or economic inefficiency, airspace (as a transport space) was acknowledged *a priori*—in the same 1919 treaty—to require joint action and a unitary, transnational system of regulation.

Relative to the other commons explored in this book, the European airspace commons unfolded very rapidly, and it displayed considerable institutional complexity almost from the outset. It required numerous support systems that either functioned interoperably or were harmonized across borders—radio communications, weather reports, maps, ticketing, identification and certification systems, legal rules, and so on. Maintaining a commercial air transport system was simply not feasible in the absence of these coordinated systems. They were established, within the first two decades of commercial air travel, through simultaneous, coordinated efforts by many organizations. The airspace commons also helped drive the expansion of other commons: it gave an enormous push to the meteorological commons, and it depended on expansion and intensification of the etherspace commons.

Despite unremitting shared effort and rapid development, however, Europe's airspace commons remained precarious. The pacification it entailed was provisional: it was not possible to fully insulate airspace from appropriation for war. It was difficult to convince states to devote fewer resources to military aviation, and, in the event of war, airspace would become unusable for commercial purposes because of the heightened risks. During World War II, Europe's international commercial air travel system ceased operation. One use precluded the other.

All told, however, the building of Europe's airspace commons in the interwar period can be seen as an exercise in cosmopolitan Europeanization. European airspace integration rivaled and in some ways exceeded the depth and reach of the postwar European Coal and Steel Community (ECSC), often hailed as the first formal step in European integration. The airspace commons involved more treaty structures, organizations, and infrastructure systems than the ECSC, including institutions with supranational decision-making authority (e.g., ICAN), although it lacked the more comprehensive institutional structure of the ECSC, which included a court, an assembly, a council of ministers, and a "High Authority."[81] The growth of the airspace governance web nevertheless had a notable effect on the development of international law in Europe, fostering significant efforts at legal harmonization.

Like all the commons investigated in this book, the history of European airspace underscores the need to consider both nature and technology, along with people and organizations, as elements of commons and their governance regimes. The geophysical characteristics of airspace affected both the need to make airspace a commons and the types of collaboration needed. Similarly, the state of aviation technology in the 1920s and the 1930s influenced governance strategies. The fact that commercial airliners of that era flew low and slow and needed frequent exit points from the sky made surveillance easy, allowing the commons to function adequately under an honor system at the transnational level, with punishment of rule breakers left to national governments. Technology shaped the commons in other ways too. An array of technical systems served to enhance safety and transform airspace into a more valuable and productive resource-space.

The history of Europe's airspace commons also highlights the role of moral economies in structuring economic activity, challenging assumptions about the primacy of market forces in this regard. The rise of European air transport was not primarily scripted by market forces, and commercial aviation was not a "free market" in interwar Europe. Airspace was state property, so its exploitation depended on government support and approval. European governments chose to support airline monopolies, a cartel arrangement for setting airline ticket prices, and pools that suppressed competition. And only governments had the resources (thanks to taxpayers) to establish, maintain, and improve aviation's costly infrastructures and support systems. Without these technical support systems and government-approved commercial agreements, an international commercial aviation market was not even possible. In addition, only governments could carry out many legal actions needed to establish and maintain international commercial aviation—everything from signing treaties to investigating aviation accidents. Governments collaborated in these activities, guided by a shared desire to open up airspace for productive use. Clearly airlines made commercial calculations, and they favored expansion of the air travel market as a route to profitability, but this vast pioneering effort took shape more through the fashioning of a moral economy that channeled and constrained economic behavior in predetermined ways than through the competitive play of market forces, particularly since the entire enterprise was subsidized from start to finish.

The history of European airspace shows how a transnational cosmo-politan commons developed as a partly self-organizing assemblage of het-erogeneous networks, organizations, technologies, and geophysical and ecological elements in which no one network or element determined the direction of the whole fabric. The analytical framework of cosmopolitan commons here reveals itself as a new way to explore questions about tech-nological agency and technological choice at the transnational scale. It clarifies how technological limits and capabilities (such as the performance characteristics of aircraft) influence regimes of collective action, and how various collectivities, in turn, steer and accommodate technological choices and manage their consequences.

Acknowledgments

The research for this chapter was supported by the Social Sciences and Humanities Research Council of Canada. I thank Nil Disco for offering inci-sive comments on several drafts of this chapter. I thank Dr. Rénald Fortier, Curator at the Canada Aviation and Space Museum, as well as the staff at the museum's library, for help in locating and gaining access to documents for this study. I thank the librarians and archivists at the International Civil Aviation Organization in Montreal for their help in navigating and using the ICAO's collections. Finally, I thank Denis Parenteau at the Air France Museum, Kate Igoe at the Smithsonian Institution, and Dominic Pisano at the National Air and Space Museum (Smithsonian Institution) for their help in gaining access to the images used in the chapter.

Notes

1. German Aerospace Center (DLR), *Annual Analyses of the European Air Transport Market, Annual Report 2007* (European Commission, 2008), p. 7.

2. Ulrich Beck and Edgar Grande, *Cosmopolitan Europe* (Polity, 2007), pp. 11–12.

3. Jörg Friedrichs, "The Meaning of New Medievalism," *European Journal of Interna-tional Relations* 7, no. 4 (2001): 475–502; Jan Zielonka, *Europe as Empire* (Oxford Uni-versity Press, 2006); J. Anderson, "The Shifting Stage of Politics: New Medieval and Postmodern Territorialities?" *Environment and Planning D* 14, no. 2 (1996): 133–154.

4. Paul N. Edwards, Steven J. Jackson, Geoffrey C. Bowker, and Cory P. Knobel, Understanding Infrastructure: Dynamics, Tension, and Design, report of workshop on History and Theory of Infrastructure, University of Michigan, 2007. Drawing on the work of Tineke Egyedi, Edwards et al. have focused on gateway technologies as means to interconnect otherwise incompatible infrastructures. The concept may

also be applied to systems that allow technologies to function between different geophysical spaces.

5. Peter Berthold, *Bird Migration: A General Survey* (Oxford University Press, 2001), p. 78.

6. "The Water Cycle: Water Storage in the Atmosphere," U.S. Geological Survey, U.S. Department of the Interior (available at http://ga.water.usgs.gov).

7. According to the U.S. National Weather Service, "almost all weather" occurs within the troposphere. See "Layers of the Atmosphere," Jetstream—Online School for Weather, National Weather Service, National Oceanic and Atmospheric Administration (available at http://www.srh.weather.gov). Recently, however, scientists have been studying the way in which stratospheric air currents interact with tropospheric weather systems. See David E. Steitz and Jana Goldman, "Whither Comes Weather? Scientists Suggest Stratosphere's Role," NASA News (available at http://www.nasa.gov).

8. B. Geerts and E. Linacre, "The Height of the Tropopause" (available at http://www-das.uwyo.edu).

9. These examples were all reported by the *Times* of London during the first year of commercial air transport between London and Paris, 1919–1920.

10. Although the treaty aspired to global universality, on a practical level it remained European. Many of the non-European states that adhered to the treaty (including Australia, Canada, India, South Africa, and New Zealand) came in as British dominions and had no voting power.

11. Convention Relating to the Regulation of Aerial Navigation Signed at Paris, October 13, 1919 (available at http://www.aviation.go.th); Oliver J. Lissitzyn, "The Treatment of Aerial Intruders in Recent Practice and International Law," *American Journal of International Law* 47, no. 4 (1953): 559–589; Peter H. Sand et al., "An Historical Survey of International Air Law Before the Second World War," *McGill Law Journal* 7, no. 1 (1960): 24–42, at 32–35; John C. Cooper, "Air Power and the Coming Peace Treaties," *Foreign Affairs* 24, no. 3 (1946): 441–452.

12. Paris Air Treaty.

13. R. E. G. Davies, *A History of the World's Air Lines* (Oxford University Press, 1964), table 51.

14. Paris Air Treaty.

15. See Eda Kranakis, "European Civil Aviation in an Era of Hegemonic Nationalism: Infrastructure, Air Mobility, and European Identity Formation, 1919–1933," in Alexander Badenoch and Andreas Fickers, eds., *Materializing Europe: Transnational Infrastructures and the Project of Europe* (Palgrave Macmillan, 2010). This article analyzes in greater detail how "hegemonic nationalism" shaped the content of the Paris

Treaty and many aspects of the development of the European commercial air transport system.

16. Albert Roper, "The Organization and Program of the International Commission for Air Navigation," *Journal of Air Law* 3, no. 2 (1932): 167–178, at 168–169; "Edward P. Warner, "International Air Transport," *Foreign Affairs* 4, no. 2 (1926): 278–293 Alfred Wegerdt, "Germany and the Aerial Navigation Convention at Paris, October 13, 1919," *Journal of Air Law* 1, no. 1 (1930): 1–32; Cooper, "Air Power and the Coming Peace Treaties," p. 449.

17. This treaty revision came into force in December of 1926, and without it, a Europe-wide commercial air transport system would not have been possible. See Manley O. Hudson, "Aviation and International Law," *American Journal of International Law* 24, no. 2 (1930): 228–240.

18. The reasons for the continued reluctance of, especially, Germany to accede to the Paris Convention were complex, but they involved the overlap between military and civil aviation. The Paris Peace Conference set restrictions on German military aviation that effectively disadvantaged its civil aviation as well, and Germany refused to accede to the Paris Treaty under these conditions, which it viewed as discriminatory. See Wegerdt, "Germany and the Aerial Navigation Convention"; Roper, "The Organization and Program of the International Commission for Air Navigation"; Oliver James Lissitzyn, *International Air Transport and National Policy* (Council on Foreign Relations, 1942); Warner, "International Air Transport"; Hudson, "Aviation and International Law," pp. 231–232; Cooper, "Air Power and the Coming Peace Treaties."

19. "L'Aéronautique au Parlement: M. Raymond Poincaré s'explique à la Chambre sur les incidents franco-allemands," *L'Air* 5, no. 98 (December 1, 1923): 15–16; "La question du survol de l'Allemagne," *L'Air* 5, no. 99 (December 15, 1923): 13–14; M. Oger, "Les relations franco-allemandes," *L'Air* 7, no. 126 (1925): 12; Cooper, "Air Power and the Coming Peace Treaties," 442–450.

20. Albert Roper, *Un homme et des ailes: Albert Roper, pionnier du droit aérien international, 1891–1969* (Officine, 2004), p. 252. The appendixes to the 1919 Paris Treaty are surveyed on pp. 21–50 of Henry Woodhouse, *Textbook of Aerial Laws* (Frederick A. Stokes, 1920). On ICAN's voting system, see William M. Gibson, "The International Commission for Air Navigation: Structure and Functions," *Temple Law Quarterly* 5, no. 4 (1931): 562–583; Lissitzyn, *International Air Transport*, pp. 369–370.

21. Albert Roper, "Recent Developments in International Aeronautical Law," *Journal of Air Law* 1, no. 4 (1930): 395–414, at 402.

22. Albert Roper, "Preface de l'Éditeur," *Revue aéronautique internationale* 1, no. 1 (1931): 1.

23. Roper, *Un homme et des ailes*, pp. 273–291.

24. "Tableau des organisations aéronautiques internationales et de leurs organismes d'études," *Revue aéronautique internationale* 9, no. 32 (1939): 2.

25. Identifying international organizations in aviation as "governmental," "official," or "non-governmental" follows the practice of ICAN Secretary General Albert Roper. In practice, the dividing line between "official" and "governmental" was blurred, and some "official" organizations later became treaty-based, international governmental organizations; among these are the International Hydrographic Bureau, the International Meteorological Organization, and the International Office of Public Hygiene.

26. Many of these treaties, organizations, and conferences are discussed in issues of *Revue aéronautique international*. See also Lucy Budd, Morag Bell, and Tom Brown, "Of Plagues, Planes, and Politics: Controlling the Global Spread of Infectious Diseases by Air," *Political Geography* 28, no. 7 (2009): 426–435; Camille Allaz, *History of Air Cargo and Airmail from the 18th Century* (Christopher Foyle, 2004), pp. 53–58.

27. John Jay Ide, "International Aeronautic Organizations," *Flight*, September 5, 1930, 1005. CAI activities were also discussed in Roper, *Un homme et des ailes* and in issues of *Revue aéronautique international*.

28. Roper, "Recent Developments in International Aeronautical Law," pp. 395–414; Roper, "The Organization and Program of the International Commission for Air Navigation," pp. 167–178.

29. Roper, "The Organization and Program of the International Commission for Air Navigation," p. 170.

30. "Freightage," *Flight*, July 22, 1937, 101. On government patronage of IATA, see S. Ralph Cohen, *IATA: The First Three Decades* (IATA, 1949), p. 21.

31. Kenneth Colegrove, "A Survey of International Aviation," *Journal of Air Law* 2, no. 1 (1931): 1–23, at 8.

32. IATA included German airlines from the outset. By 1939, nearly all of the international European airlines belonged. The German-Soviet joint airline Deruluft was a member of IATA throughout most of its years of operation (1921–1936). The IATA's activities were covered regularly in the magazine *Flight*, in *Revue aéronautique international*, and in its own technical publication, *Bulletin d'Information*, issued from 1923 through 1938.

33. This statement comes from a report of the Technical Committee, quoted from p. 58 of Cohen, *IATA: The First Three Decades*.

34. Memberships in many of the aviation organizations can be traced through the *Revue aéronautique international*. See also F. P. R. Dunworth, "Aviation in the Soviet Union," *Royal United Service Institution Journal* 80 (February 1935): 116–125; "Copenhagen Air Conference," *Flight*, December 24, 1924, 761; "The 32nd International

Aeronautical Conference," *Journal of Radio Law* 1, no. 2 (1931): 428; Ide, "International Aeronautic Organizations," *Flight*, August 29, 1930, 975.

35. Michael Heller, "The Tragedy of the Anticommons: Property in the Transition from Marx to Markets," *Harvard Law Review* 111, no. 3 (1998): 621–688, at 622.

36. As the MIT aeronautics professor Edward Warner noted in the mid 1920s, the "vast majority of the lines now operated in Europe touch at least two countries, and in some cases an intimately connected group of services makes it possible for passengers to cross half a dozen international boundaries in a single day's flight." (Warner, "International Air Transport," p. 278) Sir Frederick Handley Page, who manufactured aircraft and ran an early British commercial airline serving the London-Paris route, likewise saw European civil aviation as "essentially international." (Sir Frederick Handley Page, "The Future of the Skyways: A British View," *Foreign Affairs* 22, no. 3 (1944): 404–412, at 404.

37. Warner, "International Air Transport," pp. 284–286; Oliver J. Lissitzyn, "The Diplomacy of Air Transport," *Foreign Affairs* 19, no. 1 (1940): 156–170, at 162–163; Cooper, "Air Power and the Coming Peace Treaties," pp. 442–450.

38. United Kingdom, Air Ministry, Directorate of Civil Aviation, *Annual Report on the Progress of Civil Aviation*, 1923, Cmnd. 1900, p. 27; Davies, *A History of the World's Airlines*, table 52; Lissitzyn, *International Air Transport*, pp. 424–429; Allaz, *History of Air Cargo and Airmail*, pp. 101–113, 126–131.

39. The pools began as a Continental European phenomenon in the mid 1920s. Britain was slow to participate, and the Paris-London air route was not organized as a pool in the 1920s and early 1930s. However, by the late 1930s British airlines had begun negotiating pooling agreements with other airlines; these are discussed in articles published in *Flight* from 1934 to 1939. On airline pools generally, see "La Coopération entre les aviations civiles," *Revue générale de droit aérien* 1, no. 1 (1932): 491–521; Laurence C. Tombs, *International Organization in European Air Transport* (Columbia University Press, 1936), pp. 35–41; Ministère de l'Air, *L'Aéronautique militaire, maritime, coloniale, et marchande* (M. & J. de Brunoff, 1931), pp. 163–165, 410; Bernard Dutoit, *La collaboration entre companies aériennes: ses formes juridiques* (H. Jaunin, 1957), pp. 137–145; Walter H. Wager, "International Airline Collaboration in Traffic Pools, Rate Fixing and Joint Management Agreements," *Journal of Air Law and Commerce* 18, no. 2 (1951): 192–199.

40. Airlines treated pool agreements as trade secrets. When a committee of the League of Nations asked IATA for specific information about European airline pools, IATA responded that any inquiries should be directed to the governments (Tombs, *International Organization in European Air Transport*, p. 35). However, a generic contract, said to be a typical airline pool agreement, was published in 1932, and it included a list of facilities, equipment, and services that the contracting airlines agreed to provide each other in each other's country. See "La Coopération entre les aviations civiles," pp. 515–521.

41. Lissitzyn, *International Air Transport*, pp. 98–220; Tombs, *International Organization in European Air Transport*, pp. 39–40; A. J. Quin-Harkin, "Imperial Airways, 1924–40," *Journal of Transport History* 1, no. 4 (1954): 197–215. In the 1930s, Britain permitted establishment of two other international airlines to serve specific routes, and they participated in international pools. See "Civil Aviation," *Flight*, March 11, 1937, 234; "The Outlook," *Flight*, March 24, 1938, 272.

42. Cornelius' conclusion was paraphrased in a review on p. 327 of *Revue générale de droit aérien* 6 (1937). For an explanation of the role of Cornelius at Sabena, see pp. 48–49 and 90 of Guy Vanthemsche, *La Sabena: l'aviation commerciale belge 1923–2001: des origines au crash* (Éditions De Broek Université, 2002).

43. Quoted from Ministère de l'Air, *L'Aéronautique militaire, maritime, coloniale, et marchande*, p. 165.

44. Cohen, *IATA: The First Three Decades*, pp. 28, 42–44, 64–65. See also *L'Indicateur aérien*, IATA's airline schedule guide. Seamless air travel also demanded greater commercial security for airlines with respect to legal liability and insurance issues in international contexts.

45. Other legal issues under CITEJA's purview included compulsory insurance, air collisions, seizure of aircraft, hiring of aircraft, legal status of aircraft commanders, and damage by aircraft of property on the ground.

46. The Warsaw Treaty was signed by 21 states, all but two (China, Japan) European. CITEJA also drafted air treaties regulating damage by aircraft to third parties and seizure of aircraft by public authorities on behalf of private creditors (precautionary attachment). As World War II loomed, it was drafting a treaty "on the legal status of the aircraft commander." CITEJA coordinated with other aviation organizations, and ICAN was represented at its meetings. See "Les conferences internationals de droit privé aérien," *Revue aéronautique internationale* 1, no. 1 (1931): 11–16; Sand et al., "An Historical Survey of International Air Law Before the Second World War," pp. 24–42; John Jay Ide, "The History and Accomplishments of the International Technical Committee of Aerial Legal Experts (C.I.T.E.J.A.)," *Journal of Air Law* 3, no. 1 (1932): 27–49; Linus R. Fike, "The CITEJA," *Air Law Review* 10, no. 2 (1939): 169–186; Stephen Latchford, "Private International Air Law," *Department of State Bulletin* 12 (January 7, 1945): 11–16, 28.

47. IATA's role in ticket standardization is covered in issues of its *Bulletin d'Information*.

48. "Through Air Traffic Tickets," *Flight*, Sept. 2, 1926, 50; Cohen, *IATA: The First Three Decades*, pp. 32–35, 43, 65–70, 74–75. Airlines' use of government telecommunications facilities was initiated in the CAI.

49. "'A. Viator' on Dogs, Fogs, and Distinguished Travellers," *Flight*, November 24, 1938, 473.

50. Cohen, *IATA: The First Three Decades*, pp. 47–48.

51. W. A. Bristow, "Aerial Transport, Today and Tomorrow," *Flight*, February 16, 1922, C101.

52. Cohen, *IATA: The First Three Decades*, pp. 46–47.

53. Gordon P. Olley, *A Million Miles in the Air: Personal Experiences, Impressions, and Stories of Travel by Air* (Hodder & Stoughton, 1934).

54. European Aviation Safety Agency, *Annual Safety Review 2008* (European Aviation Safety Agency), pp. 9, 13.

55. United Kingdom, Air Ministry, Directorate of Civil Aviation, *Report on the Progress of Civil Aviation (April 1, 1926–December 31, 1926)*, Cmnd. 2844 (H.M.S.O., 1927), p. 39.

56. British statistics come from the annual reports of the British Air Ministry's Directorate of Civil Aviation.

57. Calculated from "Accident Database" (available at http://www.planecrashinfo .com).

58. "Crack-Ups Can't Hurt You in This," *Everyday Science and Mechanics*, June 1932: 623; "Daring Inventor. Dive over Precipice in 'Crash-Proof' Aeroplane," *Sydney Morning Herald Tribune*, March 28, 1932; "A 'Crash-Proof' Aeroplane: Frenchman's Daring Experiment," *Times* (London), March 26, 1932; "Inventor Pushed over a Cliff: Adventure in 'Crash-Proof' Aeroplane," *Times* (London), August 15, 1932.

59. For comparative European airline accident data, see E. D. Weiss, "A Survey of Commercial Air Transport with Special Reference to Europe, 1926–1937," *South African Journal of Economics* 6, no. 4 (December 1938): 429–451, at 451. See also the annual reports of the British Air Ministry's Directorate of Civil Aviation.

60. Clinton V. Oster, John S. Strong, and C. Kurt Zorn, *Why Airplanes Crash: Aviation Safety in a Changing World* (Oxford University Press, 1992), pp. 7–14; *Jane's All the World's Aircraft* (1935), p. 53a. The *Times* of London regularly covered interwar European air crashes and inquests after fatal air accidents.

61. Davies, *A History of the World's Airlines*, table 51; "Science and Safety in Air Travel. Fighting Fog Perils," *Times* (London), November 17, 1919; "Air Crash at Wadhurst," *Times* (London), October 20, 1925. Aircraft introduced in the mid 1930s had cruising altitudes of 10,000 feet or more, yet many older aircraft remained in service.

62. Albert Tete, Organization and Exploitation of Regular Aerial Transportation Lines, National Advisory Committee for Aeronautics Technical Memorandum no. 83 (1922) (available at http://ntrs.nasa.gov), p. 3; G. Holt Thomas, *Aerial Transport* (Hodder and Stoughton, 1920), pp. 36–48.

63. Lloyd C. Ifould, *Immortal Era: The Birth of British Civil Aviation* (Adanac, 1948), pp. 21–24.

64. Air Ministry, *Synopsis of Progress of Work in the Department of Civil Aviation 1st May, 1919, to 31st October, 1919*, Cmd. 418 (H.M.S.O., 1919), p. 6.

65. The French airline CIDNA also built wireless communication stations in Bulgaria, Romania, and Turkey and radiogoniometry stations in Bulgaria, Romania, Turkey, and Yugoslavia. See Jean Dentan, "La Flèche d'Orient de la C.I.D.N.A. en face de la crise europénne," *Revue aéronautique internationale* 3, no. 7 (1933): 57–58.

66. The archives of ICAN are housed in the Archives of the International Civil Aviation Organization in Montreal. Examples of ICAN's efforts to standardize aeronautical maps, landmarks, visual display of weather information by airports, and onboard emergency kits can be found in this collection.

67. Agreement between His Majesty in Respect of Great Britain and Northern Ireland and the President of the German Reich Relating to Air Navigation, Treaty Series no. 1, Cmd. 3010 (H.M.S.O., 1928).

68. Norman Macmillan, *Sir Sefton Brancker* (Heinemann, 1935), 332. Brancker was a founding member of IATA and served as Britain's Director of Civil Aviation.

69. Mere overflight rights (known as the "first freedom" of the air) were much easier to negotiate than landing rights and associated rights (e.g., to pick up new passengers), which in practice made more bilateralism possible.

70. Dutoit, *La collaboration entre compagnies aériennes*, pp. 137–162; Wager, "International Airline Collaboration in Traffic Pools," pp. 192–199.

71. Another example of self-interested assistance was the French airline CIDNA's contribution to financing Romania's Department of Civil Aviation. See Tombs, *International Organization in Air Transport*, p. 31.

72. See Kranakis, "European Civil Aviation in an Era of Hegemonic Nationalism," pp. 296–299; Sean Kennedy, "The Croix de Feu, the Parti Social Français, and the Politics of Aviation, 1931–1939," *French Historical Studies* 23, no. 2 (2000): 373–399.

73. For an in-depth look at the ways nationalism and militarism shaped one European nation's relationship with aviation through the interwar period, see David Edgerton, *England and the Aeroplane: An Essay on a Militant and Technological Nation* (Macmillan, 1991). This study is particularly valuable for its explanation of the way class and political ideas intersected aviation ideologies in Britain.

74. Kranakis, "European Civil Aviation in an Era of Hegemonic Nationalism," p. 295.

75. Andrew Webster, "'The Disenchantment Conference': Frustration and Humor at the World Disarmament Conference, 1932," *Diplomacy and Statecraft* 11, no. 3 (2000): 72–80. For a broader study of the way in which "liberal internationalist constituencies" sought to keep international aviation peaceful and restrain nationalist and militarist uses of aviation in the interwar period, see Waqar H. Zaidi, "Aviation

Will Either Destroy or Save Our Civilization': Proposals for the International Control of Aviation, 1920–45," *Journal of Contemporary History* 46, no. 1 (2011): 150–178.

76. Thomas R. Davies, "France and the World Disarmament Conference of 1932 to 1934," *Diplomacy and Statecraft* 15, no. 4 (December 2004): 765–780; Robert J. Young, "The Use and Abuse of Fear: France and the Air Menace in the 1930s," *Intelligence and National Security* 2, no. 4 (1987): 88–109; David Carleton, "The Problem of Civil Aviation in British Air Disarmament Policy, 1919–1934," *RUSI Journal* 111, no. 644 (1966): 307–316.

77. "Calendrier des prochaines reunions internationales," *Revue aéronautique internationale* 9, no. 32 (1939): 100.

78. Ifould, *Immortal Era*, p. 143.

79. Ibid., pp. 144–145.

80. Convention on International Civil Aviation, 1944 (available at http://www .aviation.go.th).

81. What the ECSC uniquely contributed to the process of integration was its array of formal institutions with a traditional, governmental division of powers: a High Authority, a Council, an Assembly, and a Court of Justice. In contrast, ICAN combined executive, legislative, and dispute-settlement mechanisms within a single institution, which contemporaries often likened to an international parliament.

4 Negotiating the Radio Spectrum: The Incessant Labor of Maintaining Space for European Broadcasting

Nina Wormbs

March 12, 1926

Dear Mr. Lemoine,

I deeply regret that the recent changes have not given to Stockholm and Sundsvall the improved conditions that we had striven for.

The trouble with Stockholm comes once again from Radio-Toulouse with which we have had to make several changes to get anything like order in Central and Western Europe; the trouble with Sundsvall comes from the new powerful Vienna station which seems to have a wavemeter several meters out of the standard.

I hope that by now both Stockholm and Sundsvall are free from interference. The latter should be quite free by going to 550 as Budapest has risen to 560.

Stockholm, I have placed temporarily on exactly the same wavelength as Berne, as the latter's radiation northwards appears to be small according to recent reports. There is a risk in it I admit, on the other hand the fact that you did not suffer when Rome was on the same wavelength gives hope for a solution.

If you find 437 impracticable for Stockholm, please wire me your limits of convenient movement and I will seek other alternatives.

I am looking forward to your visit in a few days. There is much to be done.
Yours sincerely,

A R Burrows, Office International de Radiophonie, Genève[1]

The preceding letter, sent to the Swedish representative to the International Broadcasting Union from the secretary of the Union, illustrates how the Union coordinated the provision of broadcasting space to stations and how it tried to ensure a service that would not be disturbed by others. Mr. Burrows' letter gives us vivid insight into how the resource of radio frequencies was handled in the early days of broadcasting in Europe. It pointed to problems with some who did not seem to abide by agreements (e.g., Toulouse disturbing Stockholm); it brought forward the issue of new players like Vienna coming into an existing system but calibrating to a faulty wavemeter, which should have been an agreed-upon standard facilitating adherence to a specific frequency. Moreover, the letter showed forcefully that broadcasting in such a remote corner of Europe as Sweden needed coordinating with broadcasting in Budapest or Berne. Mr. Burrows ended the note by welcoming his colleague to work with him, thereby illustrating that this was indeed something that needed work and collaboration to function well.

This chapter deals with the radio spectrum as a natural resource the use of which is made possible and shaped by radio technology and above all by transmission and reception equipment. In the early days of radio, the technology of using a carrier wave on a fixed frequency was introduced to allow for several users of the spectrum at the same time and to avoid intermingling and mixing of different signals. This technological solution, however, made the resource precarious and subtractable, since the number of frequencies at hand was technology dependent and was, in practice, finite. I argue that the manner in which utilization of this resource was regulated and institutionalized constitutes an example of a moral economy embedded in what we are calling a cosmopolitan commons. The empirical case to support this claim is the institutionalization of frequency plans for European broadcasting from the 1920s to the 1950s.

The reason to use broadcasting rather than any other radio service is that, owing to its direct cultural, social and political implications, it has assumed a special position among the radio services. In economic terms, until mobile phones became common, broadcasting was the communications service for which ordinary people bought personal equipment: radios and later televisions. Those purchases added up to an enormous investment in infrastructure. That, in turn, made the regulation of broadcasting a matter of public concern. Any change in resource allocation that also resulted in a station's moving its broadcasting frequency affected listeners directly. Moreover, in the case of broadcasting, whenever the commons stopped working, that became evident to everyone, since interference made listening difficult or impossible.

In the introduction to this book, cosmopolitan commons are viewed as "communities of fate" that overlap political borders, depend on technologies, and involve the actions of nature itself. As will be discussed in the next section, radio waves have certain inherent properties, and twentieth-century technology could not simply modify them to suit the needs of human communication. Instead, technological enhancements of the resource had to be complemented with organizational principles agreed upon by such a community of fate. The community coordinated beliefs about how to use the resource, how to share it, how to cooperate, how to solve conflicts, and how to punish those who failed to adhere to the system. The community was based on a moral economy in which certain kinds of behavior were accepted and others were not. Typically, behavior that threatened the quality or integrity of the resource, making it more precarious, was discouraged, whereas behavior that followed community norms was supported.

The radio spectrum bears a greater resemblance to other modern shared resources (such as airspace) than to more traditional resources embedded in grazing land, rivers, or lakes.[2] It is, moreover, instantaneously renewable and pristine when not in use. And even though it is to some extent true that using a resource almost always demands technology in one form or another (for example, to use a river for fishing or transport you also need at least a rudimentary technology), in the case of the radio spectrum this is valid in a more profound way. Without radio technology we cannot even sense the radio spectrum.

The Radio Spectrum as a Shared and Precarious Resource

In this section I will discuss the radio spectrum as a resource and address some of the features that, if unregulated, risk making it precarious and lead to suboptimal performance. In doing so I will move back and forth in history, painting a broad picture of the properties of this resource and how regulation has changed.

There are two definitions of the radio spectrum, one trivial and one non-trivial. The trivial definition is that the radio spectrum is that part of the electromagnetic spectrum that humankind has learned to use for wireless communication. It stretches from 300 hertz (Hz) to 300 gigahertz (GHz), corresponding to wavelengths between 1,000 kilometers and 1 millimeter. The non-trivial definition instead takes into account that propagation of radiation varies with frequency and place, that use and regulation have changed both in place and in time, and that the radio spectrum has a human history.[3] Clearly, the non-trivial definition is the one that interests us here.

Changes in radio technology during the twentieth century have also changed the use of the radio spectrum and can thereby illuminate this human history. Changes in technology of course also affect traditional resources—for example, new types of ships made new parts of the globe accessible, and new fishing equipment transformed fishing practices and subsequently the way fishing had to be regulated. In the case of the radio spectrum, the development of radio technology consistently extended the use of the spectrum to higher frequencies. This can be illustrated by the successive ranges of frequencies regulated by the International Telecommunication Union and its predecessors, which reveal how much of the spectrum was known, usable, and in demand. In 1912, at a conference held in London, frequencies up to 1 megahertz (MHz) were regulated. In 1927, at a conference held in Washington, frequencies up to 23 MHz were regulated. In Cairo in 1938, that was extended to 200 MHz. In Atlantic City in 1947, regulation was extended to frequencies up to 10,500 MHz. In 1959 that was extended to 40,000 MHz,[4] and in 1979 to 275,000 MHz. In the frequency map reproduced here as figure 4.1 this expansion is visible as new communication technologies occupy higher and higher frequencies. Satellite communication takes place at high frequencies; maritime communications, one of the first uses of radio technology, is still confined to low frequencies. The figure also illustrates that even though radio technology is very flexible, it also exhibits considerable inertia as an international communication system.

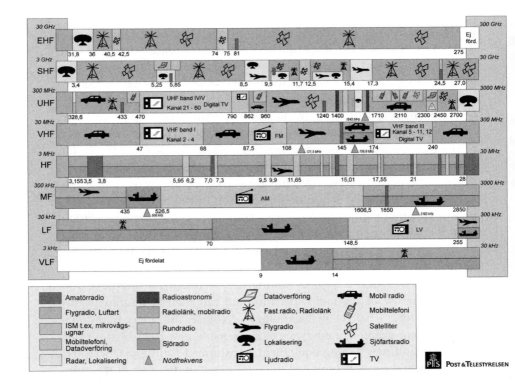

Figure 4.1
A spectrum chart from the Swedish Post and Telecom Authority. Reprinted with permission.

The radio spectrum is a spatial resource. Essentially, someone is transmitting a message and someone is trying to receive that message. Both the transmitting and the receiving are technology dependent, and the quality of each of those activities changes with changing technology. As long as only one message is transmitted at any given time, reception will be rather simple and technological development less important. However, if there are several simultaneous transmissions, they risk intermingling and making the transmission unintelligible. One way of solving this problem was to have different transmitters operate at different carrier-wave frequencies. Such a solution demanded that the frequency of each transmitter be stable, since interference is more likely if transmitters deviate from their specified wavelengths. With the advent of the use of carrier waves, after which a message was transmitted on a specific frequency that the receiver could tune in to, the radio spectrum was divided into a number of wavelength slots on which messages could be transmitted. The number of available slots depended on several things. As figure 4.1 shows, different radio services are allocated to different frequency bands, a process accomplished on an ongoing basis through international collaboration.[5] The definitions of bands vary slightly between different regions of the world. In a certain band in Europe, say for amplitude-modulation (AM) broadcasting, frequencies between 526.5 kilohertz (kHz) and 1606.5 kHz are allowed, corresponding to wavelengths of approximately 570 meters to 187 meters. Depending on the "width" of the frequency slots, a greater or lesser number of transmitting stations can be accommodated. The width of the slot, called the bandwidth, is technology dependent and is based on the amount of information that must be transmitted. If the bandwidth can be small, more slots can be fitted into the band for AM broadcasting. On the other hand, the same increase in the number of transmitting slots can be achieved if the band for AM broadcasting is enlarged to the detriment of another radio service.

Hence, one form of spatiality that is relevant to the radio spectrum for broadcasting is the number of frequencies that can be allocated to the band. However, physical geography is also an important spatial aspect, because radio waves propagate through geographical space in complex and fascinating ways.

To begin with, propagation characteristics change with frequency. In particular, the range of a transmission depends on the frequency at which it is transmitted. Some frequencies provide greater range than others (which makes them more suitable for certain kinds of communication, such as long-distance, point-to-point communication at sea); others may be more appropriate for short-range communication, such as local municipal broadcasting. The frequencies chosen by the first broadcasters had a rather long range, which meant that listeners far from the transmitting station could pick up the signal. These frequencies occupied what we often call the long-wave and medium-wave bands. However, these are historically relative terms, and their definitions have changed over time and place. The wavelengths in question were medium (from approximately 200 to 600 meters) and long (from 1,000 to 2,000 meters) in relation to shorter waves, namely those below 200 meters. In the early 1920s these shorter waves were believed to be useless and were left to radio amateurs. The amateurs, however, soon discovered that the short waves also propagated well under certain conditions and were well suited to long-distance communication.

Long and medium waves follow the curvature of the Earth's surface, and their propagation can be divided into a ground wave and a sky wave. The ground wave is always present; the sky wave is more important at nighttime when the ionospheric layers around the Earth are able to reflect the waves back to Earth again. The ground wave results in a primary reception zone, which at night is complemented by a more distant secondary zone. The sizes of these zones are, of course, also determined by the power of the transmitter and the technical quality and sensitivity of the receivers. However, propagation is affected not only by the change from day to night but also by the seasons and by shifts in solar activity. For short waves these changing propagation characteristics can be dramatic.

In addition, the propagation of radio waves depends on the conductivity of the Earth's surface, which in turn affects the reach of transmitters. Central Europe has better conductivity than Northern Europe and Scandinavia. In Sweden, which covers a considerable amount of latitude, the conductivity in the middle part of the country is a tenth of that in the southern part and twice that in the northern part. High conductivity is an asset since it helps radio waves travel. In sum, propagation of radio waves is inherent in the nature of the electromagnetic spectrum and varies with geophysical properties of the space over which it propagates. How these propagation features are handled and how the resource is used, however, are not determined by physics and geography, but by technological development and by collaborative agreement.

The spatiality of the resource thus has two dimensions: the carrier wave (which has a spatial relation to other carrier waves in the radio band) and the relation between transmitters in geographical space. Both are implicated in crowding and in interference. In the case of several simultaneous appropriators of the resource (that is, several broadcasting stations operating at once), both of these dimensions must be considered. If the carrier waves of two stations are too close in frequency, interference will make the resource subtractable. However, if these stations are located far apart geographically, their respective reception zones may not overlap but may instead remain separate and thus not interfere.

From a mathematical point of view, broadcasting over a specific area, using specific transmitting powers, and having a finite number of stations to accommodate can be seen as merely a computational problem with an optimum solution. In fact, "tiling" Europe with circular transmission zones was considered early on.[6] However, the European nations are not circular, and the population is unevenly distributed, so what might look like an optimal broadcasting solution might in fact be anything but. Clearly, some areas would want to have larger circles and others smaller ones. Moreover, by the time better coordination was required, a number of transmitting stations were already on the air, and they resisted change, but at the same time they realized that coordination would benefit everyone. This process of building a moral economy is the focus of the next section.

The Establishment of Frequency Plans for Broadcasting

Broadcasting was not one of the original uses of radio.[7] Rather, wireless, as it was initially called, originally replaced telegraph communication services, which were point-to-point rather than point-to-many. The possibility of moving the wireless telegraph out to sea was realized early on. The first international meetings to organize and regulate the radio spectrum, which took place in 1903, 1906, and 1912, hence did not include broadcasting.[8] With the introduction of broadcasting in the early 1920s, there were consequently not many institutions in place that could offer guidance on how to handle issues pertaining to the rapidly increasing numbers of both broadcasters and listeners. But as "radio fever" quickly caught on both in Europe and in the United States, different solutions to the attendant problems emerged.[9] In Europe the solutions had to be transnational and had to deal not only with the challenge of many different languages but also with the sovereignty of nation-states.

In Europe, broadcasters quickly recognized the need to cooperate in managing the worsening problem of interference as broadcasting became more widespread. In 1925, a number of European broadcasting organizations joined forces to this end in the International Broadcasting Union.[10] Among the IBU's first accomplishments was a plan to overcome the "congestion in the European ether" (which was how the situation was commonly described at the time).[11] The plan was worked out in the course of two engineering conferences held in Geneva in July and September of 1925.[12] During these meetings, an accord was reached on the guiding principles for dividing the spectrum for broadcasting in Europe. A formula was developed for calculating the fraction of the spectrum to which a country was entitled on the basis of its population, area, and economic development. A big, densely populated country that had started broadcasting early received a higher fraction than a small and sparsely populated country. This fraction, however, depended on how large Europe was deemed to be, and hence defining the borders of the region was essential.[13] This, of course, affected what work had to be done to create the community of fate to which broadcasters could belong—and to whose moral economy they would have to subject themselves.

The band at issue was what was then called the medium-wave band, comprising the wavelengths between 200 and 600 meters. The number of stations that could be fitted into this band depended on how broad each frequency slot was allowed to be and how far apart the carrier waves were nominally placed. Fairly early on, it was agreed that a separation of 10 kHz was enough—leaving 98 frequencies available for allocation, since a few were already reserved for maritime radio.[14] Of these 98 frequencies, 82 were reserved as *exclusive* wavelengths on which only one station was allowed to transmit. A station that received an exclusive wavelength would, in theory, not be susceptible to disturbance from any other station, since it alone operated on that frequency. The remaining 16 wavelengths were called *common* wavelengths and were shared by several stations, which therefore could not be located too close to one another geographically. This method of sharing frequencies provided capacity for a significantly greater number of transmitting stations than would otherwise have been possible. The allocation of long waves was postponed,[15] probably since they were fewer in number and, at the same time, more attractive thanks to their excellent propagation. Including long waves in the same arrangement might have introduced insurmountable obstacles.

In this first plan, which came to be known as the Geneva plan, exclusive frequencies were allocated to all countries in the European zone, which included the British Isles and Portugal in the West and the Soviet Union along the Ural Mountains in the East, Turkey being excepted as it was not considered part of Europe. Germany, France, and Great Britain received the most generous allocations—twelve, nine, and nine wavelengths, respectively. Sweden, Spain, Italy, and western Russia each got five exclusive frequencies; Poland got four. Austria, Belgium, Finland, Holland, and Rumania each received two. Albania, Bulgaria, Denmark, Estonia, Greece, Hungary, Ireland, Latvia, Portugal, Switzerland, Lithuania and Luxemburg, and Czechoslovakia each got only one exclusive wavelength. Even countries that had no national service in 1926, when the plan came into force, were allowed to claim and receive a wavelength.

This manner of assigning frequencies is interesting in that it illustrates an idea of fair allocation and a moral economy extending even to those yet to become users. The members of the International Broadcasting Union agreed to adhere to the plan, to keep their wavelengths stable, and to adhere to rules specifying limits to transmission power (another important factor in regulating transmission zones within the plan). Of particular interest is the generous inclusion of Germany, which at the time was still left out of most international agreements. It is possible that one difference between radio frequencies and airspace was that the potential of broadcasting for military purposes was not recognized by the community and therefore was not considered a threat, at least not by the broadcasters belonging to the community. More importantly, however, the community was better off with Germany belonging to the moral economy than it would have been otherwise.

The allocation of wavelengths to those not yet in business also betrays awareness of path dependence in broadcasting arrangements—that is, the difficulty of changing a plan once it was implemented. The engineers of the International Broadcasting Union were well aware of the pace of expansion of broadcasting, wanted their solution to be sustainable in the coming years, and therefore made provisions for new members and new stations *ex ante*.

After trials, evaluations, and ratifications by the states participating in the broadcasting organizations, the Geneva plan took effect in November of 1926. Reception improved immediately. Two factors in the success of the plan were its structure and the way it was monitored. A number of engineers who had helped to devise the Geneva plan were chosen to constitute a permanent Technical Committee of the IBU, and a Technical Centre was established in Brussels. The staff at the Technical Centre measured transmitter stability and plan adherence continuously and published the results regularly.[16] These reports functioned as a kind of public shaming for bad behavior, since any violation of the plan's provisions was advertised to the rest of the community of broadcasters. A breach of the rules undermined—but by dint of publication also reinforced—the moral economy enshrined in the plan.

Adjusting to New Regulatory Hierarchies

In 1927, a much-awaited International Radiotelegraph Conference convened in Washington. It was the first such conference since 1912. Radio technology had developed rapidly during the previous 15 years, and new services (including broadcasting) had to be attended to.[17] Representatives from 80 nations attended the conference, as did a number of non-governmental organizations. The United States, the country that organized the conference, did not invite the Soviet Union, not yet having accorded it diplomatic recognition. Hence, the Soviet Union did not ratify the resulting agreement. This, of course, affected subsequent regulatory efforts in Europe, as we will see. However, the Soviet Union took part in the European preparations for the Washington conference, and in this roundabout way its suggestions were taken into consideration.

Several important agreements were reached at the Washington conference, among them an agreement to establish an International Radio Consultative Committee (which was to have a central role in standardizing procedures for years to come).[18] However, a proposal to divide the spectrum into national bands, supported by some European delegations but opposed by the United States, was not implemented. Instead, a priority rule was established on the basis of the principle of national sovereignty. It allowed any nation to start any kind of service on any frequency as long as it did not interfere with another service already in place. Moreover, it was decided that regional agreements were a permissible way to regulate frequency plans for broadcasting, which accommodated the procedure already in place for Europe.[19] Of particular importance for broadcasting in Europe, the nominal range of the medium-wave band was narrowed to 550–1,500 kHz from its previous range of 500–1,500 kHz (corresponding to wavelengths from 200 to 545 meters instead of from 200 to 600 meters). This meant that the number of possible broadcasting frequencies decreased relative to what had been available under the Geneva plan.

The Washington conference challenged the moral economy put in place by the Geneva plan in several ways. First, the newly restricted broadcasting band put additional strain on the allocation system for broadcasting frequencies in Europe. Second, the priority rule could be interpreted as undermining the Geneva plan since it sanctioned the establishment of new broadcasting stations outside of the plan as long as these disturbed no one else. However, the priority rule could also be seen to reinforce the moral economy, inasmuch as it insisted that any newly established station manage its operations so as not to interfere with the existing Geneva arrangements.

Since the broadcast band now had a smaller frequency range, the Geneva plan had to be revised to accommodate the same number of stations within a smaller space. This became possible after the International Broadcasting Union's European Conference of Wireless Engineers, held in Brussels in 1928, at which channel separation was reduced from 10 to 9 kHz. The adjustment enabled the IBU to maintain the common resource at an adequate level. At the same time, it placed higher demands on radio technology and therefore made the resource more precarious. Despite this, the number of exclusive wavelengths was increased at the cost of common wavelengths. In the 1928 Brussels plan, 90 frequencies were designated as exclusive (up from 82 in the Geneva plan) and only 10 were left as common frequencies (down from 16 in the Geneva plan). This was a more restrictive solution, since fewer common frequencies also meant less flexibility in allowing new stations to share existing allocations. Another big change was that the Soviet Union—still called "Russia (west)" in the archival material—was left out of the new plan, a major gap that would require attention in the future. Moreover, the long waves, which had been neglected in the Geneva plan, were now allocated to eight countries: Denmark, Finland, France, Germany, Great Britain, Holland, Poland, and Sweden. Long waves were attractive because they had a greater reach than medium waves and a station could therefore cover a large area. Parallel to the Brussels plan, preparations for yet another European frequency meeting were in full swing. In Washington, it had been decided that regional agreements were to be allowed, though the Washington Convention's wording on this point was vague. The Czechoslovakian delegation had announced in Washington that it would extend an invitation for a conference to draw up a regional plan, which it did.[20] The invitation by the Czechoslovakian Telegraph Administration was extended to the national post, telegraph, and telephone administrations (PTTs) of European countries. The IBU was invited to take part in the conference, and was granted the status of expert body. In one sense this decision sidelined the IBU, because it meant that the resulting plan represented an agreement between national PTTs rather than between the private or state-owned broadcasters that constituted the IBU. The resulting plan therefore had a different juridical status, since it was now an agreement between government bodies. As we shall see below, however, it is not clear that this modification of the juridical status of the plans was of any real significance for the moral economy, as agreements in any case were ratified by governments.

In fact there was little difference between the Brussels plan and the Prague plan. The one big exception was the incorporation of the numerous stations in the western Soviet Union into the plan. This was accomplished by situating them in between the already assigned frequencies, with a difference of only 4.5 kHz on either side, which drastically decreased the nominal gaps between frequencies. The Soviet Union also had several long waves registered, and Norway and Switzerland each received long-wave allocations that they had not had under the Brussels plan. Apart from these allocations, a number of stations (ten in the non-Soviet part of Europe and six in the Soviet part of Europe) were placed between the medium-wave and long-wave bands, space which was allocated to maritime and air traffic. This practice, called *derogation*, had also been agreed on in Washington, on the condition that the stations not disturb existing traffic.[21] Once again, this can be regarded as within the constraints of the moral economy in that the allocation of frequencies out of plan is tolerable, but only so long as it does not negatively affect stations already existing, whether within a plan or not.

The fact that the Brussels plan and the Prague plan were very much alike also reveals the resilience of the basic tenets of the moral economy in the face of organizational change. This had to do with the fact that, even though PTTs were now central to the planning of the European broadcasting frequencies, the personal composition of the underlying community of fate did not change greatly. This was because many individuals were actually members of this community irrespective of whether they represented PTTs or broadcasting firms. This, in turn, had at least something to do to with the fact that broadcasting in Europe was largely monopolized by governmental broadcasting services and with the fact that those involved on this level were members of a relatively small group with the common aim of making European broadcasting work.

Maintaining Achieved Order

In the 1930s, broadcasting flourished as a cultural, a social, and an economic phenomenon. The procedures put in place by the International Broadcasting Union, with its Technical Centre in Brussels, seemed to hold major problems at bay. Interference, according to measurements done in Brussels, was declining, and the stability of transmitters was improving.[22] Transmitter stability is no longer an issue today, but in the early days of broadcasting it was. Unstable transmitters were fairly common, and they threatened to corrupt the entire system as they floated in and out of their frequency slot. Not only was listening to such a station far from pleasant; it also hampered listening to stations on nearby frequencies—a neat and doubtless effective synergy between the common interest in avoiding interference and the specific interest in cementing the loyalty of listeners.

In 1932, the International Telegraph Union merged with the International Radiotelegraph Union to form the International Telecommunication Union (ITU). It was agreed that there was a need for a plan that allowed for national broadcasting for all European countries. This was hammered out in Lucerne the following year.[23] The need for a new plan stemmed from the expansion of broadcasting. By 1933 the number of broadcasting stations in Europe had increased to 257, more than double the 123 stations registered in 1926. The underlying logic of the Prague plan could no longer accommodate broadcasting at this new scale. In preparation for the Lucerne meeting, the Technical Committee of the International Broadcasting Union gathered information on existing and planned stations and their broadcasting power, confirming that the existing space wasn't adequate to meet current broadcasting needs. After much ado, 27 of the 35 countries that met in Lucerne in late spring 1933 ratified a new plan that made room for more stations. Because of the plan's necessarily greater reliance on shared frequencies, the IBU was left with the task of sorting out any problems that arose from stations operating on the same frequency.[24]

The number of frequency plans for the European zone in the interwar period is striking. On several occasions, a plan lasted only a few years before being replaced by a new one. Why were so many frequency plans needed? A big part of the answer was the permanent revolution in radio technology and the increased use of broadcasting for nation building, with all that entails. The number of stations increased, as did their importance as vehicles for information, news, entertainment, and education. Continuous revision was a way of trying to maintain a functional, equitable, and therefore sustainable international order for a resource that was proving to be very important and yet was changing fast. From this work of permanent revision, a moral economy emerged based on a common understanding of what was needed and what might be expected from each participant in the commons. This was built not just on formal documents but also on informal and personal relations,[25] which sometimes lasted for decades. The letter reproduced at the outset of this chapter is but a sample of the correspondence, which, over time, also became more personal, reinforcing a sense of belonging and community.

World War II: the Collapse of the Commons and Enduring Tension

During World War II, chaos reigned in the ether. Not only were jamming and propaganda a burden on more ordinary broadcasting, but transmission itself was sometimes compromised for lack of spare parts. Even neutral countries had a hard time maintaining normal service. In Sweden, for example, interference from military communication over the Baltic Sea disturbed transmissions over the island of Gotland, which could only really be alleviated by the expedient of wired radio. An attempt by Sweden to acquire four new vacuum tubes from England for its long-wave transmitter in the

summer of 1940 eventuated in one tube being lost when the vessel carrying it sank in the North Sea, shipment of the other three tubes to Finland,
transport by truck and train, and a mistake at a shunting yard 20 miles from
the transmitter that caused the remaining three tubes to break. Not until
1943 was another attempt made, and not until 1944 could the transmitter
be operated at full power again.[26]

Soon after the war, efforts were made to restore some order to the airwaves. The mainspring of these efforts was the convening of the International Telecommunication Conference in Atlantic City in 1947.[27] It has
been argued that the International Telecommunication Union acquired its
present form and operational style at that conference.[28] One creation of the
conference was the International Frequency Registration Board, whose mission entailed recording (and thereby validating) frequency assignments and
providing expert advice on the allowable number of stations.[29] The International Frequency Registration Board was an important element of the commons because it legitimized the rights of established users and of would-be
appropriators of the precarious resource. If a registration was accepted by
the Registration Board, it became a validated claim that restricted the rights
of others to make use of that frequency or those nearby. Also important was
the fact that both the medium-wave band and the long-wave band were
expanded. The medium-wave band was redefined as 525–1,605 kHz and
the long-wave band as 150–285 kHz. This directly augmented the number
of stations that could operate in those bands.

After the Atlantic City meetings, a regional conference for Europe was
scheduled to be held in Copenhagen in 1948. The idea of delegating regional
arrangements to regional bodies was an ITU modus operandi. Already
established in the interwar period, it continued after the war. It eased the
pressure on the international meetings and made for a smaller community
of fate advantaged by greater mutual knowledge and a longer common history. A "committee of eight" consisting of representatives from Belgium,
France, the Netherlands, the United Kingdom, Sweden, Switzerland, the
Soviet Union, and Yugoslavia met in Brussels during the spring to prepare
for the Copenhagen meeting. As in 1925, the committee polled each country about its frequency requirements.[30] Germany was not allowed to request
any frequencies; in its place, the occupying powers could request frequencies for their sectors of the country. Moreover, the IBU was not involved in
preparations for the conference, since the community of broadcasters was
also torn asunder by the ideological and economic polarization that followed the war. The IBU was unable to bridge the emerging political tension
between East and West, and its membership followed suit. It soon broke

up into a Western and an Eastern organization: the European Broadcasting Union (EBU) for the West and the Organisation International de Radiodiffusion (OIR) for the East. Even though a few countries were members of both organizations, the expert role that the IBU had played before the war was never regained by either the EBU or the OIR.[31]

One of the biggest controversies in the committee of eight was how far to separate the stations. From the original separation of 10 kHz in Geneva, the nominal separation had been decreased to 9 kHz at Lucerne and Montreux.[32] In order to limit interference, the Soviet Union and Yugoslavia wanted to go back to a separation of 10 kHz. The others wanted to maintain the separation of 9 kHz. Even though interference would decrease, a greater separation would mean fewer channels in an already overcrowded part of the spectrum. As no compromise could be reached, the committee of eight presented two proposals to the member countries. On the basis of the responses, the committee again tried to work out a mutually acceptable plan, but without success. Hence two proposals were presented to the Copenhagen meeting, which took place from June 2 through September 15, 1948.

The political situation during the years leading up to the spring and summer of 1948 is pertinent to the remaining story and argument. Tensions between the Allies and the Soviet Union mounted and their respective positions gradually hardened. Europe was divided into two, and Germany into four pieces, with Berlin equally divided and, moreover, isolated within Soviet-occupied Eastern Germany. In what became the Eastern Bloc, Soviet-friendly governments were installed. The United States responded with efforts to reach the peoples of these nations, legitimized by what became known as the Truman Doctrine in March 1947, which advocated support to countries believed to be threatened by communism. The Marshall Plan, an effort to support the now weak and war-torn economies of Europe, was presented in June 1947, but was never implemented in Eastern Europe. During the spring of 1948, the occupying Western powers moved toward forming what in 1949 came to be West Germany. On June 20, 1948, the occupying powers replaced the ruined Reichsmark with the Deutsche Mark. Four days later, the Soviet Union stopped road and rail traffic to West Berlin. The Berlin Blockade lasted for a year. This episode became the first crisis of the Cold War.

The situation in the airwaves reflected and reinforced the tension. In March 1946, the British Broadcasting Company started broadcasting in Russian, followed by the Voice of America in February of the following year. By the spring of 1948, the Soviet Union was jamming those two stations.[33]

At the same time, the United States began broadcasting in German in their zones of occupation, partly in an effort to rebuild the German broadcasting system. Likewise, a service called RIAS (Rundfunk im Amerikanischen Sektor) was established in Berlin, and it had a counterpart in the Soviet-run Radio Berlin.[34] The Copenhagen meeting opened just a day after the beginning of the Berlin Blockade, just at the outset of the propaganda war across the Iron Curtain—a war in which radio transmitters proved to be redoubtable weapons.

It would have been surprising had these circumstances not affected deliberations in the Danish capital. Nonetheless, the difficulties that came up were not only due to the increasing tensions between east and west, but also to conflicting ideas on how to use spectrum efficiently—as we will see below.

In the first week of the Copenhagen conference, six committees were appointed to carry out as many different tasks. Among these, committee no. 4 was to establish the guiding principles for a frequency plan and committee no. 5 was charged with the task of drawing up the actual plan on the basis of these principles. However, by mid July, though committee no. 4 had not yet come up with any directives, committee no. 5 commenced work anyway. A month later their efforts gained momentum, but still without any guidance from committee no. 4. After yet another month during which different frequency plans were drawn up and considered, a final plan was accepted by 25 of the 32 states present.[35]

The final plan organized both the medium-wave and the long-wave band. The separation of channels was 9 kHz, apart from a section of the medium-wave band where the separation was 8 kHz. This meant that the resource was expanded relative to the earlier period (or perhaps more densely occupied). With these separations, the long-wave band could encompass 15 channels, of which 13 were exclusive, and the medium-wave band 121 channels, of which 48 were exclusive. The plan listed frequencies for a total of 415 stations, with another three in derogation.[36] Sweden, Egypt, Luxemburg, Austria, Syria, Iceland, and Turkey abstained from signing the European Broadcasting Convention that was drawn up at Copenhagen. Lebanon was not present.[37] One reason for not accepting was for example that a specific station was not granted the frequency at which it was already operating, but rather a new one, perhaps in derogation. Interestingly enough, the entire Eastern Bloc signed the Convention, and the Soviet Union neglected to object to the separation guideline of 9 kHz, which had been a major obstacle prior to the meeting. However, the Soviet

Union had a number of other complaints about the plan, of which the status of the Baltic countries was one. In view of the way this cosmopolitan commons actually functioned in the following year, it is noteworthy that signing the Copenhagen treaty in itself was not—as we shall see—a token of acquiescence to a moral economy.

Though the overwhelming majority of 25 states signing might give the impression that Copenhagen managed to solve the problem of realigning the use of medium and long waves in Europe, that was far from the case. The problem was, of course, Germany, which at the time was not a sovereign state, and—oddly enough—not even formally represented by its occupying powers.[38] As was noted above, countries submitted their frequency requirements to the committee of eight in preparation for the Copenhagen conference. The occupying powers requested a total of 34 frequencies for occupied Germany: 15 American, 6 British, 4 French, and 9 Soviet. The United States argued that its needs were greater, since, by contrast with the other occupying powers, it had no frequencies of its own in the area. Moreover, the US had three different needs: to counter Soviet propaganda, to reach US forces (through the American Forces Network), and to rebuild civilian German broadcasting. To this last end, the US envisioned—and had already started to implement—decentralized systems that would be more impervious to a feared new totalitarian regime. Organizationally, the system resembled the locally based broadcasting system on the American continent, a system that required more frequencies for its deployment than the standard European system, thus making inordinate demands on a resource that had been scarce in Europe since the late 1920s. Hence, the US requests reflected heterodox views on spectrum economics and broadcasting organization. The rebuilding of civilian broadcasting in the other occupational zones, on the other hand, was organized according to the established European practice of centralization.[39]

The committee of eight found the US request to be "excessive," and the two proposals presented to the conference in July incorporated only a fraction of the original request.[40] During the conference, the US tried very hard to argue for more assignments but also to reach a compromise.[41] That the USSR was unwilling to meet the demands was hardly surprising, but efforts to win over the French and British were also in vain. The reasons for this were partly that France and the UK wanted to attend to their respective national needs and were anxious to reach an agreement on European frequency allocation, two goals that were closely connected. France and the UK belonged to the same community of fate and had a history of belonging to the same moral economy, in which the frequency plan for Europe as a whole was deeply rooted in the frequencies allocated to individual countries. This can be interpreted as an illustration of the fact that the US was not part of the old community of appropriators, which had functioned so well before the war, and did not share its moral economy. However, it was not clear who should be regarded as part of the community at that point, since the map of Europe had been redrawn and the old appropriators had also changed. The underlying question was what rights Germany had. Here it seems as though the US looked after German interests better than its European allies did, albeit in a wholly non-European fashion.[42] This was a challenge to the moral economy established by European broadcasters.

The final plan denied Germany any exclusive frequency in either the long-wave or the medium-wave band. Instead, each of the occupying powers was allocated two shared frequencies for its occupational zones and one extra frequency for broadcasting to its troops.[43] This was a significant cut, not only relative to actual use at the time, but even in comparison with the Lucerne plan. In response, the US stated that it would not adhere to the plan. The other occupying powers also considered the plan inadequate to their needs, and they reserved the right to make any necessary adjustments themselves if other parties rejected its terms.[44] Hence, even before the Copenhagen Conference adjourned, it was clear that the allocation regime for Germany had failed.

Ignoring the Copenhagen plan when it was put into effect in March 1950, the US began broadcasting on frequencies that were either not assigned at all or assigned to others. This resulted in severe interference for many neighboring countries and a partial breakdown of the frequency commons. In order to overcome this interference, the power of the affected station was often raised, a measure which, while it improved reception in the transmitter's primary zone, also made for yet another breach of the plan.[45] Progress could be made and some of the mutual interference avoided, even within the constraints of the plan, by *ad hoc* mutual tuning, a pragmatic way of revitalizing the moral economy. And as was evident from the letter quoted in the beginning of this article, such *ad hoc* pragmatism had been essential from the very beginning. In the chaotic postwar situation, it allowed for better reception of the Voice of America in certain parts of Europe, for example.[46] Nonetheless, the consensus was that the Copenhagen plan was a failure. In 1952, the European Broadcasting Union reported that, of the 675 stations operating within the long-wave and medium-wave bands, as many as 40 percent operated in conflict with the plan.[47] The situation appeared hopeless, and there were no efforts to reach new agreements until 1975.

Nevertheless, broadcasting continued as a service on those wave bands, and it still does in parts of Europe. Even though the European Broadcasting Union identified increasingly severe breaches of the plan and regulators agreed that it was a failure, radio communication did not break down entirely. One reason it didn't break down was improved tuning. More sensitive and stable radio receivers—the development of which was facilitated by the incorporation of transistor circuitry in the 1950s—made listening easier. But *ad hoc* bilateral agreements on frequency adjustments also made transmissions less susceptible to interference. So, despite formal dysfunctionality, the system could be made to work well enough. Essential to get around this dysfunctionality was the belief by members of the community of fate that cooperation could provide mutual benefits.

However, I believe the major reason for the system's survival was the overall reduction of demand for long-wave and medium-wave frequency slots in the 1950s. This had an exogenous origin in technological advances that made it possible for larger and larger parts of the spectrum to be used for communication services. Most salient for broadcasting was, of course, the move to frequency modulation at Very High Frequencies, which made the resource in the medium-wave and long-wave bands less scarce and hence less precarious.

Conclusion

This chapter has argued that radio frequency plans for Europe were the formal expression of a moral economy for dividing the European radio-frequency spectrum. This moral economy undergirded a cosmopolitan commons regime that, at least in the interwar period, regulated national claims to frequency slots in an equitable and sustainable way. However, in the emerging Cold War, as new actors began to violate the established moral economy, the commons began to crumble and the resource again became precarious. Why did that happen?

Let us first reflect on the status of the formal institutions, i.e., the frequency plans. The Geneva plan (1926) was negotiated by members of the International Broadcasting Union and was then ratified by the governments of the appropriating countries, which made it, *de facto*, legally binding. The same was true for the Brussels plan that followed two years later. The Prague conference (1929) was a convention of PTT administrations, which served as national representatives, something that was also true for the meetings in Brussels and Montreux.[48] In Copenhagen (1950), governments were directly represented at ministerial levels, giving the conference a plenipotentiary character. The result was again a convention to be signed, just as in Brussels in 1926. Though there is no sharp break between Geneva and Copenhagen, we do see increasing formalization of the status of the negotiators, moving from industry participants to government agency representatives to government-appointed representatives with plenipotentiary powers. Yet this "upgrading" does not seem to have made plans for long-wave and medium-wave allocation more enduring. What this suggests is that the moral economy of European spectrum management depended on factors beyond the formal agreements.

Resource supply and demand issues, which in turn were affected by technological and organizational changes, appear to be one relevant factor in accounting for the level of success of cooperative spectrum-management plans. Substantial improvements in transmission and reception technologies during the interwar years helped to control interference and increase the efficiency of spectrum use. The long-wave band was also mobilized early on, increasing the number of frequencies available. Moreover, the minimal separation between assigned frequencies was reduced, allowing more stations to occupy a given frequency range. Nevertheless, demand for frequencies grew rapidly, as is evident, for example, in the fact that the number of European broadcasting stations increased from 123 to 257 over a span of seven years, from 1926 to 1933. And expansion not only reflected a growth in the number of stations in the original appropriating countries but also included new radio nations. By 1950, spectrum expansion could no longer keep pace with the expansion of demand, but only a few years later high-frequency FM broadcasting began to alleviate this problem.

But why did the system function well before World War II and poorly thereafter? As I argued above, the behavior of the occupying powers was decisive. The war totally changed a hitherto functioning community and distorted its existing moral economy. A central player, Germany, was suddenly stripped of its rights and was not even represented in Copenhagen when the new frequency plan for postwar Europe was forged. There, the Allied forces occupying Germany showed themselves eager to protect their own interests. To complicate things even further, the United States advocated a decentralized broadcasting system that worked well enough on the American continent but which grated with the European approach of allocating frequencies among autonomous nations speaking different languages. The US approach required too many frequencies, in view of European conditions and past practices. Accorded less than it claimed to need, the US unilaterally abandoned the agreement, corrupting the moral economy and leading others—though certainly not all—to do the same.

A final question is whether the plans tell us something else about viewing this resource as a common European patrimony. One relevant point here is the way in which the plans progressively abandoned exclusive frequencies in favor of shared frequencies. The change was most notable in the medium-wave band but also impinged on the long-wave band. In the 1926 Geneva plan, 82 frequencies were exclusive, corresponding to 83 percent of the total number of channels available. In the 1950 Copenhagen plan, 48 frequencies were exclusive, which was 40 percent of the available number of channels. If one instead counts the number of stations, the statistics are even more striking. Almost 80 percent of the European stations allocated in the Copenhagen plan worked on frequencies shared with one, two, or even three other stations. The idea of exclusivity and therefore of a certain form of property gave way to a regime in which it was deemed necessary to share and yet not disturb others using the same resource, a regime where a moral economy was in fact a precondition for success. In practice this required more extensive use of bilateral negotiations and adjustments.

Even though it might be considered simplistic to attribute the chaotic situation to the great tension of the emerging Cold War, the political situation took its toll and corrupted agreements that had been in place earlier. The final solution was to move elsewhere and explore new territory in the radio spectrum. When FM broadcasting on higher frequencies proved to solve some of the conflicting needs in Europe after World War II, the pressure on the lower frequencies eased slowly and gave way to new challenges, with waves that had different features altogether. Interestingly enough, West Germany was one of the countries in which FM technology made early headway.

The shift to FM underscores a larger point: few resources have multiplied as the radio spectrum did during the last century. When one band became full and overcrowded, a move to the next one was always possible. The move from medium and long waves to very high frequencies proved sustainable, and most broadcasting in Europe today makes use of these higher frequencies. In fact, some countries have abandoned the medium-wave and long-wave bands altogether. The practice of solving congestion problems by moving upward in the spectrum cannot continue indefinitely, however. Perhaps we have now come to the limit of spectrum expansion and will have to consider again how we can best use the spectrum in common. If that is the case, analysis of the possibilities should take into account more than just markets and auctions, solutions that have been in vogue lately. It should also revisit the idea of constructing commons that are guarded by moral economies.

Notes

1. Letter from Burrows, UIR Office, Geneva to Siffer Lemoine, Swedish PTT, Stockholm, March 12, 1926. Svenska Televerket Archive (hereafter STA), Radiobyrån F VIIIa, volume 17.

2. In an authoritative review of the commons literature, the radio spectrum has been identified as one understudied area. See Elinor Ostrom, Thomas Dietz, Nives Dosak, Paul C. Stern, Susan Stronich, and Elke U. Weber, eds., *The Drama of the Commons* (National Academy Press, 2002), 477. However, the debate on the commons properties of the spectrum is not new. Harvey J. Levin touches on it in *The Invisible Resource: Use and Regulation of the Radio Spectrum* (Johns Hopkins University Press, 1971). Recently this discussion has gained momentum through the advent of new technology which could possibly transform allocation altogether. See, for example, Yochai Benkler, "Some Economics of Wireless Communications," *Harvard Journal of Law and Technology* 16, no.1 (2002): 25–83; Stuart Buck, "Replacing Spectrum Auctions with a Spectrum Commons," *Stanford Technology Law Review* 2 (2002): 1–59; Kevin D. Werbach, "Supercommons: Toward a Unified Theory of Wireless Communication," *Texas Law Review* 82 (March 2004): 863–973. Elsewhere I have argued that the institutions put in place for broadcasting in Europe in the 1920s meet Ostrom's design principles for commons. See "Technology-Dependent Commons: The Example of Frequency Spectrum for Broadcasting in Europe in the 1920s," *International Journal of the Commons* 5, no. 1 (2011): 92–109.

3. In an interesting essay, the radio historian Hugh Aitken suggested a research project with the threefold aim of studying the discovery, exploration, and allocation of the spectrum ("Freeware: The Spectrum Project," *Antenna*, April 1994: 8–10). Alas, Aitken published only one article on the subject before dying a year later: "Allocating the Spectrum: The Origins of Radio Regulation," *Technology and Culture* 35, no. 4 (1994): 686–716. To my knowledge, his suggestions have not been followed, and the study of the use and regulation of the spectrum has attracted little interest among historians. A few notable exceptions are the following: John Tomlinson, *The International Control of Radiocommunications* (J. W. Edwards, 1945); Christian Henrich-Franke, *Globale Regulierungsproblematiken in historischer Perspektive: der Fall des Funkfrequenzspektrums 1945–1988* (Nomos, 2006); Nina Wormbs, *Genom tråd och eter: Framväxten av distributionsnätet för radio och TV* (Stiftelsen Etermedierna i Sverige, 1997); Suzanne Lommers, *Europe—On Air: Interwar Projects for Radio Broadcasting* (Amsterdam University Press, 2012).

4. Levin, *The Invisible Resource*, p. 241.

5. Tomlinson, *International Control*.

6. Siffer Lemoine, "Fördelningen av de europeiska rundradiostationernas våglängder," *Tekniska Meddelanden från Kungl. Telegrafstyrelsen*, no. 10 (1926): 73–79.

7. The invention of radio (or wireless, as it was called) has been thoroughly discussed and analyzed, as has its subsequent appropriation. Writings on Marconi, his precursors, and his contemporaries are numerous. Two books by Hugh Aitken offer a good point of entry to this literature: *Syntony and Spark: The Origins of Radio* (Wiley, 1976) and *The Continuous Wave: Technology and American Radio, 1900–1932* (Princeton University Press, 1985). Also useful is Susan Douglas' monograph *Inventing American Broadcasting, 1899–1922* (Johns Hopkins University Press, 1987).

8. On the early workings of the International Radiotelegraph Union, a predecessor of the International Telecommunication Union, see George Codding, *The International Telecommunication Union: An Experiment in International Cooperation* (Geneva University, 1952).

9. For a discussion of the US case, see Aitken, "Allocating the Spectrum." For a discussion of the European case, see Jennifer Spohrer, "Radio Interference, 'National service' and the Development of European National Broadcasting Systems in the Interwar Period" (forthcoming). See also Hugh Richard Slotten, *Radio and Television Regulation: Broadcast Technology in the United States, 1920–1960* (Johns Hopkins University Press, 2000) for the US case.

10. The formation of the Union Internationale de Radiophonie (later Radiodiffusion) is retold in the organization's own history, *Twenty Years of Activity of the International Broadcasting Union* (1945), and is treated in Lommers, *Europe—On Air*, in Jennifer Spohrer, Ruling the Airwaves: Radio Luxembourg and the Origins of European National Broadcasting, 1929–1950 (PhD dissertation, Columbia University, 2008), and in Andreas Fickers, "In Search of the 'Greenwich of the Air': Techno-Political Diplomacy in European Frequency Allocation and Control (1925–1952) (under review for *History and Technology*). I have also pointed to the fact that the story has been told differently depending on perspective—see Wormbs, "'A Simple Arithmetic Task': Technopolitics in Early Frequency Planning for Broadcasting in Europe" (work in progress).

11. International Union of Broadcasting Organisations (pamphlet, 1926, UIR Archives, Geneva), p. 11.

12. On the development of the Geneva plan, see Nina Wormbs, "Standardising Early Broadcasting in Europe: A Form of Regulation," in *Bargaining Norms, Arguing Standards*, ed. J. Schueler, A. Fickers, and A. Hommels (STT Netherlands Study Centre for Technology Trends, 2008). I expand the argument and also deal with the other conferences during the 1920s in "'A Simple Arithmetic Task.'"

13. See Lommers, *Europe—On Air*.

14. Today the term *allocation* is used for the division of bands for different services, whereas *allotment* is the term used for national distribution. I am using the historical terminology.

15. "Note on the proposed plan for the allocation of wavelengths between the European Broadcasting Stations working on waves between 200m and 600m," 12/12 1925, box 94, UIR.

16. Fickers, "In Search of the 'Greenwich of the Air.'"

17. On the Washington conference, see Codding, *The International Telecommunication Union* and Tomlinson, *The International Control of Radiocommunications*.

18. Codding, *The International Telecommunication Union*, pp. 121–122.

19. Spohrer, Ruling the Airwaves, pp. 129–132; George A. Codding, *Broadcasting without Barriers* (UNESCO, 1959), pp. 90–92.

20. Spohrer, Ruling the Airwaves, p. 132; Codding, *Broadcasting without Barriers*, pp. 93–94.

21. Protocole Final de la Conférence radioélectrique européenne de Prague, 1929, Radiobyrån F IV:3.

22. Fickers, "In Search of the 'Greenwich of the Air.'"

23. Spohrer, Ruling the Airwaves, pp. 138–140; Wormbs, *Genom tråd och eter*, p. 59.

24. Codding, *Broadcasting without Barriers*, pp. 94–96.

25. Christian Henrich-Franke, "Cookies for ITU: The Role of Cultural Backgrounds and Social Practices in Standardization Processes," in *Bargaining Norms, Arguing Standards*, ed. J. Schueler, A. Fickers, and A. Hommels (STT Publications, 2008).

26. Wormbs, *Genom tråd och eter*, p. 64.

27. Much has been written on the Atlantic City conference and the resulting radio convention. The most extensive treatment that I have found is in Codding, *The International Telecommunication Union*. See also Spohrer, Ruling the Airwaves; Henrich-Franke, "Cookies for ITU." The conference was also prepared through meetings, one of the best known of which is the meeting of the five powers in Moscow in 1946. To the confusion of historians, there were actually three conferences taking place in Atlantic City that summer. First there was the administrative conference, during which new frequency plans were to be decided and changes in the radio regulation were to be made. In July, the plenipotentiary conference was held, with delegates representing the participating states and having the power to change the radio convention. There was also another administrative conference, held with the aim of establishing a plan for short-wave allocation. See Artur Onnermark, "De internationella fjärrförbindelsekonferenserna i Atlantic City 1947," *Tekniska Meddelanden från Kungl. Telegrafstyrelsen*, no.1 (1948): 1–9.

28. See Henrich-Franke, *Globale Regulierungsproblematik in historischer Perspektive*.

29. Codding, *Broadcasting without Barriers*, pp. 98–101.

30. R. Stephen Craig, "Medium-Wave Frequency Allocations in Postwar Europe: US Foreign Policy and the Copenhagen Conference of 1948," *Journal of Broadcasting and Electronic Media* 34, no. 2 (1990): 119–135.

31. On the twists and turns of these events, see Spohrer, Ruling the Airwaves; Lommers, *Europe—On Air*; Leo Wallenborn, "From IBU to EBU," *EBU Review*, January and March 1978.

32. Hans Heimbürger and Karl Väinö Tahvanainen, *Svenska telegrafverket*, volume 6: *Telefon, telegraf och radio 1946–1965* (Televerkets centralförvaltning, 1989), p. 504; Ernst Magnusson, "Sommarens radiokonferenser i Stockholm och Köpenhamn," *Tekniska Meddelanden från Kungl. Telegrafstyrelsen* no. 4 (1948): 177–186.

33. Michael Nelson, *War of the Black Heavens: The Battles of Western Broadcasting in the Cold War* (Brassey's, 1997), pp. 10, 20.

34. Craig, "Medium-Wave Frequency Allocations."

35. Magnusson, "Sommarens radiokonferenser," pp. 182–184. On the expansion of Europe with the addition of Turkey, Syria, Lebanon, and Egypt, see Lommers, *Europe—On Air*. The total exclusion of Germany is treated below.

36. Magnusson, "Sommarens radiokonferenser," p. 184.

37. Heimbürger and Tahvanainen, *Svenska telegrafverket*, volume 6; European Broadcasting Convention, Copenhagen, 15 September, 1948, pp. 34–35, Radiobyrån F IV:53.

38. European Broadcasting Convention, pp. 39–40. It is not entirely clear from the text what status the occupying countries had with respect to Germany, but France states in her reservation that the French Delegation "is not legally authorized to represent the part of Germany under French occupation." The other occupying countries make no remark in that particular respect, but it seems fair to think that the situation was the same for the UK and the USSR. The US had no right to be a signatory.

39. Craig, "Medium-Wave Frequency Allocations," pp. 122–125.

40. Ibid., pp. 124–125.

41. For a detailed discussion of the ways in which the Department of State tried to maneuver, see Craig, "Medium-Wave Frequency Allocations."

42. Craig, "Medium-Wave Frequency Allocations," pp. 121, 125–128.

43. The literature differs as to how many frequencies were actually allocated. Craig ("Medium-Wave Frequency Allocations," pp. 124, 128) says that the Van der Pol suggestion was followed, which entitled every zone to two frequencies. However, in the Convention (p. 39) it is stated that the US is not "prepared to implement any allocation Plan which envisages only one programme per zone in Germany with

minimal frequencies for the purpose." Heimbürger and Tahvanainen (*Svenska tele-grafverket*, volume 6, p. 506) state that each occupational power got three frequencies, of which one was for troop broadcasting.

44. European Broadcasting Convention, pp. 39–40.

45. Heimbürger and Tahvanainen, *Svenska telegrafverket*, volume 6.

46. Craig, "Medium-Wave Frequency Allocations," p. 130.

47. Codding, *Broadcasting without Barriers*, p. 97.

48. Codding (*Broadcasting without Barriers*, p. 93) argues that the Prague conference was "more official than previous meetings." This conclusion is based partly on the fact that the delegates represented administrations rather than companies and the fact that the USSR was present. Codding (*The International Telecommunication Union*, p. 157, note 115) regards the Lucerne plan as the first one created under the auspices of the ITU, which is simply another hierarchical level. By this standard, not even Prague, which was called by PTTs and initiated in Atlantic City, falls under the auspices of the ITU. Codding states (p. 159) that the result of the Lucerne Conference was a European Broadcasting Convention with the plan attached. However, he also states (pp. 96–97) that the Copenhagen conference was different in that, contrary to those at Prague, Lucerne, and Montreux, it was attended by representatives of countries rather than administrations. One would think that it is the resulting treaty that is of importance. In Sweden, the status of the conference hardly influenced who was sent as a representative.

5 Conflict and Cooperation: Negotiating a Transnational Hydropower Commons on the Karelian Isthmus

Kristiina Korjonen-Kuusipuro

This chapter examines the genesis of a transnational hydropower commons across a barrier of ideological division and hostility. This hydropower commons is located on the upper reaches of the 150-kilometer Vuoksi River, which runs across the Karelian Isthmus. From the date of Finland's independence from Russia in 1917[1] until 1940, the Vuoksi lay entirely within Finland's borders. Its awe-inspiring rapids and mythical forests earned it a place in the Finnish epic poem, the *Kalevala*, and hence the Vuoksi figured prominently in Finnish identity. Yet the Winter War of 1939–40 between Finland and the Soviet Union resulted in Finland losing all but 14 kilometers of the river to its foe.[2] The river had been an important source of hydropower for Finland since the late nineteenth century. But because of the new border, two out of the four Finnish hydropower plants that had been built along the Vuoksi (one unfinished) lay within the territory ceded to the USSR, and the small nation's hydropower output was reduced by 30 percent. The loss of these resources and the traumatic upheaval of the war fomented deep and long-lasting animosity. Yet the new border created a new interdependency: using the river effectively for hydropower demanded cross-border cooperation, most notably to control water levels that affected hydropower output.[3]

This chapter seeks to understand what was needed to govern the Vuoksi's newly born transnational flow, and how a cooperative framework could be built and made to succeed despite the divisions that pervaded Finnish-Soviet relations. Border areas, which have been described as spaces of flow across a dividing line,[4] form a unique environment, precarious in many ways. According to Häkli and Kaplan, cross-border cooperation may take place on various levels, ranging from the international to the local, and may involve several kinds of actors, from national governments to regional networks and institutions, to local authorities, to companies or individual actors.[5] Therefore, in analyzing Russo-Finnish cooperation to manage the Vuoksi, we must take a multi-level approach.[6]

The case of the Vuoksi offers an important perspective on transnational commons making because of the difficulties it posed. The Soviet-Finnish border was a hostile, polarized divide. In Finland, the border with the Soviet Union/Russia (which Finns, since gaining their independence in 1917, have simply called "The Border") not only marked the edge of the national state but also stood as a symbol of independence.[7] In view of the divergent historical meanings of this border for Finland and the USSR, and the many social and cultural differences between the two states, it is not surprising that negotiation processes were slow and complex. Moreover, these negotiations had to cope with fundamental asymmetries stemming from the power differential between Finland and the Soviet Union, as well as from the Vuoksi's direction of flow from Finland (upstream) to the Soviet Union (downstream). Nevertheless, and in spite of the seemingly incommensurable needs and priorities that separated the two sides, effectual bilateral extraction of hydropower from the Vuoksi demanded a unified, shared system of governance that could regulate the Vuoksi as an "organic machine," a technologically monitored and controlled energy flow system.[8]

The Vuoksi River and Its Development as a National Hydropower Resource

The Vuoksi runs from Lake Saimaa (Finland) toward Lake Ladoga (Russia). Its catchment area is 68,500 square kilometers—almost the size of the Republic of Ireland. Part of the largest lake system in Finland, the Vuoksi is the only outlet for Lake Saimaa (the fourth-largest lake in Europe, with a surface area of 4,400 square kilometers). Economically, the Vuoksi is the most important transboundary river between Finland and Russia because of its four hydropower plants—Tainionkoski (62 megawatts); Imatra (170 MW); Svetogorsk, previously known as Enso-Vallinkoski (100 MW); and Lesogorsk, previously known as Rouhiala (100 MW).[9]

The Vuoksi has something of a dual nature. Its first stretch contains rapids that make it particularly valuable for industrial hydropower. With its headwaters situated 75 meters above sea level, the water level falls more than 60 meters over the first 25 kilometers, passing through a series of rapids that became hydropower sites: Tainionkoski, at the headwaters of the river (a 6-meter fall), Imatra (an 18-meter fall), Svetogorsk/Enso-Vallinkoski (a 9-meter fall), and Lesogorsk/Rouhialankoski (a 7.5-meter fall). The lower part of the river, toward Lake Ladoga, descends in a more leisurely fashion and was thus more suited for navigation and recreational purposes than for industrialism and hydropower production.[10]

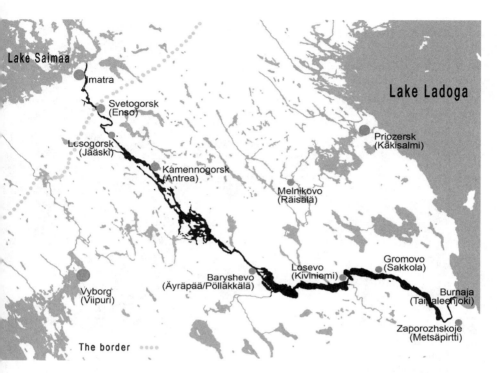

Figure 5.1
Map of the Karelian Isthmus and the Vuoksi River. Source: South-Karelian Institute,
Lappeenranta University of Technology.

Up to the early twentieth century, the Vuoksi River was seen as part of
a pristine "Finnish" landscape, but after independence in 1917 it became
a centerpiece of Finland's quest for economic independence and industrial
modernity. To earlier generations, the valley of the Vuoksi had represented
a kind of Eden—a part of the mythical Karelia, where the Finns were said
to have originated. The Vuoksi's rapids, known for their beauty and their
spectacular power, were already important tourist sites in the eighteenth
century, further cementing the area's symbolic link to Finnish identity.

After gaining independence in 1917, Finland embraced industrialism
both as a necessary support for independence and as a preferred route
to modernization. Given the state of the art and Finland's abundance of
fast-flowing watercourses, Finnish industrialization came to depend on
hydropower. The embrace of hydropower technology, dam building, and
electricity production radically transformed the Vuoksi over the ensuing

Figure 5.2
Akseli Gallen-Kallela, *Imatra in Winter*, 1893. In Finland, this painting by Akseli
Gallen-Kallela became a symbol of strong and forceful nature. Courtesy of Finnish
National Gallery.

decades. At first the power of the river was utilized locally as industries were
established near the sources of hydropower. But the growth of electricity
networks increasingly made the Vuoksi's power available nationwide.

A complex relationship between private ownership and national control shaped the evolution of hydroelectric power along the Vuoksi. Before
independence, the Finnish Senate repeatedly rejected proposals to tap
the Vuoksi's hydropower that were perceived as too eagerly serving non-
Finnish interests—notably projects aimed at transferring electric power
beyond Finland to nearby Saint Petersburg.[11] With independence, the new
Finnish state established a hydropower committee (Koskivoimakomitea)
that was tasked with developing plans to utilize the nation's hydropower
resources.[12] Yet Finnish hydropower development also involved private
interests. According to Finnish law, an individual who owned land along
a riverbank also owned the right to use the river's power at that location,
e.g. by installing hydraulic machinery where the river abutted the property. However, the rapids along the Vuoksi were initially deemed to be the

communal property of the adjacent farming communities. This provision made it difficult but not impossible to establish private hydropower plants. Such undertakings required a "time-wasting and troublesome process of acquiring all the shares of water rights for the ownership or use of the developer."[13] Two large Vuoksi rapids (Rouhialankoski and Ollikkalankoski) came to be privately owned in this way by a power company that built the Rouhiala hydropower plant in collaboration with a consortium of six industrial companies, completing it in 1937. The state viewed this private project with some suspicion, granting approval only after several years of debate and evidently only because the state was a majority owner of one of the companies involved (Enso-Gutzeit).[14]

The remaining Vuoksi power plants were partly or entirely owned by the Finnish state. Imatra, built from 1922 to 1929 and the most important of the Vuoksi hydropower plants, was owned and operated by the state.[15] At the time it was put into operation, Imatra was one of the biggest hydropower plants in Europe, and Imatran Voima, the government-owned power company that managed it, became a symbol of progress and national pride.[16] The state of Finland was indirectly involved in the building of two other hydropower plants, Tainionkoski and Enso-Vallinkoski, not only because it had to approve construction of the plants but also because it owned a majority share in the companies that built and operated them: Tornator and Enso-Gutzeit. Tornator began construction of the Tainionkoski plant in 1921, after completion of a dam at the site where the power plant would be built. Enso initiated construction of a further power plant, Enso-Vallinkoski, in 1938, but the project ground to a halt during the Winter War.[17] At first privately owned, Enso became majority owned by the Finnish state in 1918.[18] Thus, through its regulatory power as well as through its ownership of company shares, the Finnish state's role and interests in hydropower development became predominant. Indeed, it would be fair to say that hydroelectric power in Finland was mostly nationalized, even though state control was sometimes exercised indirectly and in cooperation with private enterprises.[19]

Though full of promise for Finnish national development, the goal of harnessing Vuoksi hydropower raised two distinct problems. The first was technical and operational: the power plants were built so close together that altering the water level at one plant affected the efficiency and power output of the next plant upstream. The amount of energy that can be extracted at any one point on a river is proportional to the available "head" (the difference in the river's height just upstream of the power plant and just downstream of it). "Head" can be manipulated at each power plant by using a moveable weir to raise the water level upstream of the plant to increase the potential energy. But when power plants are very close together, raising the water level above a downstream plant also raises the water level below the next plant upriver, hence reducing its total head and its potential power output. The Vuoksi power plants thus formed an interlocked system that demanded an overarching system of regulation in order to function optimally for all parties. In practice this meant that the operators of the plants had to agree on water levels and on sharing of information in order for the network as a whole to remain maximally productive. When the state of Finland had a majority interest in all the plants, achieving this kind of overarching regulation was not a problem, but it became a significant issue when two of the plants became the property of the Soviet Union.

The second problem was a security risk resulting from the proximity of Vuoksi hydropower installations to the Soviet border. Finns were proud of their industrial modernization. During the 1930s, the upper Vuoksi valley was presented to foreign visitors as the "Finnish Ruhr" and was touted as American-style development. The region seemed to have a glorious future. Yet the Finnish Defense Forces considered it dangerously close to the Russian border, and their fears proved to be well founded. Indeed, the expansion of Finnish hydropower projects made the upper Vuoksi valley more attractive as a target of conquest.

From National to Transnational Resource, 1940–1945

After the Winter War and the ceding of two hydropower plants to the Soviet Union as a result of the 1940 Moscow Peace Treaty, the first talks on the management of the newly transnational space were held in the context of thorny border negotiations. These discussions initially concerned a problem of disputed ownership of the important Enso industrial area along the Vuoksi, an industrial complex and community that included mills, factories, offices, and workers' residential buildings.[20] The dispute was caused by carelessness. The new border had originally been drawn on a map with a thick pen. The thick line passed directly through the Enso industrial area, and it was not clear on which side of the border the mills, factories, offices, and housing were supposed to belong. Naturally, because of the economic importance of the complex, both Finland and the Soviet Union wanted it to be on their side of the borderline. Finnish negotiators claimed the Enso industrial complex and community on the basis of the desirability of keeping economically integrated areas together. The Soviet side agreed on the general principle, but wanted it to be applied in the opposite way: they demanded ownership of the entire Enso industrial area.

The Enso dispute shows that hydropower was an important issue in the Soviet-Finnish border talks, and the location of the new border made further negotiations on this issue inevitable. Early discussions were confusing and vague, and the atmosphere was hostile.[21] One incident, recounted below, shows how different kinds of issues were connected to the negotiation processes and how volatile the situation was after the war. In March of 1940, Finnish authorities received a telegram from Finnish workers at the Rouhiala hydropower plant stating that Soviet soldiers who had occupied the plant were holding them hostage. The Soviet Union wanted to keep the new hydropower plant operating because the nearby city of Leningrad desperately needed more energy.[22] Yet there were not enough skilled Soviet workers available to run the plant. Finnish workers were needed to secure an uninterrupted flow of power. In order to defuse tension and facilitate the ongoing border negotiations, Finnish authorities persuaded Rouhiala's chief engineer and ten of his workers to remain in what had now become a part of the Soviet Union to keep the plant running. The Soviet Foreign Minister, Vjatšeslav Molotov, guaranteed that the Finnish workers would be allowed to go home as soon as new Soviet workers were ready to take over.[23]

The last of the hydropower plants in the Vuoksi valley, Enso-Vallinkoski, was under construction at the time the Enso industrial area was ceded to the Soviet Union, and Soviet engineers and builders had to complete it. This raised the difficult question of getting hold of the construction plans. The plans were owned by Enso-Gutzeit, which refused to hand them over to the Soviet Union on the ground that it was a private Finnish company, even though partly owned by the state.[24]

Joint water management also became an issue in 1940. The Soviets announced that they wanted to raise the water level in their part of the Vuoksi in order to gain more power. That would entail great losses for Finland's own hydropower production. It was clear that each country wanted to maximize its own benefits. The conflicting desires led to a heated debate about the ownership of the water in the Vuoksi, and about possible compensation for costs incurred by Finland if it were to lose power because of Soviet actions.

The mutual desire to tap the Vuoksi's hydropower potential thus brought the two countries into a situation of forced interdependence. In international politics, *interdependence* refers to circumstances characterized by reciprocal effects among countries. In this particular situation, Finland and the Soviet Union potentially suffered what Keohane and Nye have termed "reciprocal (although not necessarily symmetrical) costly effects of transactions."[25] There was no easy way out of the dilemma caused by the one-sided decision of the Soviet Union to raise the water level in its section of the river. Rejecting the Vuoksi's immense industrial and hydropower potential would have been too costly for both countries, yet it became clear that interdependence did not automatically lead to mutual benefits.[26] The question was whether Finland and the Soviet Union could overcome this impasse and find a sustainable and mutually beneficial solution for water management. Unfortunately, as a new war began in 1941 (the Continuation War), Finland and the Soviet Union did not come to terms on how to deal with the Vuoksi, nor did the war enable Finland to reclaim any of the territory it had lost. In the end, the border remained where it had been set in 1940. And by 1945, the entire river valley was in sad shape after the two wars. The Karelian Isthmus had been a major battle scene, and the Vuoksi itself was part of the defensive Mannerheim Line.[27] Final agreement on the border meant that hundreds of thousands of Finns from the Karelian Isthmus had to relocate to other parts of Finland as the Soviets resettled the region with residents from central parts of the Soviet Union. To top it all off, Finland had to pay crippling war reparations.

The Finns' experience of war with the Soviet Union, and their loss of symbolically and economically important territory, can only be described as cultural trauma. Cultural trauma occurs when members of a group feel that they have been subjected to a shocking, overwhelming event that leaves an indelible effect on their group consciousness, marking their memories forever and changing their future identity in fundamental and irrevocable ways. Cultural trauma is a threat to the existence and to the future of a collectivity, and it is sometimes associated with vengeful behaviors.[28] An informative example of such behavior occurred when upstream Finns dumped great quantities of wood and paper refuse into the Vuoksi. Not surprisingly, this irritated downstream residents. Soviet authorities lodged official complaints and demanded compensation. The head of the Finnish local police in Imatra accordingly issued a warning that no litter should be thrown into the river, and officers were detailed to patrol the riverbanks and monitor the condition of the water. Although that measure seems to have stopped the littering, it did not erase people's negative feelings toward the Soviet Union. The locals turned their backs on the border, and the river on the other side of the border disappeared from the Finnish public imagination.

The Period of Selfish Profit Seeking, 1945–1959

During the 1940s and the 1950s, animosity and mistrust held sway and a jointly managed transnational hydropower commons remained a distant prospect.[29] Yet the importance of the issue for both countries compelled them to attempt negotiations immediately after the wars. A primary concern for Finns was regulation of the water level of Lake Saimaa, which was possible only by means of the Vuoksi. The Finns wanted to protect Lake Saimaa riparians from damage due to periodic flooding and to balance this aim with the needs of power production and navigation. Regulating the level of Lake Saimaa was not a new issue; the possibility had been explored in Finland since the 1920s, but implementation had been postponed by the wars.[30] Downriver, however, the Soviet Union wanted primarily to maintain ideal conditions for hydropower production: a constant water level with minimal changes in the river's flow. Conflicting interests rooted in the river's geomorphology, directional flow, multiple uses, and widely varying flow levels thus made it difficult for Finland and the Soviet Union to reach agreement on a common plan to regulate the Vuoksi. The calculus of costs and benefits differed for each side, and there was no easy way to develop a common system of cost and benefit calculations that would be agreeable to both parties.

In 1949, the Finns rebuilt the Tainionkoski dam (originally constructed in 1921) to enable it to be used for regulation of the level of Lake Saimaa.[31] The Tainionkoski site was strategically important because it was situated close to the head of the river and because the Tainionkoski dam, after its reconstruction in 1949, blocked the whole span of the river. The reconstruction made it possible to impede the Vuoksi's flow, which would cause the level of Lake Saimaa to rise and would simultaneously decrease the flow to downstream power plants.

A year before the dam's rebuilding, the Soviet Union had proposed an agreement for joint regulation of the river, but Finland had rejected it because the Soviet Union had refused to pay $8 million of compensation demanded by Finland, including rent for use of part of the Vallinkoski power plant's potential on the Finnish side of the border ($5 million), damages to Finland caused by damming the waters of the Vuoksi on the Soviet side ($900,000), disturbance of the Imatra hydropower plant ($300,000), and compensation for construction work done on the Finnish side of the border in order to regulate Lake Saimaa ($1.5 million).[32] Finland's rejection stopped all negotiations. It was in this context, in early 1949, that Finland went ahead with the Tainionkoski dam reconstruction and started to regulate Lake Saimaa's water level without negotiating with the Soviet Union. Finland merely provided advance notice of its planned discharges once a month. The discharges, carried out partly in response to large natural variations in the Vuoksi's flow, generally decreased during the summer months and increased during winter months.

It is important to view negotiations over the Vuoksi not only with respect to local issues but also within the larger context of the structure of foreign relations between Finland and the Soviet Union and the unique relationship the two countries established during the Cold War.[33] The "official" relationship was defined by the Agreement of Friendship, Cooperation, and Mutual Assistance (also known to Finns as the YYA treaty) in 1948. This treaty helped to stabilize Finnish-Soviet relations. It committed both parties to respect the existing border and to guaranteeing they would not engage in or condone any activities that threatened either nation.[34] The underlying relationship between the two countries naturally had complexities not captured by treaty formalities. In the West, the term "Finlandization," which came into use in the 1960s, evoked the way power inequalities affected Finnish-Soviet relations. For Finland, maintaining good relations with the Soviet Union was seen as "a must" and as intelligent politics. For its own security, Finland needed to guarantee a peaceful relationship with its mighty neighbor.[35]

Finnish-Russian relations were also influenced by the increasing porosity of the Iron Curtain: information and knowledge increasingly passed through it along several paths, and a growing range of interactions affected thinking on both sides.[36] Fresh ideas together with new connections and forms of cooperation helped push both Finland and the Soviet Union to see and pursue their mutual benefit in new ways. For example, after Stalin's death in 1953, the focus of the Cold War shifted to scientific and technical competition between the East and the West across multiple areas and disciplines—biology, computer science, mining, space exploration, and so on. Closer scientific-technical ties across the Iron Curtain were fundamental in order to succeed in this competition, and the Kremlin's political elite developed a more open strategy in this regard. Finland became an important intermediary between the Soviet Union and the rest of the West, and on August 16, 1955 the two states signed an Agreement on Finnish-Soviet Scientific-Technical Cooperation.[37]

Despite political developments that assisted mutual understanding, however, this period was characterized by the fact that both countries wanted to gain maximum benefit from Vuoksi hydropower whatever the cost. On the

Finnish side, this attitude was evident in the reconstruction of the Tanion-koski dam. On the Soviet side, an analogous attitude manifested itself when the Soviet power company Lenenergo unilaterally raised the water level above the Enso-Vallinkoski plant in order to increase its power output. That action decreased the head and thus diminished the hydropower potential of the Imatra plant on the Finnish side of the border. In Finland, the power company Imatran Voima tried to compensate for this loss of power in several ways (for example, by installing a new turbine at the plant), but the deficit was not entirely overcome. Between 1950 and 1953, Imatran Voima also reconstructed part of the Vuoksi in order to lower the water level below Imatra and thereby increase the plant's energy output. A huge excavator, a Marion 7600, was transported to the Vuoksi; islands were moved and the river rerouted. This expensive project successfully increased hydropower production in Finland, yet it did not entirely compensate for the losses caused by the Soviets' actions.[38]

At that point, mutual reciprocity and equitable sharing of a common power resource were still just distant dreams. In practice, each party seemed to feel that the other was profiting at its expense. Both parties were reluctant to work together for common goals or to make compromises needed to maximize the river's hydropower production overall. Each tried to solve its problems within its own national framework.

Building New Institutional Foundations for Cooperation and Reciprocity, 1960–1990

After the 1950s, Finland and the Soviet Union began to negotiate cooperatively to develop an integrated regulatory regime for the frontier waters they shared.[39] Beyond the increasing porosity of the Iron Curtain, there were two more specific reasons for the shift to a more conciliatory approach. First, Finland's prime minister (and future president) Urho Kekkonen met with the Soviet leader Nikita Khrushchev in 1955, and the two developed an immediate rapport that evolved into a deep friendship, which in turn helped launch a broader pattern of Finnish-Soviet rapprochement.[40] The relationship between Kekkonen and Khrushchev helped buttress a new political doctrine, the idea of peaceful coexistence, which served as an ideological basis for the opening up of the Soviet Union toward the West.[41] Second, Finland wanted negotiations for the regulation of Lake Saimaa to proceed because these negotiations were connected to efforts to increase Finland's overall energy production. During the 1940s and the

1950s Finland had built a number of big hydropower plants in two northern rivers, the Oulujoki and the Kemijoki—one at Isohaara (1949), one at Merikoski (1954), one at Pälli (1954), one at Nuojua (1955), one at Utanen (1957), one at Petäjäskoski (1957), and one at Pirttikoski (1959). Although Lake Saimaa was not directly connected to these rivers, it played a backup role in their management because it served as a massive water reservoir for the entire Finnish water system.[42]

The first breakthroughs occurred in 1959 and 1960. In 1959, Finland heralded the shift toward cooperation by terminating its unilateral system of flow regulation, which the Soviet authorities had always opposed.[43] The following year, Finland and the Soviet Union finalized a general agreement concerning the "Regime of the Finnish-Soviet State Frontier and the Procedure for the Settlement of Frontier Incidents." Part II of this agreement pertained to the governance of frontier watercourses. It defined frontier watercourses in a rather simple way, and it set rules for navigating in frontier water areas. It included a provision to keep frontier waters clean by prohibiting pollution and other kinds of deliberate damage. The treaty also established rules for regular exchange of information between the two parties concerning water levels, water volumes, ice conditions, and any other factor needed to govern frontier waters in a mutually advantageous fashion.[44] Even though this agreement was a huge step forward and offered some clear guidelines for both countries, a more detailed agreement was still needed. In particular, the treaty did not establish rules of cooperation for hydropower production.

In February of 1961, Finland's Ministry of Trade and Industry took a big step toward greater cooperation when it appointed a committee to draft a bilateral agreement on the management of transboundary waters. Its members, Eero Manner, Simo Jaatinen, Aarno Karhilo,[45] and Pentti Saarikko, were all either engineers or jurists. In addition to reviewing written sources, the committee based its work on findings from a field trip to the Saimaa area and the hydropower plants.[46] The committee carefully examined the local circumstances of all frontier waters between Finland and the Soviet Union. It also reviewed international precedents for transboundary water-management agreements and based its draft treaty partly on earlier examples.[47] This approach was intended to insure that any Finno-Soviet agreement would not conflict with accepted international jurisprudence. In drafting the agreement, the Finnish party drew inspiration in particular from transboundary water-management agreements made between the Soviet Union and Czechoslovakia, between Finland and Norway, and between Norway and the Soviet Union.[48]

The committee submitted its final report and a draft treaty in December of 1962. The Soviet Union accepted this as a point of departure and declared itself ready to begin negotiations in Helsinki in February 1964. The final stage of the negotiations took only three weeks, and the issues negotiated were hydroelectric power, flood control, fisheries, and water pollution. On April 24, 1964, the Agreement between the Republic of Finland and the Union of the Soviet Socialist Republics Concerning Frontier Watercourses was signed in Helsinki. It was soon followed by the establishment of a joint Finnish-Soviet Commission on the Utilization of Frontier Watercourses.[49]

The 1964 treaty applied to all Finnish-Soviet transboundary watercourses. The decision to pursue an agreement of this scope followed discussion and debate about the diversity of transboundary watercourses and the most appropriate level at which to frame such agreements. The working group preferred a more encompassing agreement because they felt that local agreements would be more likely to be based on circumstances that might change rather quickly, which might in turn threaten long-term cooperation. They believed that a more encompassing framework for cooperation would bring greater stability to Soviet-Finnish relations in the realm of watercourse management.[50]

The 1964 agreement provided a foundation for the handling of a variety of issues concerning the management of surface waters. The first chapter defined terms and set out the guiding principles. It defined transboundary watercourses and prohibited measures that would harm the other party's waters, fisheries, land, structures, or other property. It also mandated that a party causing harm, damage, or losses pay compensation. Both parties were obliged to keep main fairways free for transport, for the floating of timber, and for the movement of fish. Interestingly, this provision echoed the traditional rule known as the "king's vein," which included a definition of the main fairway as one-third of the width of the watercourse at the deepest point at mean water level.[51] The fourth article of the first chapter stipulated that the signatories would jointly monitor the water quality of the frontier watercourses. The second chapter established procedural formalities and defined the rules for future cooperation. The third chapter concerned logging and water transport (which were not immediately relevant to the Vuoksi), and the fourth fish stocks and fisheries. The last chapter dealt largely with adjudicating conflicts.[52] At first glance, the agreement appeared to say very little about the utilization of hydropower. However, if we look at the negotiators and the parties that were influential in the background on the Finnish side, we see the names of familiar hydropower or industrial companies, among them Imatran Voima and Enso-Gutzeit. In similar fashion, the Soviet Union's delegation was intertwined with Lenenergo and with the Hydraulic Structures Planning Institute joint-stock company Gidroproekt-Institute.

The 1964 agreement made a conceptual shift toward an innovative water regime that treated frontier waters as a shared resource to be managed for mutual benefit.[53] After the signing of the agreement, a permanent Finnish-Soviet Commission on the Utilization of Frontier Watercourses was established. Each state appointed three members and three alternates to the commission, and both states agreed to make experts available to the commission. Decisions had to be unanimous and were then binding on both sides. The agreement provided a three-stage procedure to solve disagreements. Disagreements at the first stage were to be settled by the commission. If that proved impossible, the conflict would move to the second stage and a joint board would be established to resolve the dispute. If that failed to produce an agreement, the differences would be settled at a still higher level, through diplomatic channels. So far no dispute has moved up to the third stage.[54]

The commission's mandate was to examine and deal with matters concerning utilization of all frontier waters between Finland and the Soviet Union. It first met in 1966, and its first task was to make a comprehensive list of transboundary waters between Finland and the Soviet Union. That task was accomplished in 1971, and 448 transboundary waters were identified.[55]

The commission has met once a year since 1966, and between meetings the work has been carried on by four scientific-technical thematic working groups (an integrated water management group, a water protection group, a fisheries group, and a border guards group) and by an umbrella group consisting of the chairmen of all the working groups and the Russian and Finnish chairmen of the commission. The annual meeting is the formal decision-making body. The members of the commission are Russian and Finnish authorities, representatives of hydropower companies, and experts on water issues. The border guards of both countries also contribute to the commission's efficiency. They act as messengers, but they also provide assistance in fulfillment of some practical tasks, such as taking samples for the purpose of monitoring water quality.[56]

The water quality of the Vuoksi has been monitored since the establishment of the commission in the 1960s. The Vuoksi's catchment area is densely populated, and the heaviest polluters are the towns and industrial plants located in the area. Phosphorous and nitrogen loads in the industrial effluents and agricultural pollution cause eutrophication of the water system. The water quality deteriorates in the downstream direction. Yet the quality of the water in the Vuoksi has improved as a result of transbound-ary activities. Over the years, the transboundary commission has discussed issues concerning municipal wastewater purification plants and wastewater purification for industrial plants. Monitoring has increased knowledge and has helped the two sides to make common plans to improve water quality in the future.[57]

Even though the transboundary commission created a hole in the Iron Curtain through which communication was possible, in practice cross-border activities remained difficult. Miscommunication was frequent. At the most basic level, this was due to language barriers: Finns did not speak Russian and Russians did not speak Finnish. Moreover, owing to the cul-tural traumas of earlier Russification and war, Finns were unwilling to learn Russian. And because of strict Soviet ethnic policies, those citizens on the Soviet side who were able to speak Finnish (perhaps because they were of Karelian or Ingrian origin) could not do so publicly.[58]

Communication, especially face-to-face communication, is an effective means of improving cooperation, but it requires time, effort, and commit-ment.[59] In the case of the Vuoksi, difficulties in communication were visible at the local level, in efforts to coordinate flows through the hydropower plants, but also in the work of the commission, which always required the mediation of interpreters. At the local level, technological and orga-nizational changes improved communication. Telephone service between Imatra (in Finland) and Svetogorsk (in the Soviet Union) was initiated in 1964, but in practice telephone communication was difficult because of the language barrier. In order to overcome this problem, workers at the Imatra power plant developed a specific code language for situations that required quick responses when interpreters were not available. The code language consisted of simple sentences that were transformed into numbers and Rus-sian names. For example, the sentence that in English would be "Please receive a telephone message today at 11 o'clock" was "adin Tamara"[60] in the code language. Workers in Finland also invested in learning Russian, but the code language worked better and made it possible to communicate more quickly and reliably by telephone. Teleprinters became an important means of communication in 1979, when a 400-kilovolt power line between Finland and the Soviet Union was opened and teleprinters were connected to it. The teleprinters overcame acute language problems and were easy to use.[61]

One of the commission's most important practical achievements, after years of discussion, negotiation, and preparatory research, was the Discharge Rule of Lake Saimaa and the Vuoksi River. The origins of this rule go back to a 1973 Russian proposal to develop a system of continuous regulation of the Vuoksi's flow. Because of the many complexities in reaching a common system of flow regulation agreeable to both sides, the proposal took many years of study, debate, and negotiation to bear fruit.[62] Not until 1989 did the two sides reach a mutually agreeable solution. A new treaty signed in October of that year took effect in 1991. It stipulated that the Vuoksi was to be left to its "natural flow" unless water levels went above or below a "normal zone," defined as being within 50 centimeters above or below the mean level for each season. If Lake Saimaa's water level rose above this zone, discharge volumes from the lake were to be increased; if the water level was expected to fall below the normal zone, the volume of discharge would be reduced.[63] The adoption of this rule helped to prevent severe flooding. At the same time, it outlined a financial compensation procedure for losses of energy output suffered by Russia as a result of altered discharge levels. If the discharges from Lake Saimaa proved exceptional, Finland had to compensate Russia for the concomitant loss of energy output suffered by its power plants. Such occurrences have been rare. Low water levels for Lake Saimaa were raised in 2002 and 2003, and high water levels required special discharges in 1992, 1995, 1998 and 2005. Despite having to provide compensation when unusual discharges occurred, Finnish authorities consider the system helpful and flexible. Some Finns have expressed strong opposition to the idea of having to compensate Russian power companies for power losses incurred to prevent flooding in Finland.[64] Yet it has been estimated that for every euro paid by Finland in compensation for Russian power losses, Finns have saved ten euros in averted flood damages.[65]

Overall, the period from the 1960s to the end of the 1980s can be described as a constructive one for building a transnational (in this case a bi-national) system of governance for the Vuoksi. One factor that helped raise the level of cooperation was a growing sense of trust that developed within the commission. Individual members served on the transboundary commission for many years, which helped to build a greater sense of mutual understanding and respect, commitment to a shared mission, and personal friendship. Simo Jaatinen, who served as chairman of the Finnish part for more than twenty years, has written that personal relationships developed through many field trips the members of the commission took together. "Because of these trips," he has noted, "it has become easier to look at things from the other side as well. The Finns have learned how things are managed in nations with strong centralized governance."[66] The Russian chairman, K. S. Kornev, has remarked that "frequent unofficial meetings with Finnish colleagues have made open exchange of views possible, and this has increased mutual understanding."[67] The positive

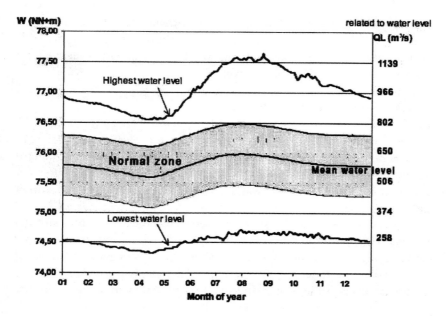

Figure 5.3
Graph illustrating the principle of the Discharge Rule of Lake Saimaa and the Vuoksi River. Finnish Environment Institute SYKE (modified by the author; original obtained from http://www.rajavesikomissio.fi).

atmosphere persisted despite changes in personnel within the commission. Greater trust also developed through the many positive outcomes of the commission's formal and informal meetings. Repeated demonstrations of the ability to solve different kinds of problems quickly and successfully created a mutual sense of accomplishment and increased the participants' commitment to the organization's shared mission, which in turn strengthened mutual understanding at the transboundary institutional level as well as the local level. In practice, all members sensed that they were working for mutual benefit, rather than merely pushing their own interests.[68]

The importance of the establishment of a solid institutional foundation to implement and oversee common rules for governance of the Vuoksi watershed cannot be overemphasized. Elinor Ostrom and other commons scholars have pointed out the importance of the role played by institutions in successful and enduring governance of shared resources,[69] and the case of the Vuoksi lends further support to this hypothesis. The case of the Vuoksi shows that appropriately designed, shared institutional frameworks helped overcome a deep tradition of hostility, mistrust, and divergent thinking that, moreover, involved a flow resource with upstream-vs.-downstream inequities and asymmetries. And the case of the Vuoksi shows that such an institutional framework simultaneously improved relations at the local level as well as the international level.

Cooperation for Mutual Benefit since 1991

In 1991, when the Soviet Union collapsed, Russia became its assignee in matters pertaining to relations with Finland, such as the 1964 transboundary waters treaty. Among the many changes that occurred on the Russian side as a result of the breakup was a growing debate over the ownership of natural resources. The first conflicts on this score were not between (potential) private owners and the state, but between the federal center and the regions. Gradually, however, advocates of private ownership of natural resources also became more vocal, with some consequences for the Vuoksi.[70] Vuoksi waters in Russia are still owned by the state, but private ownership of land is now possible. This change has become visible in the new summer houses that are continuously being built on the river's banks.[71] And both of the Vuoksi hydropower plants located in Russia are now owned by the Lenenergo power company, which was privatized in 1992.[72]

Despite profound political and economic changes in Russia, however, the regime of transboundary governance of the Vuoksi remained intact, and Russian-Finnish relations concerning the Vuoksi continued to improve. Collaboration became more frequent. Finnish workers visited Russian hydropower plants more often and vice versa. And one disagreement that extended back to the 1980s was revisited and resolved. "Until the 1990s," one Finnish insider commented, "we had a sort of quarrel with the Soviets about water levels: they said we keep the water level too high and we said they keep the water level too low. In 1993 the hydropower companies and the commission together decided that this quarrel must end, and we stopped all power plants except the first one for eight hours and then started them all again. In this way we could justify the water levels in all power plants. We even found out that on the Soviet side they had misaligned their water-level instrument and it was because of this they had wrong measuring results."[73] For years these faulty measurements had raised questions and doubts at the Imatra power plant in Finland. Since the 1980s they had observed a decline in their plant's efficiency (i.e., its power output relative to the available head of water). This diminished efficiency suggested that water levels at the downstream Russian plant differed from the levels being reported, but it was not until all the Russian and Finnish plants were stopped simultaneously in 1993 that the inadvertent cause of the problem was discovered.

The collapse of the Soviet Union also had another major effect on Vuoksi governance: it opened the door to Finland's becoming a member of the European Union. When that occurred, in 1995, the Vuoksi, now straddling an external border of the EU, became not merely an issue of Finnish-Russian relations but also an issue of EU-Russian relations. Membership in the EU brought new rules, including the Water Framework Directive, but it also brought new financing possibilities for cooperative projects. In the Vuoksi area, the Russia-Finland commission launched two separate projects involving local authorities from both countries. The VIVATVUOK-SIA project of 2001–2003 investigated the sustainable use of the Vuoksi, focusing particularly on its shore areas. It also examined negative effects of hydropower production, such as erosion. The VUOKSIAGAIN project of 2004–2006 encouraged local actors such as the Regional Council of South Karelia, the South East Finland Regional Environmental Centre, and the Priozersk District Administration to work together in new ways in order to strengthen protection of the Vuoksi as a common resource and a holistic ecosystem. One of the goals of the VUOKSIAGAIN project was to jointly establish a protected zone along the banks of the Vuoksi, and to restrict home construction and the use of pesticides and fertilizers inside that zone so that flooding would not add toxic pollutants to the river or damage property built too close to its banks. In practice these projects were examples of the "neighboring area cooperation" encouraged and supported by the EU; they were also examples of how joint management of a shared resource, pursued over decades by high-level authorities, can encourage cooperation at multiple levels.[74] As results of these two projects, the idea of common regional planning has been developed further, new collaborative networks have been established involving increased public participation, and new plans have been established for long-term environmental protection of the Vuoksi and its shorelands.[75]

Conclusion

The first steps in transboundary water interaction between Finland and the Soviet Union in the Vuoksi valley brought both conflict and reconciliation. At the outset, the idea of sharing water and cooperating seemed to be an insurmountable challenge because of the trauma of the war. However, both sides needed the Vuoksi's hydropower, and authorities in both countries felt the need to cooperate despite deep feelings of mistrust. A new commitment toward greater collaboration took root in the 1960s, and the results have proved successful. The 1964 treaty emphasized the building of a long-term, equitable relationship and an institutional structure to encourage systems of joint governance for all Finnish-Russian transboundary water resources, rather than a solution that required separate agreements and institutions for each case. Admittedly, this agreement constituted a top-down management system with a central role for the nation-states and for national interests. And the agreement was not tailored to the utilization of hydropower, so additional agreements had to be worked out in the case of the Vuoksi. Yet the 1964 treaty and its institutions have, on balance, operated successfully for nearly half a century, and indeed have promoted deeper cooperation, as shown by the important Lake Saimaa discharge rule. Joint governance has helped the two sides optimize the benefits each side gains from the Vuoksi hydropower commons, including protection against flooding, maximizing overall power output, financial compensation for power losses, improved water quality, and better environmental protection.

Yet differences between the Finnish and the Russian cultural, political, and social systems affected transnational water management. Vladimir Kotov has argued that Russian water management is regulated not only by law but also by the behavioral norms and informal practices of the Russian system. Kotov sees the conflict between informal practices and formal regulations as one of the main characteristics of the Russian water-management system.[76] In the case of the Vuoksi, these cultural differences can also be seen in negotiation styles. The Russians tended to be more formal and hierarchical than the Finns. However, in the case of the Russian-Finnish Transboundary Waters Commission, the generally long terms of service of its members, and their shared professional commitment, helped overcome differing cultural norms. Commission members came to know one another in both official and unofficial settings, and relationships of trust developed. This trust has both interpersonal and institutional dimensions.[77] Members of the commission not only have come to trust one another; they also trust the institutional framework of the cooperation, as witnessed by the histories they have written.

Participants in Finnish-Soviet/Russian water cooperation have consistently emphasized the smoothness, openness, and equality of relations in the transnational commission rather than conflicts or negativity. In the two histories the commission has published, there is almost no mention of severe problems. However, we might well question whether these official accounts tell the whole story and whether some aspects of cooperation don't deserve more careful consideration. At the local level, an attitude has prevailed toward the Vuoksi as a transboundary resource that has not been entirely conducive to realizing a cosmopolitan commons. Finns still long for their lost territory, and the waters of Lake Saimaa, which feed the Vuoksi, are considered to be national property, and that has led to negative feelings toward the issue of paying financial compensation to Russia for power losses when these have occurred as a result of planned discharges from Lake Saimaa. On a broader level, some might interpret Finland's cooperation in transboundary water management as merely a manifestation of "Finlandization." And it is certainly true that the power disparity between Finland and the Soviet Union affected the cooperative relationship in important ways. Finlandization or not, however, effective transboundary water management could not occur without a level of shared commitment and authority sufficient to produce concrete yet broadly equitable results, and this is what has occurred in the case of the Vuoksi.

Rather than as primarily a case of Finlandization, the Transboundary Water Commission is better viewed as an epistemic community of professionals committed to working efficiently toward common goals on the basis of shared esoteric knowledge and good interpersonal relationships.[78] However, it is important to emphasize that the commission is still a rather closed community of authorities, and local people or members of non-governmental organizations do not always feel that they have sufficient access to the organization or its working groups. There have been some positive signals that the commission is becoming a more open community. For example, it has launched its own website, and Timo Kotkasaari, the current chairman of the Finnish part of the commission, asserts that the organization now handles initiatives proposed by individual citizens and by non-governmental organizations.[79] Still, the commission's website concentrates on history rather than on current issues, and proceedings of its meetings and lists of the participants still are not available to most outsiders. It is possible to obtain proceedings by request, but they are not routinely made available to the general public.

Looking back, we can also say that technological development has affected the management systems of the Vuoksi hydropower commons. In everyday communication across the border, new ways of communicating have replaced the old ones. Weekly notices of scheduled discharges are now sent by email, and all email messages are both in Finnish and in Russian. Because of the language barrier, communication by telephone is used only in unexpected situations that require quick responses. In communication by telephone, an interpreter is still needed. Also, both of the Finnish hydro-power plants are automated and operate almost without personnel, and this situation has promoted the use of coded and formal communication. In a similar way, the diffusion of information has changed dramatically. Even though the Vuoksi's water level has been regularly monitored since 1904, using pretty much the same measurement procedures, recent advances in communication technology have produced dramatic changes in the rate and frequency of information flows. The Internet, for example, now gives everyone real-time access to important data concerning the Vuoksi. Though the river's water level is still controlled by the Tainionkoski dam, it is now possible to follow the daily discharges on the Internet.[80]

Cross-border governance of the Vuoksi has been, above all, a dynamic and always changing process. Finnish-Soviet governance has broadened to become a more complex system of EU-Finnish-Russian governance, which, moreover, now involves greater participation by local organizations and NGOs. The formal rules of governance have evolved considerably, and they have been constantly re-evaluated and periodically renegotiated at the macro level as institutional practices and at the micro level as informal and cultural practices. Moreover, the tasks of Vuoksi governance have expanded, as evidenced by the EU-supported VIVATVUOKSIA and VUOKSIAGAIN projects. Analyzed over the span of more than half a century, the emergence and the evolution of Vuoksi governance as a cosmopolitan commons show that even unequal, ideologically opposed enemies can resolve conflicts and build successful cooperation.

Acknowledgments

I am most grateful to the editors of this book, Nil Disco and Eda Kranakis, for their valuable comments and help during the writing process. In the last phase, Eda's comments helped me to revise the manuscript once more. I also express gratitude to Veikko Puska at Imatra; he has helped me with numerous details by using his excellent network of experts. I would also like to thank the two anonymous MIT Press referees for their comments.

Notes

1. From 1809 until its independence in 1917, Finland was an autonomous Grand Duchy of the Russian Empire. Its history as a Grand Duchy of Russia can be divided into three separate phases. The first phase (1809–1869) can be described as a phase of consolidation of a new ruler. The capital was moved from Turku to Helsinki, and connections to St. Petersburg were strengthened both physically (navigation routes

and railroads) and ideologically. The second phase, from 1860 on, can be characterized as a period of increased independence. The reforms of the 1860s and the 1870s freed the Finnish economy and society from the tight restrictions of an earlier age and helped lay foundations for an independent state. Finnish became an official state language, and the Duchy gained its own government and parliament, as well as greater economic and financial independence. Legislative developments also laid the foundations for a modern economy, and economic policy became more pro-industrial. This period ended in 1899, when Russia began imposing policies of Russification in everything from currency to language and religion. Widely opposed and resisted by the Grand Duchy's overwhelmingly Finnish population, Russification could not stop Finland's move toward full independence. See Osmo Jussila, *Suomen suurruhtinaskunta 1809–1917* (WSOY, 2004), pp. 771–795; David Kirby, *A Concise History of Finland* (Cambridge University Press, 2006), p. 113; Jouko Vahtola, *Suomen historia. Jääkaudesta Euroopan Unioniin* (Otava, 2003), pp. 257–289.

2. In Russia, the Winter War is called Sovetsko-Finskaja Voina (Советско-финская война). Finland actually fought three separate wars during World War II: the Winter War (1939–1940) against the Soviet Union, which resulted in the Moscow Treaty wherein Finland ceded territory to the USSR; the Continuation War (1941–1944), in which Finland fought as an ally of Germany against the Soviet Union; and the Lapland War against Germany (1944–1945). The border mandated by the Moscow Treaty was reaffirmed at the end of the Continuation War in 1944 and remains the border today.

3. The empirical material for this study consists of documentation from the Archives of the Ministry for Foreign Affairs (MFA 109/E 12 Enso-Vallinkoski, MFA 12/L Venäjä: Suomen ja Neuvostoliiton yhteinen rajavesisopimus I b, MFA 12/2 Venäjä) and Fortum (former Imatran Voima) of Finland and interviews I have conducted for my forthcoming doctoral thesis, The Common Vuoksi—Human-Environment Interaction as Cultural Process in the Vuoksi River Valley from 1800 to Present. I have interviewed local authorities and workers of the Imatra power plant and people who live or have lived in the Vuoksi valley.

4. Barbara J. Morehouse, "Theoretical Approaches to Border Spaces and Identities," in *Challenged Borderlands: Transcending Political and Cultural Boundaries*, ed. V. Pavlakovich-Kochi, B. Morehouse, and D. Wastl-Walter (Ashgate, 2004).

5. Jouni Häkli and David H. Kaplan, "Learning from Europe? Borderlands in Social and Geographical Context," in *Boundaries and Place: European Borderlands in Geographical Context*, ed. D. Kaplan and J. Häkli (Rowman & Littlefield, 2002), p. 11; Anssi Paasi, "Place, Boundaries, and the Construction of Finnish Territory," in ibid., pp. 179–181.

6. Mark Zeitoun and Naho Mirumachi, "Transboundary Water Interaction I: Reconsidering Conflict and Cooperation," *International Environmental Agreements* 8 (2008): 297–316. Zeitoun and Mirumachi suggest the need to study "transboundary water

interaction" in a way that simultaneously takes into consideration different forms and levels of conflict and cooperation.

7. Maria Lähteenmäki, "Dreaming of a Greater Finland. Finnish-Russian Border Demarcations from 1809 to 1944," in *The Flexible Frontier: Change and Continuity in Finnish Russian Relations*, ed. M. Lähteenmäki (Aleksanteri Institute, 2007), p. 145.

8. Richard White, *The Organic Machine: The Remaking of the Columbia River* (Hill and Wang, 1995).

9. Tainionkoski and Imatra are on the Finnish side of the border; Svetogorsk (previously named Enso) and Lesogorsk (previously named Rouhiala) are now on the Russian side.

10. Ossi Seppovaara, *Vuoksi. Luonto ja ihminen vesistön muovaajina* (SKS, 1984); V. G. Drabkova et al.," Ecological State of the Vuoksi," in *Karelia and St. Petersburg: From Lakeland Interior to European Metropolis*, ed. E. Varis and S. Porter (Joensuu University Press, 1996).

11. Jaakko Auer and Niilo Teerimäki, *Puoli vuosisataa Imatran voimaa. Imatran Voima Oy:n synty ja kehitys 1980-luvulle* (Helsinki, 1982), pp. 15–17.

12. Malgosia Fitzmaurice, Olufemi Elias, and Vaughan Lowe, *Watercourse Co-operation in Northern Europe—A Model for the Future* (T.M.C. Asser, 2004), p. 3; Timo Herranen, *Valtakunnan sähköistyskysymys. Strategiat, siirtojärjestelmät sekä alueellinen sähköistys vuoteen 1940* (Suomen Historiallinen Seura, 1996), pp. 53-56; Timo Myllyntaus, *Electrifying Finland: The Transfer of a New Technology into a Late Industrializing Economy* (Macmillan, 1991), pp. 169–170. Fitzmaurice, Elias, and Lowe have stated that many dam projects are organized and undertaken as local initiatives. This was the case in the Vuoksi valley during the late nineteenth century and the early twentieth centuries, but after Finnish independence harnessing the Vuoksi became a national goal.

13. Myllyntaus, *Electrifying Finland*, p. 163.

14. Kristiina Korjonen-Kuusipuro, "Voimaa Vuoksesta," *Tekniikan Waiheita* 3 (2007): 5–15; Myllyntaus, *Electrifying Finland*, p. 70.

15. The first of these hydropower plants, the Linnankoski plant, built in 1898–1900, never actually worked and was destroyed when the Imatra power plant was finished and the water level was raised.

16. Auer and Teerimäki, *Puoli vuosisataa Imatran voimaa*, pp. 30–31; Myllyntaus, *Electrifying Finland*, p. 165.

17. Herranen, *Valtakunnan sähköistyskysymys*, pp. 96–103; Myllyntaus, *Electrifying Finland*, p. 72; Kristiina Korjonen-Kuusipuro, "Voimaa Vuoksesta," p. 13. Enso-Gutzeit acquired Tornator in 1931.

18. The firm was renamed Enso-Gutzeit in 1927.

19. Timo Myllyntaus has also argued that harnessing the Vuoksi's rapids can be seen as a national measure for several reasons. Beyond the observation that Finland's Senate, before independence, rejected plans to harness the rapids submitted by international investors or companies, he notes that two years after independence Finland passed a law prohibiting the selling of energy abroad. At the same time, Finland rapidly developed plans to harness the important rivers when the Committee for Hydropower (Koskivoimakomitea) was established in 1917, including the state-sponsored Imatra hydropower plant. See Myllyntaus, *Electrifying Finland*, p. 82.

20. In the 1930s, Enso, in Finland's Jääski Parish, was an industrial community of more than 10,000 people. At that time, Enso-Gutzeit Oy was one of the largest industrial companies in Finland and the third-biggest pulp producer in Europe. In Enso, Enso-Gutzeit had pulp and paper mills and sawmills that utilized Vuoksi river hydropower. Antti O. Arponen and Reijo Miettinen, *Enso. Jääsken pitäjän teollisuuskeskuksen elämää 1800-luvulta syksyyn 1944* (Karjalan kirjapaino Oy, 2002), pp. 16–24.

21. Kristiina Korjonen-Kuusipuro, "Critical Water: Negotiating the Vuoksi River in 1940," *Water History* 3, no. 3 (2011): 169–186.

22. Some scholars have assumed that the causes of the Winter War were partly economic. This is still a debated issue. It is in any case clear that the Soviet Union was aware of the economic importance of the upper Vuoksi area, because Russia was involved in the planning of hydropower production at the end of nineteenth and the beginning of the twentieth century—i.e., before Finnish independence. See, e.g., J. Kilin, "Rajaseudun väki kahdesti panttina 1939–1940," *Historiallinen Aikakauskirja* 3 (1993): 209.

23. MFA 109/ E 12 Enso-Vallinkoski.

24. Korjonen-Kuusipuro, "Critical Water."

25. Robert O. Keohane and Joseph S. Nye, *Power and Interdependence*, third edition (Longman, 2001), pp. 7–9.

26. Situations of this kind are typical in interdependence relationships. According to Keohane and Nye, there is always an issue of how benefits are shared between multiple (both governmental and non-governmental) actors. See Keohane and Nye, *Power and Interdependence*, pp. 8–9.

27. The Mannerheim Line was a defensive fortification line built by Finland against the Soviet Union. The first plans for the defensive line were made in the spring of 1918. The Mannerheim Line was constructed in two phases during the periods 1919–1924 and 1932–1939. It consisted of small and large bunkers as well as parts of the Vuoksi.

28. Jeffrey Alexander, Ron Eyerman, Bernard Giesen, Neil Smelser, and Piotr Sztompka, *Cultural Trauma and Collective Identity* (University of California Press, 2004), pp. 1, 10.

29. Simo Jaatinen, "Rajavesikomissio: Hyvä yhteistyöelin," in *Finnish-Russian Commission on the Utilization of Frontier Water Courses: Cooperation on the Frontier Watercourses during Thirty Years* (Painatuskeskus, 1995), p. 13; Memorandum 27 Dec. 1962, MFA 12/L.

30. Memorandum 27 Dec. 1962, MFA 12/L.

31. In Finland most hydropower plants are built as part of a dam, and in the Vuoksi all hydropower plants included a dam. The dams are all rather small. Among these, the Imatra dam (20 meters high) is one of the largest. See Myllyntaus, *Electrifying Finland*, pp. 169–172.

32. Memorandum 27 Dec. 1962, MFA 12/L, 11–12.

33. Tatiana Androsova, "Economic Interests in Soviet-Finnish Relations," in *Flexible Frontier: Change and Continuity in Finnish-Russian Relations*, ed. M. Lähteenmäki (Aleksanteri Institute, 2007), p. 142.

34. Agreement of Friendship, Cooperation, and Mutual Assistance (YYA Treaty), 17/1948 (available at http://www.finlex.fi).

35. Androsova, "Economic Interests in Soviet-Finnish Relations," pp. 135–137; Tatiana Androsova, "Economic Interest in Soviet Post-War Policy on Finland," in *Reassessing Cold War Europe*, ed. S. Autio-Sarasmo and K. Miklóssy (Routledge, 2011), pp. 33–48.

36. Sari Autio-Sarasmo and Katalin Miklóssy, "Introduction: The Cold War from a new perspective," in *Reassessing the Cold War Europe*, ed. Autio-Sarasmo and Miklóssy, pp. 2–3.

37. Mikko Kohvakka, "Science, Technology and Changing Power Relations: The negotiation process of the agreement on Finnish-Soviet Scientific-Technical Cooperation, 1955," *Scandinavian Journal of History* 36, no. 3 (2011): 349–370.

38. Osmo Korvenkontio, "Svetogorskin-Imatran padotussopimus," in *Finnish-Russian Commission on the Utilization of Frontier Water Courses: Cooperation on the Frontier Watercourses during Thirty Years* (Painatuskeskus, 1995), p. 17.

39. According to many leading commons researchers, including Elinor Ostrom, resilient institutions are of major importance for governing commons and especially for cross-border water-management cooperation. See Elinor Ostrom, *Governing the Commons: The Evolution of Institutions for Collective Action* (Cambridge University Press, 1990); Meredith A. Giordano and Aaron T. Wolf, "Sharing Waters: Post-Rio International Water Management," *Natural Resources Forum* 27 (2003): 163–171. See also Fitzmaurice et al., *Watercourse Co-operation in Northern Europe*, pp. 61–62, 119.

40. Esa Seppänen, *Itäsuhteiden kolmiodraama. Kekkonen-Brežnev-Kosygin 1960–1980* (Ajatuskirjat, 2007), pp. 23–29.

41. Autio-Sarasmo and Miklóssy, "Introduction: The Cold War from a New Perspective," p. 5.

42. Memorandum 27 Dec. 1962, MFA 12/L, 42.

43. Memorandum 27 Dec. 1962, MFA 12/L, 10–12; Lasse K. Kivekäs, "Saimaan vedenjuoksun ja pinnankorkeuden muuttamisyrityksistä," *Rakennustekniikka* 3 (1990): 21–35; Eino Seppänen, "Saimaan säännöstelyn historiaa," *Rakennustekniikka* 4 (1981): 259–263.

44. Agreement Concerning the Regime of the Finnish-Soviet State Frontier and the Procedure for the Settlement of Frontier Incidents, signed in 23 June 1960, 379 United Nations Treaty Series; see also Fitzmaurice et al., *Watercourse co-operation in Northern Europe*, p. 111.

45. Aarno Karhilo was later transferred abroad and replaced by Tero Lehtovaara.

46. Memorandum 27 Dec. 1962, MFA 12/L.

47. The members of the committee drew upon the models provided by the general agreements signed in Madrid in 1911 and in Barcelona on 20 April 1921 concerning the usage of general waterways, which were at the time almost the only existing agreements on transnational waters. They also carefully considered the recommendations of the International Law Association and the Institut de droit international and emphasized the importance of the United Nations Water Resources Development Centre and the Economic Commission for Europe. However, the members of the committee made no mention of using the Rhine River treaties as guiding models.

48. Memorandum 27 Dec. 1962, MFA 12/L, 70–73.

49. Agreement between the Republic of Finland and the Union of Soviet Socialist Republics concerning Water Courses, signed at Helsinki on 24 April 1964 (available at http://www.rajavesikomissio.fi); Timo Kotkasaari, "Transboundary Cooperation between Finland and Its Neighbouring Countries," in *Management of Transboundary Rivers and Lakes*, ed. O. Varis, A. Biswas, and C. Tortajada (Springer, 2008), p. 133.

50. Memorandum 27 Dec. 1962, MFA 12/L, 74.

51. The king's vein was an open mainstream that was to be kept open at the deepest part of each river. The free mainstream guaranteed the movement of migratory fish and the possibility for ship and boat traffic, and later also timber floating. The first notions of the king's vein can be found in the fifteenth century. The principle became part of Swedish legislation in 1734. At that time (and until the early nineteenth century) Finland was part of the Kingdom of Sweden. See Pekka Hallberg, "Vesioikeuslaki 100 vuotta," *Vesitalous* 5/2002: 7–8; Erkki J. Hollo, "Vesioikeuslaki 100 vuotta. Kehitystä ja mukautumista," *Vesitalous* 5/2002: 9–13. Eva Jakobsson, *Industrialisering av älvar. Studier kring svensk vattenkfartutbyggnad 1900–1918* (Göteborg, 1995), pp. 120–121.

52. Agreement 1964, no. 7804.

53. Compare Fitzmaurice et al., *Watercourse co-operation in Northern Europe*, p. 111; Jaatinen, "Rajavesikomissio—hyvä yhteistyöelin," pp. 11–15.

54. Agreement 1964, no. 7804, article 19.

55. The first Finnish chairman was E. J. Manner (1965–1968). He was succeeded by Simo Jaatinen, who remained chairman for more than 20 years. The first Russian chairman was K. S. Kornev (1965–1980); he was succeeded by V. P. Loginov. See Yhteinen suomalais-neuvostoliittolainen rajavesistöjen käyttökomissio, *20 vuotta suomalais-neuvostoliittolaista rajavesistöyhteistyötä = 20 let sovetsko-finljandskomu sotrudničestvu po pograničnum vodnym sistemam* (Valtionpainatuskeskus, 1987).

56. Sirkka Haunia, "Trends in Information Exchange," presented at St. Petersburg, 2005 (available at http://iwlearn.net); author's interview of P.H.; *Finnish-Russian Commission on the Utilization of Frontier Water Courses: Cooperation on the Frontier Water Courses During Thirty Years*, pp. 53–55.

57. Drabkova et al., "Ecological State of the Vuoksi," pp. 129–141; Yhteinen suomalais-neuvostoliittolainen rajavesistöjen käyttökomissio, *20 vuotta suomalais-neuvostoliittolaista rajavesistöyhteistyötä*, pp. 16–22.

58. On cultural trauma, see, e.g., Alexander et al., *Cultural Trauma and Collective Identity*, pp. 1 and 10; author's interview of V.P.

59. Elinor Ostrom and James M. Walker, "Communication in a Commons: Cooperation Without External Enforcement," presented at conference in 1989 (available at http://dlc.dlib.indiana.edu); Steven Hackett, Edella Schlager, and James Walker, "The Role of Communication in Resolving Commons Dilemmas: Experimental Evidence with Heterogeneous Appropriators," *Journal of Environmental Economics and Management* 27, no. 2 (1994): 99–126.

60. "Adin" is a way to pronounce the number one in Russian.

61. Author's interview of V.P.; "Codes to be used in telephone conversation between Enso and Imatra 1 May 1971," Archives of Fortum.

62. Osmo Korvenkontio, "Svetogorskin—Imatran padotussopimus," p. 24.

63. Discharge Rule of Lake Saimaa and the Vuoksi River, 91/1991.

64. Author's interview of P.H.; Kotkasaari, "Transboundary Cooperation between Finland and Its Neighbouring Countries," p. 136.

65. Marku Ollila, "Joint Risk Management Planning and Implementation Case Study: River Vuoksi," presented at workshop on Transboundary Flood Risk Management, Geneva, 2009 (available at http://www.unece.org).

66. Jaatinen, "Rajavesikomissio—hyvä yhteistyöelin," pp. 12–13.

67. T. Kotkasaari, N. N. Mihejev, and K. S. Kornev, "Rajavesiyhteistyötä jo kolmen-kymmenen vuoden ajan," in *Finnish-Russian Commission on the Utilization of Frontier Water Courses: Cooperation on the Frontier Water Courses during Thirty Years* (Paina-tuskeskus, 1995), p. 8.

68. Kotkasaari, Mihejev, and Kornev, "Rajavesiyhteistyötä jo kolmenkymmenen vuoden ajan," pp. 5–9; Jaatinen, "Rajavesikomissio—hyvä yhteistyöelin," pp. 11–15.

69. Ostrom, *Governing the Commons*. On the role and significance of institutions in the governance of shared resources that transcend borders, see also Giordano and Wolf, "Sharing Waters," p. 165; Gary D. Libecap, "The Conditions for Successful Collective Action," in *Local Commons and Global Interdependence: Heterogenity and Cooperation in Two Domains*, ed. R. Keohane and E. Ostrom (Sage, 1995), pp. 161–162; Oran R. Young, "The Problem of Scale in Human/Environment Relationships," in *Local Commons and Global Interdependence*, ed. Keohane and Ostrom, p. 37.

70. According to Stephen Wegren, land reform is still an ongoing issue in Russia and will remain so for the foreseeable future. He argues that use rights and land ownership are crucial elements of power and economic development. Even though private ownership of land is possible, Maria Tekoniemi has calculated that in 2007 more than 90% of the land was still owned by the state or municipalities and only 7% by private owners. Stephen K. Wegren, *Land Reform in Russia: Institutional Design and Behavioral Responses* (Yale University Press, 2009), pp. x–xi; Maria Tekoniemi, "Yksityinen maanomistus Venäjällä—monet ongelmat rajoittavat järjestelmän toimivuutta," *Focus/opinion* 1 (2007) (available at http://www.suomenpankki.fi).

71. Everyman's rights is a traditional legal concept in Finland and other Nordic countries that allows the use of and access to natural resources (e.g., berries, mush-rooms, lakes, rivers) despite private ownership of the space. In Russia, there is an equivalent for "everyman's rights," by which people have a right to use riverbanks and forests if they are not privately owned. See Soili Nystén-Haarala, *Russian Law in Transition* (Kikimora Publications, 2001), pp. 36–39. Information about Russian summer homes along the Vuoksi stems from the author's interview of P.H.

72. The Russian energy company Lenenergo was founded in 1886. Today Lenenergo is the largest grid company in the Northwest Federal District and one of Russia's system-forming enterprises. Lenenergo's main shareholders are MRSK Holding (50.31% stake) and the St. Petersburg administration (25.16%). A history of the Lenenergo company is available at http://www.lenenergo.ru.

73. Author's interview of V.P.

74. Author's interview of R-S.W; Kotkasaari, "Transboundary Cooperation between Finland and Its Neighbouring Countries," pp. 137–138.

75. Riitta-Sisko Wirkkala, Ljubov Smirnova, and Anna-Riikka Kohonen, "Recom-mendations to be Applied in the International River Basin Management of the River

Vuoksi," in *River Basin Analysis and Management of the River Vuoksi in the Context of the Russian Legislation and the EU's Water Framework Directive* (25.7.2008) (available at http://www.ymparisto.fi), pp. 38–40.

76. Vladimir Kotov, "Russia: Historical Dimensions of Water Management," in *The Evolution of the Law and Politics of Water*, ed. J. Dellapenna and J. Gupta (Springer Science and Business Media, 2009).

77. Andrew C. Wicks, Shawn L. Berman, and Thomas M. Jones, "The Structure of Optimal Trust: Moral and Strategic Implications," *Academy of Management Review* 24, no. 1 (1999): 99–116; Nan Lin, "Social Capital," in *International Encyclopedia of Economic Sociology*, ed. J. Beckert and M. Zafirosvki (Routledge, 2006); author's interviews of V.P., P.H., and R.W.

78. On the role of epistemic communities in water management, see Paula Garb and John M. Whiteley, "A Hydroelectric Power Complex on Both Sides of the War: Potential Weapon or Peace Incentive?" in *Reflections on Water: New Approaches to Transboundary Conflicts and Cooperation*, ed. J. Blatter and H. Ingram (MIT Press, 2001), p. 231.

79. Kotkasaari, "Transboundary Cooperation between Finland and Its Neighbouring Countries," p. 134. Allowing greater input from outside organizations into the workings of the commission would be a way to bring a wider range of stakeholders into the process of negotiating and analyzing diverse interests and develop action plans for the Vuoksi. See John Kerr, "Watershed Management: Lessons from Common Property Theory," *International Journal of the Commons* 1, no, 1 (2007): 89–109.

80. Discharge information is available at http://wwwi3.ymparisto.fi.

II Protecting Humans and Nature

6 Predicting the Weather: An Information Commons for Europe and the World

Paul N. Edwards

Weather affects virtually everything people do: where and how we live, what we eat, what we can and cannot do on any given day. A shared resource when it brings sunlight, warmth, and water, it is a shared risk when it brings floods, droughts, or extremes of temperature. Weather affects agriculture, urban planning, government, insurance, and much else. It even gets inside us. We routinely describe moods, sensations, and relationships as "stormy," "foggy," "cold," or "sunny." We can't change the weather (at least not on purpose), so to escape its tyranny we go indoors. Architects design buildings that protect us from it, yet even their best designs too often succumb to hurricanes, tornadoes, floods, snow, or heat. One thing we *can* do about the weather is to try to predict it, enabling us to prepare for its worst effects and to take better advantage of its best ones.

Weather prediction is an ancient pursuit, the province of sailors, farmers, and shamans long before it became an object of science, and of science long before the advent of practical forecast technology. In this chapter, I will focus mainly on weather forecasting in the modern era. Today's weather prediction system collects weather data from countless sources and blends them into coherent data images via computerized data analysis. Using computer simulations, it then creates weather forecasts for large areas from those data, and it broadcasts both the forecasts and the treated data. I call this form of forecasting "technoscientific" in order to signal that it binds devices (computers, weather instruments, satellites), large-scale information and communication infrastructures, and scientific understanding (meteorological theory, simulation modeling).

I begin by outlining a case for viewing the global weather forecasting system as an information commons—an interpretation that is partially at odds with the more common picture of forecasting as a public good. I then briefly survey the history of the international weather prediction system before moving on to the chapter's main subject: a history of the European

Centre for Medium-Range Weather Forecasts. The ECMWF opened its doors at Shinfield Park, England, in 1974 as a joint project of sixteen European nations. Its primary goal—at the time a highly ambitious one—was to supply credible weather forecasts for a period four to ten days in the future: the "medium range." By the mid 1980s, the ECMWF had already achieved remarkable success. Within a decade after its first operational forecast in 1979, it had built a reputation as the world's most important weather forecasting center, supplying not only Europe but also the rest of the world with data, analysis, and forecasts and contributing substantially to the science of climate change. This episode highlights the technical and political challenges of building a cosmopolitan information commons. Throughout the chapter, I point to tragic possibilities—ways in which the commons can be unintentionally disrupted or destroyed by the withdrawal of data or the privatization of public resources.

Weather Information Systems as Cosmopolitan Commons

Elsewhere I have described the technoscientific weather forecast system as a global knowledge infrastructure comprising the physical, organizational, and knowledge elements that underlie the practice of forecasting.[1] In this chapter, I want to look at it from another angle, thinking of it instead as a cosmopolitan information commons. This perspective complements a widely held and largely correct view of weather forecasting as a pure public good.

Early conceptions of public goods (as discussed in chapter 2 of this volume) defined them as resources that are both non-excludable (i.e., people cannot be prevented from consuming them) and non-subtractable or "non-rivalrous" (i.e., people can consume them without leaving less of the resource for others).[2] National defense, lighthouses, and scientific knowledge were classic examples. By contrast, common-pool resources were described as non-excludable *but subtractable*. This combination of properties was held to be the basis for "tragedy of the commons" effects—that is, overconsumption, such as the depletion of ocean fisheries or grazing lands. More recent scholarship has emphasized, however, that common-pool resources have rarely been entirely non-excludable in practice. Each chapter in this volume, including this one, describes ingenious ways—from physical barriers to customs and legal systems—that people have found to restrict access to and use of common-pool resources. It can be argued that this also applies to public goods. Even national defense and scientific knowledge can be rendered excludable under certain conditions. For example, when the Confederate army defended the seceding states of the antebellum American

South, the African slaves living within the Confederacy could hardly be considered its beneficiaries. Similarly, the "patent thickets" that effectively privatize numerous scientific and technical innovations have been called a "tragedy of the anticommons" in reference to the deleterious effects of the labyrinth of conflicting rights and rights holders they establish.[3]

What about subtractability, often held to distinguish public goods from common-pool resources? Unlimited numbers of people can certainly "consume" (read, view, hear) a weather forecast without diminishing its value for anyone else; it is even the case that the more people consume some kinds of weather forecasts (such as those for hurricanes or tornadoes), the *greater* their value for everyone.[4] Indeed, the head of the World Meteorological Organization once called public weather forecasting "the ultimate example of a pure public good."[5] Yet this chapter will show that the weather data and computer modeling on which forecasts are based depend on contributions from many quarters, which can be, have been, and still are threatened by rivalrous alternatives. Considered as a cosmopolitan commons, technoscientific weather forecasting comprises (a) a shared resource-space, (b) a set of "tragic" possibilities that could diminish or destroy the commons, and (c) a moral economy that governs both contributions to knowledge production and consumption of its outputs. Weather forecasts are much more than costless, abstract information that could be produced anywhere by anyone. Instead, they are inextricably tied to the physical phenomena they predict, and they require data from all over the world. Acquiring knowledge of these phenomena demands that information creators, including people, equipment, and institutions, span numerous international boundaries. As a result, the technoscientific forecasting system is inherently enmeshed in political structures and choices whose vicissitudes make weather forecasts a precarious resource at certain times.

Consider the idea of a "resource-space" (presented in chapter 2). Cosmopolitan commons theory contends that the scale at which and the ways in which resource-spaces are organized and exploited depend not only on the geophysical characteristics of the spaces (the radio frequency spectrum is a very different kind of resource-space from the North Sea, for example), but also on available technologies, techniques, and theories, and on how these are organized into functioning commons. A weather forecast describes physical phenomena for a certain area over some time period. Because weather moves, the longer the time period, the larger the physical volume the forecast must assess. Today, with satellite imagery available at the click of a mouse, we see quite readily that weather systems are huge (often larger than most European nations) and that they move swiftly (often traversing the entire European continent in a day or two). Yet as recently as the 1950s, these systems could not be seen directly. Before high-altitude photography from rockets and satellites, images of weather systems had to be constructed by painstakingly plotting individual data points on maps, then connecting those points according to principles that began as visualization techniques before they gained support from scientific theory.

In seeking to make better, longer-term weather forecasts, the weather forecast infrastructure—the resource-space that is organized and exploited to produce weather forecasts—has expanded from essentially local dimensions to regional and national scales, and in the last thirty years to a global scale. Thus, weather forecasting now is carried out within a *human-constructed* resource-space, rather than within a naturally existing resource-space such as the Rhine or airspace. In fact, weather forecasting is carried out within a constructed "resource- space/time," we might say, since the speeds of data transmission and processing determine how much data can be collected and used, while the quality of computer models determines how far into the future usable forecasts can look. Obviously weather itself occurs in geophysical space, whether we predict it or not. The "resource-space/time" I am talking about is, rather, the space/time of weather knowledge—i.e., the space/time that is exploited to produce modern weather forecasts, and the space/time of those forecasts' validity, which today runs to about 10 days in the northern hemisphere.

Weather forecasts are, of course, based on data. A moral economy of freely shared, widely disseminated data originated in the seventeenth and eighteenth centuries, when such data were not only inexpensive to acquire but also entirely devoid of economic value. Remarkably, this moral economy remained in place even after the advent of computers and satellites, which multiplied the costs of forecasting (and its power) many-fold.

The "tragic possibilities" that attend the weather information commons stem from challenges to this moral economy. For example, each of the world wars resulted in a temporary collapse of global data sharing. In a more recent example, movements to monetize publicly produced weather data and to privatize major elements of forecasting arose in the 1990s. Some monetization and some privatization did in fact occur. Yet, thanks to defensive action by the World Meteorological Organization and others, basic weather data and forecasts remain freely shared public-domain resources. We will return to each of these points below. First, though, let us look briefly at the historical trajectory of technoscientific weather forecasting since its earliest days.

Topographies of Weather Information: Weather Telegraphy

Meteorology is among the oldest examples of scientific internationalism. This fact is due at once to the large physical scale of weather systems and to the technological means required to produce information and knowledge about those systems. Italian scientists developed thermometers and barometers, the basic instruments of weather observation, in the 1640s. Seeking an understanding of how weather moves, scientists established observing networks almost immediately. From 1654 to 1667 the Accademia del Cimento oversaw a pan-European network consisting of ten weather stations ranging from Florence to Paris, Warsaw, Innsbruck, and Osnabrück. James Jurin published European data (and some data from the British colonies in North America) in the British Royal Society's *Philosophical Transactions* from 1724 to 1735. In the late eighteenth century, the Societas Meteorologica Palatina, based in Mannheim, organized a network of 37 weather stations scattered across Europe and the United States.[6]

Before the advent of the telegraph, however, these observing networks could share data only long after the weather in question had passed. Such forecasting as there was took place locally and was based chiefly on the barometer, whose imprecise predictions were valid only for 12–24 hours. Dependent on patronage and unable to offer forecasts of much practical value, these early networks all collapsed after a decade or two. Surprisingly, until the nineteenth century they almost never used maps, relying instead on tables that mixed instrument readings with qualitative descriptions that lacked a standard vocabulary. Still, these early observing networks established the idea that wide-area data analysis might reveal patterns that could be used for prediction.

Even the earliest meteorologists dreamed of an ability to "see" the weather of the entire world. Scientific understanding of Earth's atmosphere as a global system dates to Edmund Halley's articulation of the mechanism of the trade winds in 1686, which was accompanied by a map of those winds across the Atlantic, the Indian Ocean, and part of the Pacific.[7] The Prussian meteorologist Heinrich Dove mapped global temperature averages from the equator to the middle latitudes in 1853.[8] By 1856, the American William Ferrel, using a combination of theory and observation, had produced a remarkably modern diagram of the large-scale global atmospheric circulation.[9] However, all of these were climatological features derived from data recorded over many years, and were of little use in forecasting weather in the immediate future.

At an 1839 meeting of the London Meteorological Society, John Ruskin spoke of meteorology's desire "to have at its command, at stated periods, perfect systems of methodical and simultaneous observations; it wishes its influence and its power to be omnipresent over the globe so that it may be able to know, at any given instant, the state of the atmosphere on every point on its surface."[10] The arrival of the telegraph in the 1840s seemed to bring Ruskin's goal of meteorological omniscience within reach. For the first time, meteorologists could share weather data over large areas within hours of making observations.

In 1849, Joseph Henry of the Smithsonian Institution established an American weather telegraph network, with government support. Henry secured the agreement of commercial telegraph companies to transmit weather data free of charge. This agreement brought a commercial enterprise and a new communications medium into the moral economy of forecasting under the same arrangement as meteorology's previous no-cost data-sharing regime, such as it was.[11] Europe soon followed suit. In 1854, during the Crimean War, a disastrous storm destroyed a French fleet near Balaklava on the Black Sea. Since observers had seen the same storm moving across the Mediterranean the previous day, they realized that advance warning (by telegraph) might have prevented the debacle. As a result, in 1855 France established a national weather telegraph service and an international meteorological center. Other European nations followed suit, arranging national weather telegraph networks within their borders. By 1857 Paris was receiving and forwarding daily telegraph reports from Russia, Austria, Italy, Belgium, Switzerland, Spain, and Portugal.[12] This transformed the resource-space of forecasting from an essentially local topography to one largely coextensive with telegraph networks. In Britain, in the United States, and elsewhere, many of these networks used railway lines as convenient rights-of-way, leading to overlapping topographies of large technical systems.

Weather telegraphy permitted meteorologists to map the weather over very large areas within a few hours of observations. The resulting "synoptic" maps functioned like snapshots.[13] They charted pressure, temperature, and other weather conditions at each observing station. Wind direction and speed told which way the weather was moving, and how fast. By the end of the nineteenth century, weather maps looked much like those in use today, showing isolines of temperature and pressure and indicating the direction of motion of weather systems. These maps provided a basis for at least a rational guess at what would happen next, and where. Yet the complex motion of the atmosphere—a turbulent, chaotic, global system—means constant change. Meteorologists struggled in vain to find consistent patterns in the behavior of weather at local and regional scales. In the absence of an understanding of atmospheric dynamics and an observing system in the vertical dimension, synoptic maps added little to the quality of forecasts beyond 24–36 hours. Their principal benefit lay in advance warning of the approach of major storms.

By 1865, some twenty European nations had formed the International Telegraph Union to promote and develop technical standards for international telegraphy. Weather services created simple telegraph codes for reporting basic data using a minimal number of characters. The telegraph agencies generally agreed to transmit these weather bulletins at no cost, as a public service. Certain special characteristics of weather data made this information-commons approach sensible at a time when telecommunication was quite expensive. First, basic weather data are quite simple and compact, requiring only a few words of telegraph code. For example, the entire report from a New York City station in the 1870s, coded for telegraph transmission, read simply "York, Monday, Dead, Fire, Grind, Himself, Ill, Ovation, View." Expanded, this report translates into the following.

York: New York (station)
Monday: 30.07 (barometer corrected)
Dead: 29.90 (corrected barometer for temperature and instrumental error)
Fire: 70° (thermometer)
Grind: 75 per cent (humidity)
Himself: west, fair (wind and weather)
Ill: 6 miles (velocity of wind)
Ovation: 1/2 cirrus clouds, calm (upper clouds)
View: 67° (minimum temperature during night)[14]

The compactness of the data, further reduced by ingenious codes, made it feasible for the telegraph operators to carry weather reports without charge. Had many more variables been required, this might not have been possible. A second unique characteristic of weather data is that their value decays rapidly. Before World War II, data more than a day old were of no use in forecasting. A third characteristic is that a *collection* of weather data, suitably mapped, has vastly greater value than individual data points alone. In general, all parties gained by sharing data, and none benefited from keeping them secret—except in certain circumstances, as we will see below. These special characteristics of weather data and forecasting structured both the push to enroll telegraph operators and the operators' acceptance of that task on a no-cost basis.

Meanwhile, the topology of telegraph networks—which rapidly became national systems, in some cases under government control—helped shape the forecasting system that emerged in the second half of the nineteenth century. That system's fundamental unit was the national weather service.[15] National weather services provided both verbal forecasts and weather maps, which could be viewed at weather stations or even, starting in the 1880s, could be transmitted as crude facsimiles over telegraph lines and published in newspapers—a practice that was routine in some places by 1900.[16] As both system builders and network users, the national weather services experienced conflicting pressures. Because they were answerable to their governments, they gave the highest priority to improving their services within their respective countries' borders. Yet as forecasting techniques improved, all nations needed data from beyond their own borders. This was especially true in Europe, with its relatively small nations. Even countries on Europe's western edge, such as Britain and Norway (whose weather comes primarily from the North Atlantic), needed data from Canada, from the United States, and from ships at sea. The heads of European and American national weather services joined forces in the 1870s to form an International Meteorological Organization that was oriented toward standardizing weather observations and telegraph codes for international data sharing. By 1900, the IMO had articulated the goal of a Réseau Mondial—a worldwide network—that might one day report weather data from across the globe in hours over the telegraph network that by then covered much of the world.

Elsewhere I have described the International Meteorological Organization and its successors as engaged in "infrastructural globalism." This phrase emphasizes the deliberate project of global infrastructure building in meteorology. The well-defined and persistent focus on the planetary scale of weather, especially after 1950, created a clear trajectory for the sociotechnical systems needed to monitor the weather and to model its processes.[17] Cosmopolitan commons theory complements the idea of infrastructural globalism. Nineteenth-century meteorology explicitly conceived weather forecasting as a public good, properly supplied by national governments to their citizens, and made weather data freely available to forecasters. Meteorologists saw global data sharing as desirable not in some abstract or political sense, but because the physical phenomena in question are global in scale. The resource-space of the weather information commons began to grow.

Weather Information, World War, and the Growth of a Global Network

The long-term goal of a global weather information infrastructure, the moral economy of freely shared data, and the notion of weather forecasts as public goods were widely accepted by professional meteorologists. Yet until after World War II the institutional basis for an internationalized forecasting system remained extremely weak, and it conflicted with the political structures on which meteorology depended. Each national or imperial weather service established its own standards for data formats, telegraph codes, units of temperature (Fahrenheit vs. Celsius), units of pressure (pascals vs. millibars vs. pounds per square inch), observing hours, and so on. As a result, there were many thousands of slightly different national standards. In the pre-computer era, converting one format, code, or unit into another required significant work. This chaotic state of affairs constrained forecasters, in practice, to use only a small subset of the available data.

A major goal of the International Meteorological Organization was to standardize data, formats, codes, units, and so on. But the IMO had no governmental status or authority. From 1871 to 1939 its activities consisted of little more than occasional meetings of the heads of major national weather services and the publication of suggested standards. Between meetings, the IMO did very little. It didn't acquire a permanent secretariat until 1926, and even then the secretariat's annual budget never exceeded $20,000. The IMO did make slow progress on international standards. But getting dozens of national weather services to agree on and conform to common standards and techniques often cost more in time, money, and aggravation than it seemed to be worth. Therefore, as in many situations where national sovereignty conflicts with internationalist or globalist goals, the national weather services often behaved in contradictory ways, sometimes guarding their existing standards and systems against "outside" interference and at other times urging the adoption of international norms. The one thing on which all agreed was that the national weather services retained sovereign rights to work as they wished. This tension between meteorological nationalism and internationalism severely limited the IMO's potential well into the Cold War era.

The two world wars did nothing to improve the situation, and in fact illustrate the precarious nature of the weather knowledge commons. During each war, the combatants did their best to cut off each other's access to the global flow of weather data in order to reduce their enemy's forecasting capability. As soon as hostilities ceased, access was restored. The world wars inhibited data sharing, but they also advanced meteorological science in numerous ways. The armies of World War I made meteorology central to military affairs, promoted the growth of dense observing networks and upper-air observation by aircraft (largely in order to overcome the loss of data that otherwise would have been shared), and led indirectly to a conceptual revolution in weather forecasting spearheaded by the Norwegian

Vilhelm Bjerknes and his so-called Bergen School.[18] The World War II story of the weather forecasts for Operation Overlord (the landing at Normandy) makes for exciting reading. The invasion succeeded in large part because Allied forecasts showed a brief period of calm weather in the midst of a relentless series of storms in the English Channel. German forecasts had failed to predict the lull.

Military meteorology underscored the national orientation of weather forecasting infrastructures. Nations could not necessarily rely on other nations for the weather information their armed services needed in wartime, when that information suddenly ceased to be seen as a common resource and became a tool of conflict instead. As a result, the military forces of the major world powers established their own, largely separate meteorological services, which established independent networks of weather stations around the world. In peacetime these networks shared data with the civilian weather services, but they stood ready to operate independently in the event of war. These separate infrastructures demonstrate a flaw in the public-good perspective on forecasting, which sees forecasts only as an abstract knowledge product. By contrast, the cosmopolitan-commons perspective highlights the sociotechnical systems that shape the character and the extent of commons. In the case of forecasting, the commons can be damaged by withholding of data (as in wartime), and forecasts can also be created by restricted, non-public systems (such as military weather services).

The vicissitudes of the interwar period prevented major change in the structure of international meteorology, except in one important area. At the 1919 Paris Peace Conference, signatories adopted a Convention relating to the Regulation of Aerial Navigation (discussed in detail in Eda Kranakis' chapter in this book). This convention laid out the legal basis for international air traffic, but it also specified guidelines for international meteorological data exchange, to be carried out several times daily by "radiotelegraph." Finally, the Convention established an International Commission for Air Navigation (ICAN), a body with supranational decision-making authority (based on a system of qualified-majority voting) charged in part with implementing these meteorological standards.[19] This put ICAN several steps ahead of the International Meteorological Organization, which could claim neither a legal mandate nor any official governmental status. The IMO staked a competing claim to "aeronautical meteorology," establishing its own Technical Commission for the Application of Meteorology to Aerial Navigation. But participating governments officially recognized only ICAN, not the IMO's technical commission. By 1935, this led the IMO to transform its technical commission into an International Commission for Aeronautical Meteorology (known by its French acronym, CIMAé) with members appointed by governments. CIMAé was the first, and until after World War II the only, IMO entity to acquire official intergovernmental status. In the event, most CIMAé members also sat on ICAN, so it functioned more as a liaison than as an independent organization.[20]

Meanwhile, in the first half of the twentieth century the telegraph-based weather data network rapidly became an amazingly complicated web. New technologies arrived in rapid succession. The weather network had to integrate not only new instruments, such as radiosondes (weather balloons carrying radio transmitters), but also new communications media, such as telex and shortwave radio. By the 1920s, both aircraft data and maritime (ship) data provided new data streams. New airports became observing stations, increasing the density of the network.[21] Radio eliminated the need for fixed cables, permitting cheaper, faster data exchange both within and among nations. Radio also mattered enormously in distributing weather forecasts, both as broadcasts to the public and as targeted forecasts for aviation and shipping. During this period, most of the development was driven by the internal system-building dynamics of national weather services. Even though the national services of Europe depended on each others' data, rationalizing *international* data networks remained a secondary priority. IMO standards served only as guidelines, routinely violated yet producing considerable convergence.

The interwar phase of technology transfer and growth resulted in numerous different systems, some linked and others not, all governed by a very loose patchwork of sometimes conflicting national, regional, and international standards. The pre–World War II network made rudimentary worldwide data *available* to forecasters nearly in real time by the 1920s. But forecasters' ability to *make use* of those data remained extremely limited, largely because of the extreme difficulty of sorting out the numerous formats and standards used by various national weather services.[22] Describing the conditions of commercial air travel in the interwar period, Eda Kranakis' chapter in this volume brings home the somewhat primitive quality of forecasting at that time, when international flights were routinely interrupted by "unexpected" weather conditions along the way. The reason for this low quality was that before the computer age forecasting lacked a strong scientific basis, relying largely on experience and intuition instead.

As information commons, mid-twentieth-century weather forecasting systems exhibited a fragmented and unstable character. From the point of view of those who made forecasts, data were widely available and freely shared but often were too difficult to use, owing to a lack of well-established standards. Nevertheless, a shaky, multi-modal, but basically functional data-sharing network existed over land in most of the northern hemisphere, and in a good part of the southern hemisphere too. (See figure 6.1.)

Each national weather service created its own forecasts. Since the value of weather data decays rapidly, the speed of available information and communications infrastructures made international collaboration on forecasting extremely difficult. In the interwar period, then, the prevailing moral economy of weather forecasting held that both data and forecasts should be generated by government agencies as a public service and treated as public goods—except in wartime, when governments halted the trans-border flow of weather data and military services treated weather forecasts as secret assets.

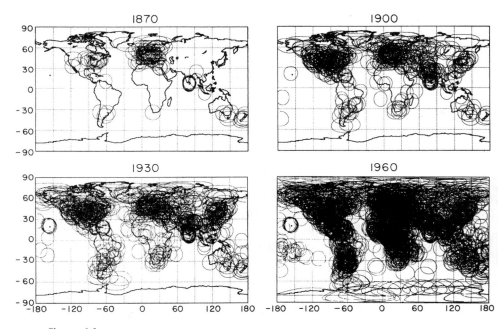

Figure 6.1
The evolution of the surface observing network. Top: Evolution of coverage by surface stations in WMSSC, principally based on World Weather Records and Monthly Climatic Data for the World. Coverage shown as a 1,200-kilometer radius around each station. Bottom: surface stations included in the Goddard Institute for Space Studies version of World Monthly Surface Station Climatology as of 1987. Grid cells demarcate regions of equal area. Numbers in each cell represent the date on which coverage began (top), total number of stations in that region (middle), and a grid cell identifier (bottom right). Reproduced from J. Hansen and S. Lebedeff, "Global Trends of Measured Surface Air Temperature," *Journal of Geophysical Research* 92, no. D11 (1987): 13, 345–372.

Consolidation: The World Meteorological Organization and the World Weather Watch

A unifying theme of this book is the complementary role of nature and technology as agents in the creation and maintenance of commons. Meteorology presents an excellent case in point. As has already been noted, diagrams of the global atmospheric circulation had appeared by the middle of the nineteenth century. Scandinavian meteorologists employed a hemispheric conception of weather in forecasting during World War I, when Vilhelm Bjerknes developed the idea of the "polar front" (an explicitly military metaphor) to describe the interaction of polar and mid-latitude circulatory cells that governs northern Europe's weather. By the time of the International Geophysical Year (1957–58), the "single physical system hypothesis," as it was called in IGY documents, dominated the research goals of meteorology.[23]

A single physical system implied a single, unified weather information system. Consolidation of the pre-World War II network began around 1950 and continues into the present. On the institutional side, the International Meteorological Organization acquired intergovernmental status as a specialized agency of the United Nations, renaming itself the World Meteorological Organization. National sovereignty remained a touchy issue, and Cold War politics sometimes created headaches for an entity that sought to integrate all the world's weather services but had to function under UN rules. Still, the WMO brought considerable new authority to develop and propagate standards and systems.[24]

On the technical side, consolidation was driven by the arrival of computer models for weather forecasting, first used operationally in Sweden in 1954. Instantly perceived as a vastly superior technique, computer modeling brought with it a voracious appetite for data. Weather forecasters began with regional models of the North Atlantic, North America, etc., but the major weather services were already transitioning to hemispheric models by the late 1960s, and to global models by the mid 1970s. These computer models required huge quantities of data, demanding prodigious efforts in standardization, communication systems, and automation. These trends in science and technology shaped the resource-space of forecasting, shifting it to ever larger scales. By 1965, there was no doubt among leading meteorologists that the world weather information infrastructure would eventually be global in scope.

In the 1960s, meteorologists eagerly awaited weather satellites, which can view the entire globe every twelve hours from their polar orbits. Satellites would, they hoped, complete the fully global observing system of which many had dreamed. In private correspondence, the American president, John F. Kennedy, and the Soviet premier, Nikita Khrushchev, agreed to promote weather satellites through the United Nations. The UN General Assembly received Kennedy's proposal for "further cooperative efforts between all nations in weather prediction and eventually in weather control" with great enthusiasm when he presented it in September 1961. In the same speech, Kennedy also announced cooperative efforts in telecommunications satellites. These efforts were major elements of a concerted push to prevent the militarization of outer space, and to characterize it as a global commons. That characterization entailed limits on national sovereignty, since one plausible (even if unenforceable) view of sovereign rights would be that they extend to an unlimited height above a nation's soil.[25]

By 1962 the WMO was directing its principal energies toward systems, standards, and institutional mechanisms for a World Weather Watch—an integrated system of satellites, surface-based observing systems, aircraft, and radiosondes (weather balloons) that would produce coherent global data images through computer processing. Although its name suggests a top-down organization, in practice the World Weather Watch's planners never expected to replace the existing patchwork. Instead, they relied on improved standardization and greater cooperation among national weather services.[26] Today the World Weather Watch remains the WMO's principal activity. It instantiates the realization of Ruskin's dream of a "vast machine" for "methodical and simultaneous observations" all over the world.[27]

The sharing of data from weather satellites and the concept of the World Weather Watch grew directly out of Cold War politics. Both were heavily promoted as counterweights to military and ideological tensions. The principal planners of the World Weather Watch were the US Weather Bureau scientist Harry Wexler and his Soviet counterpart, Viktor A. Bugaev. The International Geophysical Year had established "world data centers" for the participating geophysical sciences. World Weather Watch planners adopted a similar framework, envisioning three World Meteorological Centers: one in the United States, one in the Soviet Union, and one in Australia. These centers were to communicate with six Regional Meteorological Centers—one of which would be in Europe—which would organize communication among National Meteorological Centers.[28]

In the 1960s, the world's most advanced meteorological research centers were in the United States (the US Weather Bureau, the National Center for Atmospheric Research, several NASA facilities, the meteorology departments of the University of Chicago, the Massachusetts Institute of Technology, and the University of California at Los Angeles) and in the Soviet

Union (Guri Marchuk's laboratory at Novosibirsk). In Sweden, an International Meteorological Institute—established by Carl-Gustav Rossby (the foremost meteorologist of the mid twentieth century) in 1955—served as a European hub for meteorologists, but many European scientists preferred the US and Soviet laboratories. The US laboratories, in particular, featured more powerful computers than most of their counterparts in Europe.

The World Meteorological Centers were initially conceived mainly as data centers and communication hubs. They would collect and consolidate global data, then forward relevant subsets of those data to the Regional Meteorological Centers. The idea was to reduce the redundancy that characterized meteorological communication. At the time, each national weather service in a region (Europe, for example) had to collect bulletins from every other country, a process that took hours on the slow communication channels of the day. The data problem, however, was only one piece of the puzzle. In addition, integrating the data would require forecasting tools—computer models—capable of making use of them. Therefore, World Weather Watch planning soon led to calls for a Global Atmospheric Research Program (GARP) that would develop ways to link data systems, to automate communication, and to build computer models capable of analyzing data for the entire world. For several years in the mid 1960s, WWW and GARP planners discussed the idea of establishing a world meteorological research center, probably in Europe. However, it proved impossible to assemble a politically feasible combination of location, computer technology, funding, and leadership for such a center, which would have had to supersede or else compete with the already strong US research system, and the project was abandoned.

A European Meteorological Computing Center

The major European weather services (in Sweden, the United Kingdom, West Germany, and France) had all established computer forecasting centers by 1960. These centers faced financial strains as rapid advances in modeling techniques required ever faster, larger supercomputers. Weather models are essentially simulations of the atmosphere, represented in the models as a grid of points containing numerical values for temperature, humidity, motion, and other conditions in the vicinity of that point. The models recalculate values for every point on the grid at a "time step," typically 10–15 simulated minutes. Reducing the distance between grid points generally improves the simulation, for the same reason that a computer or television screen with more pixels per square inch displays a higher-quality image—but halving the distance between grid points multiplies the amount of computation required by a factor of at least 8 (2^3). The earliest forecast models used a grid spacing of about 600 kilometers. By 2000, global forecast models typically had a one-kilometer resolution and twenty vertical levels, amounting to roughly 1.3 million grid points. Scientists also improve models by simulating additional physical processes, further increasing the computational demands. Finally, in a forecast model, incoming weather observations—data—are injected into the simulation to correct it. These data are not simply dumped into the model, but must be checked for quality, interpolated to grid points, and analyzed in a variety of other ways beforehand. This analysis process also requires computer time. Hence, forecasters' appetite for computer power has been virtually insatiable.

Supercomputers were extremely expensive in the pioneering days of computer forecasting. In 1970, the most advanced supercomputers cost between 8 million and 16 million dollars (roughly equivalent to 47 million and 93 million 2012 dollars), and owing to the rapid improvement of computer technology their effective lifetime was only four or five years. But the machine was only the beginning. Running a supercomputer center required large additional sums for electric power, peripherals, and smaller computers for preparing programs and data. Centers also needed highly trained professional staff to program, operate, and maintain the machines.

In the early 1960s, the absence of a strong indigenous computer industry became a policy concern for major European governments. The principal industrial strategy turned initially on promoting "national champions"— state-supported firms such as ICL in the United Kingdom, Bull in France, and Siemens in West Germany. By 1970 this strategy had failed to make any headway against IBM and Control Data Corporation, which had emerged as the leader in the niche market for advanced supercomputers. The next European approach to regaining market share in computing was a collaborative, government-supported program known as Unidata,[29] begun in 1972 by CII-Bull (France) and Siemens (West Germany) and joined in 1973 by Philips (Netherlands). Unidata proved no more successful than the national champions, however, and it collapsed in 1975.

The Unidata joint effort reflected a general trend toward collaborative technology projects. In 1967, faced with an ongoing "brain drain" and withering competition from American technology firms, the Committee for Medium Range Economic Policy of the European Economic Communities had initiated studies to determine the feasibility and cost-effectiveness of cooperative efforts in science and technology. Included in early lists of possibilities were "longer-range weather forecasts" and "influencing weather."[30] The committee accordingly established an Expert Group on

Meteorology under the direction of Erich Süssenberger, president of the Deutscher Wetterdienst (German weather service). In consultation with the weather services of all six European Economic Community countries, the Expert Group generated a long list of potential meteorological collaborations, including an ozone-monitoring network, air-pollution studies, and the manufacture of weather balloons. Süssenberger himself emphasized the meteorological "frontier" of medium-term (four-to-ten-day) forecasts. Achieving usable forecasts at that range would require enormous progress in modeling, supported by the most advanced supercomputers. This emphasis stemmed in part from Süssenberger's involvement in the World Weather Watch program and the Global Atmospheric Research Program, then in the earliest phases of conception and planning.[31]

The European Centre for Medium-Range Weather Forecasts

The Expert Group on Meteorology recognized the extreme difficulty of estimating the precise value of weather forecasts. Still, it noted, an American cost-benefit analysis had calculated that an accurate five-day forecast would be worth about $6 billion per year, which suggested that Europe might anticipate annual savings of several billion dollars. Such an effort might also help Europe to catch up scientifically with the United States and the Soviet Union, which led the world "principally due to the high level of their technology," i.e., computing.[32] Finally, the Expert Group noted that no computer with sufficient capacity to address the problem of medium-term prediction existed anywhere in Europe.

Accordingly, the Expert Group proposed a "European Meteorological Computing Centre" in 1969. Later that year, the Council of Ministers of the European Communities agreed to open the proposal to nine non-EC countries: Austria, Denmark, Ireland, Norway, Portugal, Spain, Sweden, Switzerland, and the United Kingdom. Of these, Sweden and the UK were especially important owing to their historical strengths in meteorology; in fact, the UK initially reacted skeptically to the EMCC idea, apparently viewing its own programs as superior to those of the other countries.[33] Meanwhile, the absence of Eastern Europe from the proposal—as from all 49 other EC cooperative technology projects—reflected the political realities of the Cold War.

By 1971 all other concepts for cooperation in meteorology had been swept aside by the idea of a computing center. Planning then proceeded rapidly. First, the Expert Group commissioned an economic study of the potential benefits of more accurate medium-term forecasts, which might reduce weather-related losses in agriculture, shipping, air travel, construction, and a myriad of other weather-dependent industries and activities. Lacking standard numerical data and operating under tight time constraints (only a few months were allotted), the economic study group employed an unusual method, interviewing 156 experts in fifteen countries about potential benefits in a variety of economic sectors. From these interviews and some basic data for various economic sectors, the economists estimated that reasonably accurate four-to-ten-day forecasts for Europe would produce a benefit of at least 200 million u.a. (units of account, a standard measure of value) per year—at least 25 times the center's projected operational costs of 7.5 million u.a. The report concluded that, despite uncertainties in its calculations, "the mere money value obtained is so considerable that no more than a partial realization of the benefits expected would largely justify the creation of EMCC."[34]

As with the World Weather Watch, relations between the proposed EMCC (renamed European Centre for Medium-range Weather Forecasts in late 1971) and the national weather services required a delicate touch in order to avoid the perception that some pan-European monolith would take over the national weather services' functions. In a bow to the meteorological nationalism described earlier in this chapter, initial discussions emphasized that the ECMWF would not provide forecasts directly to consumers, but would instead hand off its forecasts and data to the national weather services, which then would process them further and provide national forecasts. By then, the EC program Cooperation in the Field of Scientific and Technical Research program (abbreviated COST) had received financial authority to implement its recommendations, so planning proceeded swiftly. Süssenberger noted the unusual nature of this situation: "Normally, meteorologists develop programmes and then ask the Finance Minister for funds, which usually leads to complex discussions. In the case of the ECMWF, however, the financial means were made available first with a request to plan appropriate projects."[35]

The decision to create a European center entailed the thorny question of where such a center would be located. COST explicitly considered the technical, the economic, and the human factors, but implicitly considered political ones as well. The technical and economic issues of computer power and telecommunications links played a very considerable role in this choice.

One way to acquire sufficient computer power would have been to co-locate the ECMWF with a national meteorological center, where it could share computer facilities and thus possibly offset some operating expenses. COST roundly rejected this idea, noting that "approximately 10 hours computing time are needed on a 50 MIPS [million instructions per second]

computer to carry out a 10 day forecast. Since these tremendous requirements exceed by far the total capacity of all European meteorological centres (only the British centre will reach 10 MIPS in the near future, the others
ranging in the order of 1 MIPS or less), the possibility to place EMCC within
a national centre is not a feasible solution."[36] Another solution might have
been to use computer time from many centers over a network—an idea
much like today's concept of "cloud computing." COST concluded, however, that "it is not at present possible to aggregate computing power from
different sources in a liaison network and to concentrate this accumulated
capacity on one single problem,"[37] and that only a dedicated, stand-alone
computer center could achieve the intended goal.

Furthermore, despite a nod to its sister "Working Party on Data Processing" (tasked with improving the competitive position of the European
computer industry), COST did not expect any European computer supplier
to be able to handle the ECMWF's needs until 1980 or later. The group suggested a Control Data machine and estimated the likely operational cost
of computers, software, and associated maintenance at 4.8 million u.a. per
year.[38] Furthermore, the sophisticated computer equipment would require
maintenance. This implied a need for proximity to technical support from
the computer manufacturer, most likely to be available in or near a major
European capital.

The telecommunications issue involved how raw meteorological data
would reach the ECMWF, and how the latter would, in turn, distribute
processed data and forecasts. The cost of telecommunication depended on
prices in the eventual host country: "a central location of EMCC will generally minimize the network costs . . . ; a peripheral location of EMCC could . . .
raise the annual network costs from u.a. 650.000 to u.a. 1.000.000." Cost
was not the only consideration, however. The amounts of data required for
regional and global forecasting were so large that the capacity of data transmission lines had to be taken into account. In the 1970s—long before the
Internet's many-to-many connectivity became the norm—a few major data
and forecasting centers (hubs) were connected by dedicated long-distance
"trunk lines" with the highest available data rates (then around 2,400 bits
per second). Smaller centers were connected to a regional hub via spoke
lines with lower data rates, at lower costs. For the proposed EMCC, this
implied a trunk connection:

[EMCC] requires a considerable amount of input data for its operations to be obtained mainly from European WMO centres situated on the 'Main Trunk Circuit' of
the WMO Global Telecommunication System, i.e. London, Frankfurt, Paris. In order
to speed up and to secure the operational data inflow direct and relatively short

connections between EMCC and these 3 centres are desirable; this condition suggests a central location for EMCC.[39]

Planning thus focused on the London-Frankfurt-Paris "triangle," but other
locations were not ruled out.

Human factors also entered into the placement decision. The plan envis-
aged regular, long visits (months to years) by considerable numbers of scientists from all over Europe, and they would need available, affordable,
high-quality housing, schools for their children, and easy transport to and
from their home countries. Urban amenities and a pleasant environment
would also make it easier to attract top scientists.

The various national delegations volunteered sites in Belgium, Denmark,
Germany, the Netherlands, Italy, and the United Kingdom. The eventual
choice of Shinfield Park (near Reading, and less than 50 kilometers from
London) reflected political factors as well as other factors described above.
Shinfield Park lay near Britain's Met[eorological] Office College and to the
Met Office's headquarters at nearby Bracknell, but it lacked international
schools and other amenities. However, the COST negotiations in 1972
coincided with Britain's buildup, under Prime Minister Edward Heath,
to membership in the European Economic Community, which it joined
on January 1, 1973. According to Austin Woods' history of the ECMWF,
Heath—an amateur meteorologist—was easily persuaded of the ECMWF's
benefits, but also saw its near-term political utility:

[A] strong memorandum was sent from the Government of the UK to COST detail-
ing the technical advantages of having the Centre at Shinfield Park. It went on:
'There are also political considerations. Her Majesty's Government considers that
at the time of our entry into the EEC it is particularly important that we should be
in a position to be able to announce publicly that an important European scientific
institution is being set up in the United Kingdom.'

In the final COST vote on placement in early 1973, the UK proposal competed chiefly with one from Denmark, backed by Germany (which had
withdrawn its own bid in order to make way for Germany to obtain the
European Patent Office). The UK forced the issue, hinting strongly that it
might not participate in the ECMWF if it failed in its bid to host the center.[40] This arm twisting may or may not have influenced delegates' votes; in
any case, the UK bid succeeded and the project moved swiftly to fruition.
In 1973 fifteen nations signed the Convention establishing the European
Centre for Medium-Range Weather Forecasts.

Activity began almost immediately. The ECMWF's first director, Aksel
Wiin-Nielsen, launched the Centre in January 1974 at temporary facilities in Bracknell. Wiin-Nielsen, a Dane, had begun his career at the Danish

Meteorological Institute and moved to the International Meteorological Institute in Stockholm in 1955. Another move, this time to the US Weather Service's Joint Numerical Prediction Unit, followed in 1959. Soon afterward, he joined the US National Center for Atmospheric Research during its earliest years. In 1963 he was named the first chairman of the Department of Meteorology and Oceanography at the University of Michigan. His personal career thus embodied the frequent and international movement of people and ideas that was typical of meteorology during this period—as well as the flight of excellent meteorologists from Europe to the United States in the 1950s and the 1960s, one of the conditions that COST hoped the ECMWF might reverse.

The ECMWF leased its first computer, a CDC 6600, in 1975, and immediately began work on building a global general circulation model (GCM). The US National Meteorological Center had introduced a GCM for hemispheric forecasting several years earlier, but after periods beyond a few days this and other hemispheric models became unstable owing to their artificial methods of handling computations at the models' equatorial "edge." Using a global model would eliminate this instability, but would require advanced numerical techniques as well as much greater computer power. Rather than construct the ECMWF's global forecast GCM from scratch, Wiin-Nielsen contacted his friend Joseph Smagorinsky, leader of the Geophysical Fluid Dynamics Laboratory (GFDL) at Princeton University, and Akio Arakawa and Yale Mintz, two professors of Meteorology at the University of California at Los Angeles. Though neither group had an operational forecast model, they had the world's best-developed global GCMs. Wiin-Nielsen requested working copies of these models, and they agreed. Neither group imposed any conditions other than appropriate credit—an impressive generosity considering that each group's model had taken more than a decade to develop. Acquiring the model codes called for in-person visits. In 1975, Robert Sadourny, a French modeler who had studied with Arakawa and Mintz in the 1960s, spent four weeks at UCLA. Meanwhile, an Irish meteorologist, Anthony Hollingsworth, made his way to Princeton's Geophysical Fluid Dynamics Laboratory. Both returned to the ECMWF bearing code and documentation, as well as personal knowledge gained during the visits. After extensive comparison, the ECMWF settled on the GFDL scheme. Soon, however, the ECMWF replaced part of the GFDL model (known as the "model physics") with a new package of its own, retaining only the GFDL dynamical core (the part of the model that simulates atmospheric motion). Later this too was replaced with a spectral core coded "from scratch."[41] After four years of research and development, the ECMWF commenced operational medium-range forecasting in August 1979.

Six years later, in 1985, the ECMWF's global forecasts outdid the British Met Office forecasts for the northern hemisphere on some measures. (The ECMWF's performance in the southern hemisphere was slightly worse than that of the Met Office model.) According to B. J. Mason, former head of the Met Office, the accuracy of forecasting had advanced by a full day: "[T]he 72-hour, 500mb forecast is now as good as the 48-hour forecast was 7 years ago, and the 48-hour forecast is now as good as the 24-hour forecast was then."[42] In fact, by 1985 the ECMWF had achieved the objective imagined by the 1971 cost-benefit analysis: producing six-day forecasts of about the same accuracy as the two-day forecasts of 1971.[43] By comparison with all earlier technoscientific forecasting, in which the limit of accurate forecasting had advanced from about 24 to about 48 hours, this was an astonishing achievement.

At a seminar commemorating the ECMWF's first ten years, in 1985, Joseph Smagorinsky—a towering figure both in computer modeling and in the creation of a global weather data infrastructure—told the assembled representatives of European national weather services that the ECMWF "commands the awe and admiration of the meteorological world." By the time I began interviewing climate scientists in the early 1990s, most of my interviewees described the ECMWF as the best forecast center in the world. The ECMWF's global and regional forecasts and its thousands of "data products"—subsets of collected, analyzed and processed data—were, and are, widely used by weather services not only in Europe, but all over the world.

At the same 1985 seminar, Ernst Lingelbach, former head of the Deutscher Wetterdienst, noted that of about fifty large projects sponsored by COST, each with numerous "sub-programmes," "only one . . . has led to the establishment of a great common research institute, namely the ECMWF. All other [COST] actions are being implemented by coordinating the research efforts of individual national laboratories."[44] Concerning the role of the ECMWF in European integration, Lingelbach went on to say:

The fathers of the idea of European unification still have many things to hope for. However, in the meteorological community, their ideas have been realized with the integration of the work of seventeen European states [through the ECMWF]. The lead over other countries has been re-gained in this area and many industries in Europe have become aware of the great use that they can make of the Centre's ever improving medium range weather forecasts.[45]

With the demise of communism, the ECMWF expanded its membership. Today 34 European states support its operations.

In the late 1980s, the ECMWF's role expanded to encompass climate change analysis. The ECMWF produced a highly cited "reanalysis" of fifteen years' weather observations. This was an entirely new way to trace the evolution of Earth's climate during the period of instrumental records. Rather than the *average* temperatures, pressures, etc. pre-calculated at individual weather stations used in most climate datasets, reanalysis would begin with unanalyzed data: *every available instrument reading* for some long period. Then it would pipe those data through a state-of-the-art forecast model. The forecast model would fill in gaps in the observational data. This would overcome the serious problem of incomplete climate data, on the one hand, and a constantly changing observation and forecasting system, on the other. The result would be a continuous, consistent, and complete data image of the weather over the entire Earth.

The ECMWF completed a 15-year reanalysis in 1996, covering the years 1979–1994. In 2000 it initiated a 40-year reanalysis, then extended that to 50 years.[46] Other agencies also created reanalyses, but the outstanding quality of the ECMWF's reanalysis model garnered profound respect from the climate science community. Reanalysis has not displaced traditional data analysis in the study of climate change, and for various technical reasons it may never do so. Nonetheless, reanalysis is widely regarded as a major benchmark for understanding and measuring global climate change, and the ECMWF was among its pioneers.[47]

Conclusion

Benedict Anderson famously argued that maps contribute to the "logoization of political space."[48] He also liked to say that people encounter such "logos"—simplified maps of their own nations—most frequently while watching the weather report on the nightly news. Yet although people naturally care most about the weather at their own location, modern weather reports no longer focus only on the nation. Instead, they offer a nested series of views at various scales. Continuous reporting of weather around the world occupies entire television channels. Even in ostensibly national news, European weather reporting typically zooms out to the scale of the entire North Atlantic, then zooms in to a smaller transnational region before focusing down to the national level. An information commons built from instruments, computers, and arcane mathematics thus joins global natural systems with multiple levels of political identity—globe, region, nation, city—in the everyday consciousness of people around the world. Weather information plays a major supporting role in activities and infrastructures of the modern world: agriculture, air travel, shipping, managed waterways, and others far too numerous to mention. The advance warning we now have of storms, snowfall, heat waves, floods, and other extreme weather events saves hundreds of billions of dollars and thousands of human lives each year, yet the total annual cost of the world weather forecast system has been estimated at just $10 billion. Many people routinely check the radar on smart phones or computers before heading out the door. Weather forecasts for the next few days—long enough for weather systems to travel from central Canada to Germany (though this is not exactly what happens)—strongly influence planning of weather-related events.

Like other chapters in this volume, this one has emphasized the joint contributions of nature (weather itself, a structurally global phenomenon) and technology (observing systems and computer models) to the construction of this commons. Among other things, we have seen how building the European Centre for Medium-Range Weather Forecasts required the pooled scientific, financial, and technological resources of an entire continent, yet soon provided benefits not only to the European region but to the whole world.

The moral economy of this commons is rooted simultaneously in the natural phenomena of which it provides knowledge, in the institutions that generate weather data and forecasts, and in the scientific knowledge and technological systems that permit those forecasts to be created and shared. Weather data were freely shared first as scientific curiosities; later as nationally produced quasi-public goods, most beneficial when widely shared; and finally as products of an international enterprise, as global public goods. The ECMWF has been this chapter's primary example, but a number of other entities (including the US National Aeronautics and Space Administration and the European Organisation for the Exploitation of Meteorological Satellites) contribute equally to this widely used information commons.

Are there tragic possibilities that could affect the weather information commons? Yes—not because of overconsumption, which cannot occur, but through the withdrawal of major contributors (as in wartime), or through the privatization of weather information. Beginning in the 1980s, private companies such as Accuweather began to challenge the public monopoly on weather forecasting.

An extreme version of this challenge arose in 2005, when US Senator Rick Santorum sponsored a "National Weather Service Duties Act" that would actually have *prohibited* the US Weather Service from issuing routine weather forecasts, since such forecasts were available from commercial weather forecasters—who use data produced at public expense, and provided to the commercial forecasters at no cost, by the Weather Service. Instead, the bill would have required the Weather Service to release only forecasts of severe storms and other dangerous weather events—a task it could not possibly fulfill without creating routine forecasts as well.[49] The act garnered no co-sponsors and died in committee. Yet it was only one in a long series of similar efforts by private-sector forecasters to wrest partial control of the weather forecast infrastructure from public agencies.

This issue is more complex than it appears, however. Basic weather forecasts and severe weather alerts clearly belong in the public domain, but there is a large category of "value added" forecast products, such as targeted forecasts specially produced for individual clients, that arguably should not be the responsibility of public agencies. For example, a burgeoning industry in "weather risk management" uses publicly available data in combination with proprietary models and methods toward such purposes as "management of the financial consequences of adverse weather for [firms and organizations] with natural exposure to weather, [and] commercial trading of weather risk, both in its own right and in conjunction with a variety of commodities." The expense of these efforts is best borne by the private entities that require them and profit from them.[50]

As for weather data themselves, it is most ironic—after all I have said about Europe's pivotal role in building the global weather information commons—that the single most direct challenge to a global data commons also came from Europe. By 1995, many European national weather services had gradually introduced commercial operations, selling specialized forecasts and data in order to recoup some of their costs as well as to stave off private competition. These operations had engaged in a "gentleman's agreement" not to sell products outside their national borders—but with European integration in the then-new European Union, this arrangement could not hold.[51] A coalition of European weather services pressed the World Meteorological Organization for the right to charge fees for data. They succeeded in pushing through a compromise, WMO Resolution 40, which altered the long-standing WMO policy that weather data should be freely shared. "Recognizing . . . the requirement by some [WMO] Members that their National Meteorological Services initiate or increase their commercial activities," Resolution 40 created two tiers: "essential" data, which must be freely shared, and "additional data and products," for which fees could now be charged.[52] Today the ECMWF still provides many kinds of data and forecasts freely or at cost, especially to scientific researchers—but, like the European national weather services, it now also sells many of its data products, for fees that can run to tens of thousands of euros.[53]

Probably the single most controversial action in the history of the WMO, Resolution 40 remains severely contentious. John Zillman, the WMO's president from 1995 to 2003, strongly opposed the introduction of high fees for data, and wrote several papers and presentations on the subject. In one address, he said:

The widespread trend towards implementation of competition policy at the national level, and governments' increasing tendency to open up many former government functions to competition and the operation of market forces, represents a full-frontal assault on the most fundamental principle of meteorological service provision worldwide—its reliance on cooperation rather than competition and its dependence on free and open exchange of information and knowledge in the overall global public interest.[54]

When I interviewed Zillman in 2001, he told me this had been "his number one cause over the last decade and a half."[55]

Ironically, the practice of charging high fees for weather data produced at public expense has actually hobbled Europe's competitive position and reduced the potential value of weather data to European economies. A 2002 study found that European private-sector meteorology is only one-tenth as large as the same industry in the United States. "Given that the US and EU economies are approximately the same size," that study concluded, "the primary reason for the European weather risk management and commercial meteorology markets to lag so far behind the US is the restrictive data policies of a number of European national meteorological services."[56] The weather information commons thus remains a precarious resource, one that highlights the crossroads humanity currently faces with respect to virtually all scientific data and information: open and freely shared, common resources for all to use, or proprietary and reserved for private gain.

Notes

1. Paul N. Edwards, *A Vast Machine: Computer Models, Climate Data, and the Politics of Global Warming* (MIT Press, 2010). Some portions of the present chapter are adapted from this book.

2. P. A. Samuelson, "The Pure Theory of Public Expenditure," *Review of Economics and Statistics* 36, no. 4 (1954): 387–389.

3. Michael A. Heller, "Tragedy of the Anticommons: Property in the Transition From Marx to Markets," *Harvard Law Review* 111, no. 3 (1997): 621–688.

4. Todd Sandler, "Intergenerational Public Goods: Strategies, Efficiency and Institutions," in *Global Public Goods*, ed. I. Kaul, I. Grunberg, and M. Stern (Oxford Univer-

sity Press, 1999); J. E. Stiglitz, "Knowledge as a Global Public Good," also in *Global Public Goods.*

5. John W. Zillman, "Meteorological Service Provision in Changing Times," presented at Third Technical Conference on Management of Meteorological and Hydrological Services in WMO Regional Association V (South-West Pacific), Manila, 2001.

6. D. C. Cassidy, "Meteorology in Mannheim: The Palatine Meteorological Society, 1780–1795," *Sudhoffs Archiv* 69 (1985): 8–25.

7. Edmond Halley, "An Historical Account of the Trade Winds, and Monsoons, Observable in the Seas Between and Near the Tropicks, with an Attempt to Assign the Phisical Cause of the Said Winds," *Philosophical Transactions of the Royal Society of London* 1, no. 183 (1686): 153–168.

8. Heinrich Wilhelm Dove, *The Distribution of Heat Over the Surface of the Globe: Illustrated by Isothermal, Thermic Isabnormal, and Other Curves of Temperature* (Taylor and Francis, 1853).

9. William Ferrel, "An Essay on the Winds and Currents of the Ocean," *Nashville Journal of Medicine and Surgery* 11, no. 4–5 (1856): 287–301, 375–389.

10. John Ruskin, "Remarks on the Present State of Meteorological Science," *Transactions of the Meteorological Society* 1 (1839): 56–59.

11. Mark S. Monmonier, *Air Apparent: How Meteorologists Learned to Map, Predict, and Dramatize Weather* (University of Chicago Press, 1999), p. 41.

12. Aleksandr Khristoforovich Khrgian, *Meteorology: A Historical Survey* (Israel Program for Scientific Translations, 1970).

13. The word "synoptic" refers to charts, forecasts, or data covering a large area, typically several tens of degrees of latitude and longitude. See American Meteorological Society, *Glossary of Meteorology* (Allen, 2000).

14. W. J. Johnston, *Telegraphic Tales and Telegraphic History: A Popular Account of the Electric Telegraph, Its Uses, Extent and Outgrowths* (W. J. Johnston, 1880), p. 172.

15. Frederik Nebeker, *Calculating the Weather: Meteorology in the 20th Century* (Academic Press, 1995).

16. Monmonier, *Air Apparent*, pp. 157–168.

17. Paul N. Edwards, "Meteorology as Infrastructural Globalism," *Osiris* 21 (2006): 229–250; Edwards, *A Vast Machine.*

18. Robert Marc Friedman, *Appropriating the Weather: Vilhelm Bjerknes and the Construction of a Modern Meteorology* (Cornell University Press, 1989).

19. Convention Relating to the Regulation of Aerial Navigation, Paris, 1919.

20. Howard Daniel, "One Hundred Years of International Co-operation in Meteorology (1873–1973)," *WMO Bulletin* 22 (1973): 171–175.

21. Gordon D. Cartwright, and Charles H. Sprinkle, "A History of Aeronautical Meteorology: Personal Perspectives, 1903–1995," in *Historical Essays on Meteorology, 1919–1995*, ed. J. Fleming (American Meteorological Society, 1996); Eda Kranakis, "The 'Good Miracle': Building a European Airspace Commons, 1919–1939," in this volume.

22. Edwards, *A Vast Machine*, chapters 3 and 5.

23. Morton J. Rubin, "Synoptic Meteorology and the IGY," in *Geophysics and the IGY: Proceedings of the Symposium at the Opening of the International Geophysical Year*, ed. H. Odishaw and S. Ruttenberg (American Geophysical Union, 1958); World Meteorological Organization, *International Geophysical Year 1957–1958, Meteorological Program: General Survey* (World Meteorological Organization, 1956).

24. Edwards, *A Vast Machine*, chapter 8; Clark A. Miller, "Scientific Internationalism in American Foreign Policy: The Case of Meteorology, 1947–1958," in *Changing the Atmosphere: Expert Knowledge and Environmental Governance*, ed. C. Miller and P. Edwards (MIT Press, 2001).

25. Walter A. McDougall, . . . *The Heavens and the Earth: A Political History of the Space Age* (Basic Books, 1985).

26. World Meteorological Organization, *First Report on the Advancement of Atmospheric Sciences and Their Application in the Light of Developments in Outer Space* (Secretariat of the World Meteorological Organization, 1962).

27. Ruskin, "Remarks on the Present State of Meteorological Science."

28. Edwards, *A Vast Machine*, chapter 9.

29. The European Unidata industrial group of the 1970s is entirely unrelated to the contemporary Unidata Program based at the University Corporation for Atmospheric Research in the United States, which provides meteorological software and data to universities.

30. "Annex 1," *Report of the Working Group on Policy in the Field of Scientific and Technological Research*, Secretariat of the Committee for Medium-Range Economic Policy, European Economic Community, Brussels, 1967. ECMWF archives.

31. H. Taba, "Interview with Erich Süssenberger," *Bulletin of the World Meteorological Organization* 51, no. 2 (2002): 112ff.

32. "Projet de rapport préliminaire du groupe spécialisé 'Météorologie,'" Commission des Communautés Européennes, Groupe spécial recherche, Brussels, 22 January 1968, EUR/C/639/68 (available at http://ecmwf.int).

33. Austin Woods, *Medium-Range Weather Prediction: The European Approach: The Story of the European Centre for Medium-Range Weather Forecasts* (Springer, 2006), p. 30.

34. COST/136/71 (III/1118–71-E), "Project Study on European Centre for medium-range weather forecast (ECMWRF)," Secretariat of the program on European Cooperation in the Field of Scientific and Technical Research, Brussels. Annex 1, 26 August 1971, p. 4.

35. Taba, "Interview with Erich Süssenberger."

36. COST/136/71 (III/1118–71-E), "Project Study on European Centre for medium-range weather forecast (ECMWRF)," Secretariat of the program on European Cooperation in the Field of Scientific and Technical Research, Brussels, 5 August 1971, p. 51.

37. Ibid.

38. Ibid., p. 78.

39. Ibid., p. 81.

40. Woods, *Medium-Range Weather Prediction*, pp. 65–66.

41. Ibid., chapter 7; Anthony Hollingsworth, Austin Woods, and Robert Sadourny, interviews with Paul N. Edwards, June 1998.

42. B. John Mason, "Progress in Numerical Weather Prediction," in *Medium-Range Weather Forecasts—The First Ten Years: Proceedings of a Seminar to Commemorate the 10th Anniversary of ECMWF* (European Centre for Medium-Range Weather Forecasts, 1985), p. 68.

43. Godwin Olu Patrick Obasi, "On the Occasion of the Tenth Anniversary Celebrations of the Centre for Medium-Range Weather Forecasts," in *Medium-Range Weather Forecasts—The First Ten Years*, p. 14.

44. Ernst Lingelbach, "The Background to ECMWF and the Economic and Scientific Justifications for Establishing the Centre," in *Medium-Range Weather Forecasts—The First Ten* Years, p. 14, quoting in part "Detailed background of COST," General Secretariat of the Council of European Communities, Brussels, November 1981.

45. Ibid., pp. 25–26.

46. J. K. Gibson et al., "ERA-15 Description," *ECMWF Reanalysis Project Report Series* 1 (1997): 72; S. Uppala et al., "ERA-40: ECMWF 45-Year Reanalysis of the Global Atmosphere and Surface Conditions 1957–2002," *ECMWF Newsletter* 101 (2004): 2–21.

47. Edwards, *A Vast Machine*, chapter 12.

48. Benedict R. Anderson, *Imagined Communities: Reflections on the Origin and Spread of Nationalism*, third revised edition (Verso, 1991): xiv.

49. National Weather Services Duties Act of 2005 (available at http://thomas.loc.gov).

50. Weather Risk Management Association, "What Is Weather Risk Management?" (available at www.wrma.org).

51. Peter Weiss, *Borders in Cyberspace: Conflicting Public Sector Information Policies and Their Economic Impacts* (US Department of Commerce, National Oceanic and Atmospheric Administration, and National Weather Service, 2002); Woods, *European Centre*, p. 226ff.

52. WMO Resolution 40 (CG-XII), "WMO policy and practice for the exchange of meteorological and related data and products including guidelines on relationships in commercial meteorological activities," 1995 (available at http://www.wmo.int).

53. European Centre for Medium-Range Weather Forecasts, "Tariffs" (available at http://www.ecmwf.int).

54. Zillman, "Meteorological Service Provision."

55. John Zillman, interviewed by Paul N. Edwards, July 30, 2001.

56. Weiss, *Borders in Cyberspace*, p. 8.

7 Breeding Europe: Crop Diversity, Gene Banks, and Commoners

Tiago Saraiva

The 30-year campaign to create a World Gene Bank culminated in 2008 with the opening of a Global Seed Vault inside a mountain on Norway's Svalbard Archipelago, far north of the Arctic Circle. The precarious resource of crop diversity seems now to be safely conserved inside this technologically updated green version of Noah's Ark. It promises to conserve agricultural biological diversity for the benefit of humankind in case of major catastrophes (tsunamis, terrorist attacks, nuclear war, insurrections).[1] With a storage capacity of about 4.5 million seed samples, it serves as the equivalent of a secure bank vault for the multiple collections of crop diversity scattered around the world located at sites as turbulent as Iraq's national gene bank in Abu Ghraib. The promoters justify their undertaking by stating that humanity's food supplies would be in jeopardy without access to the crop diversity that is needed to adapt cultivated plants to changing environments.

Repeated references to the protection of humanity's common heritage also invoke images of a pre-modern world in need of protection from the overwhelming forces of industrialization and globalization.[2] The facility at Svalbard was designed to resist major disasters, which is paradoxical because the crop diversity it harbors is endangered by the very standardization of agriculture at a global scale that breeding based on seed banks has facilitated.[3] The problem of diminishing crop diversity is particularly evident in Europe. Plant breeders have been worrying about the increasing genetic uniformity of European fields since the beginning of the twentieth century. The much-lamented nostalgic world of local varieties based on the exchange of seeds among farmers—a typical case of a gift-based moral economy sustaining traditional commons—has disappeared before the onslaught of the monotonous operations of industrialized agriculture.[4] The nostalgia is based on convincing data: in the 1980s one single variety of rye accounted for half the acreage planted in Germany; Dutch farmers in 1990 were sowing close to 90 percent of their acreage with no more than three

high-yielding varieties of the main crops (wheat, barley, oats).[5] Denuncia-
tion of "genetic erosion" in European fields, and of the associated dangers
of increased vulnerability to pests or climate change, thus dovetails nicely
with the critique of "local cultural erosion."[6]

This chapter argues that the history of crop diversity is more than a
mere simple enclosure of the commons by industrialization processes. In
fact, it shows how local commons gave way to more extended "cosmopolitan commons." It explores how plant breeders, after transforming the
traditional local commons of crop diversity into a commercial and national
resource in the period between the two world wars, made it into a cosmopolitan commons in the aftermath of World War II through the medium
of coordinated gene banks. As the technological sophistication of the Svalbard vault suggests, crop diversity is much more than a "common heritage
of humankind." The meters-thick walls of reinforced concrete, the heat-
sealed, laminated, moisture-proof foil seed packages, and the standardized
descriptors of the collection all strongly suggest the technological sophistication of the new commons. The narrative argues the case for the historical
relevance of the building of this technoscientific commons for the process
of European integration. Gene banks are taken as central elements of a narrative that tries connect the history of the hidden integration of Europe
with the history of managing plant genetic resources—in other words, to
merge European history with the history of commons through the history
of technology.

The first section of the chapter discusses the role of crop diversity in modern plant breeding since the second half of the nineteenth century, emphasizing the significance of large collections of cultivated plants for European
breeders. The next sections first illuminate how Nazi and Soviet continental ambitions depended on controlling plant genetic resources, and then
examine the transnational efforts to guarantee access to the resource across
Western and Eastern Europe in the postwar years. The chapter concludes
with a discussion of different modes of ensuring crop diversity. In particular, it compares strategies based on *ex situ* facilities—the gene banks—with
initiatives relying on *in situ* conservation—the farmers' plots. The different
ways of relating to the commons and defining the community of commoners can be perceived as different ways of building Europe.

Crop Diversity and Pure Lines

Historians have profusely documented the large-scale consequences of the
Columbian exchange for agriculture worldwide, Europe included.[7] And no
scholar dealing with European imperialism ignores the role of botanical

gardens as promoters of global seed exchanges sustaining cinchona plantations in Java or cocoa estates in the Caribbean.[8] It is thus surprising that
twentieth-century gene banks, in which much of the world's crop diversity
is currently deposited, have received so little attention from historians. To
be sure, there are a few accounts of the history of these scientific institutions on which to base our narrative, but there is nothing comparable to
the volume and sophistication of the literature dedicated to botanical gardens. And most of the secondary literature was produced by scientists or
activists directly involved in crop diversity conservation, whose presentist
and specialist perspective typically ignores the more significant historical
questions. In particular, there is little elucidation of how the development
of gene banks intersects with European history. Since Soviet farm collectivization, Nazi food policies, and the European Common Agricultural Policy
are all difficult to understand without reference to such technoscientific
infrastructure, a revised view of the historical role of gene banks in accounts
of twentieth-century Europe seems justifiable.

First we must understand the importance of crop diversity in agricultural modernization. In the second half of the nineteenth-century European crop breeders successfully began to sell seeds for profit. Their success
was due in large part to the technique of pedigree selection, first developed
in the 1850s by a Frenchman named Louis de Vilmorin.[9] In contrast to
the traditional farmer's mass selection, in which seeds from the best plants
are selected and sowed together in the next year, in Vilmorin's pedigree
selection all the descendants were derived from a single individual through
self-fertilization. Following the example provided by animal breeders and
their stud books, plant breeders now used detailed records to identify the
genealogy of each individual plant that was cultivated in their plots. The
bulk of the breeder's work thus consisted of rambling around in farmer's
fields, identifying an interesting plant, reproducing it through self fer-
tilization, and carefully documenting the characteristics of the progeny.
Through pedigree selection, breeders produced pure lines—stable varieties
selected for some important feature such as pest resistance, early ripening,
or milling properties. They then combined different properties by crossing
different pure lines to obtain the hybrids that made their fortunes in the
seed market.

The Vilmorins mixed family genealogy with pedigree selection. Henry
de Vilmorin, son of Louis, scored his first breakthrough by crossing hardy
English squarehead wheats and early-maturing varieties from Aquitaine,
creating the Dattel hybrid in 1883. In 1905, by then under the leadership
of Philip de Vilmorin, the company released Le Bon Fermier (Good Farmer).
In 1927 it released the celebrated Vilmorin 27. Most French wheats today

are direct descendants of these original Vilmorin crosses.[10] From the early 1900s to the end of the century, wheat productivity increased from an average of 1.2 metric tons per hectare to 4 metric tons per hectare in Europe (over 10 metric tons per hectare in some countries), and about half of the increase is currently estimated to have resulted from selective breeding.[11]

The standardized pure lines coming out of the breeders' plots promised to accomplish the dream of converting agriculture into an industrialized activity by putting an end to the chaos of the farmers' fields cultivated with unreliable traditional crop varieties, the so-called landraces produced by mass selection and adapted to local conditions.[12] The paradox is that the breeders' primary resource was none other than these populations of landraces. From these the breeders isolated pure lines by repeated inbreeding. No successful breeder could dispense with the genetic variability provided by traditional landraces. In 1911 the Vilmorin Company had about 1,200 varieties of wheat at its disposal, one of the largest collections in Europe.[13] This diversity had been accumulated during voyages of exploration, by participation in international fairs, and, more importantly, by means of a dense national and international network of correspondents that sent local landraces to the Vilmorin headquarters near Paris.

The establishment of such networks and the formation of collections with thousands of different varieties were also central to the work of the Italian breeder Nazareno Strampelli, whose "elite races" of wheat would quickly replace local landraces all over Italy in the first decades of the twentieth century.[14] His hybrids were a combination of pure lines obtained chiefly from the Vilmorins, pedigree selections of Italian landraces (such as Rieti), and exotic material from Japan and other countries. The genetic variability of these different sources enabled the development of varieties like Ardito, which combined rust resistance with short thick stems. This last feature was a crucial characteristic of the Strampelli varieties; it allowed for generous use of chemical fertilizers without the problem of lodging (falling over) of the wheat plants that was associated with the long stems characteristic of traditional landraces.

It made no difference that the Vilmorins were a private company and Strampelli was working in a public institution; in order to sustain their operations, both were tapping into typical traditional local commons developed by farmers in the course of many generations, commons ruled by seed exchanges between neighbors. These were clearly examples of gift economies, resembling those described in the literature on traditional commons. Informal farmer-to-farmer exchanges allowed innovations to spread and be tested by others for adaptation and adoption, thus expanding the spatial and social reach of these commons.[15] The plant breeders, however, were now transforming these commons into a resource for sustaining their private or public purposes. But by replacing the diversity of the landraces with new standardized forms of life, they were destroying the very commons they relied on.

That was not the entire story, however. The exchange networks on which the breeders built their collections also included other plant breeders. Strampelli's hybrids made use of Vilmorin's wheats. Hence, another commons was being formed. Breeders, public or private, exchanged their new creations with one another, enabling colleagues and rivals to build on previous innovations. Strampelli did not have to pay royalties to use Vilmorin's pure lines. And yet this new commons was not the same as the traditional commons built on neighborly relations. The new breeders' commons was a creature of the global transportation system that emerged in the nineteenth century, enabling rapid and reliable exchanges with agents in Europe, America, or Japan. Railways, mail service, catalogs, and international exhibitions all contributed to the formation of the breeders' collections. But in the end this was still not a very sophisticated commons. All too often breeders found it necessary to visit the most famous collections in person, for the simple reason that published catalogs could never give enough information on the qualities of the varieties. One could even say that in spite of its new global reach this commons was still not very different from the one built by farmers. As we shall see, it didn't take long for this first informal breeder's commons to be replaced by nationalization of the resource.

The Soviet Empire and the Scramble for the World Genetic Resources

As Strampelli's use of Japanese seeds made clear, the genetic variability needed to produce wheat forms amenable to the techniques of modern agriculture, namely chemical fertilizers, couldn't be limited to European landraces. In 1914, Erwin Baur, who would soon become one of the most distinguished German geneticists, was already advocating the collection and preservation of cultivated primitive races, especially of wheat and barley in the Orient, Asia, North Africa, and Ethiopia.[16] And in view of the enormous costs involved, he had no qualms about arguing for state intervention in the business of plant breeding. Correspondence and occasional visits were not adequate to sustain the scale of operations demanded by industrial breeders. Only the state had the means to organize the systematic collection expeditions that could transform global crop diversity into a resource available to plant breeders.

It is thus entirely fitting that the first institution dedicated to the planned collection and preservation of landraces and their wild relatives was in the Soviet Union: the All-Union Institute of Applied Botany and New Crops, founded in 1924 and later renamed the All-Union Institute of Plant Industry (VIR).[17] Its institutional ancestor, the St. Petersburg Bureau of Plant Botany, had been surveying Russian territories since the end of the nineteenth century, documenting local landraces. Its director, Robert Regel, was responsible in particular for a comprehensive inventory of Russian barleys native to the different territories of the Empire. But it would be Nikolai I. Vavilov, director of the Bureau from 1920 on, who would transform the institution into a truly global one in order to better serve the policies of rapid industrialization launched by the Bolshevik regime.

It is hard to overestimate Vavilov's importance to the emerging understanding of crop variability. He made a name for himself in the international community of plant geneticists by proposing two theories that were crucial in understanding genetic diversity at the global scale. His "theory of the homologous series of hereditary changes" predicts how similar mutations appear in related species. Darwin had described "parallel variations," but Vavilov studied the phenomenon in a large collection of cultivated plants and proposed a genetic explanation. Homologous series of phenotypes were shown to correspond to homologous series of mutations. Vavilov himself describes the significance of his theory for gene banks in a 1920 text:

All the attention of our laboratory is being paid to the search for series in isolated genera which cannot be intercrossed. Vetch, lentil and pea cannot be crossed, but their series of variation, we may say, are almost the same. These days we have received new vetch samples from Kharkov, so all the gaps in the series are now filled. For 3 years we have been trying to do the same with other forms of cultivated plants. . . . We have not got enough examples. It is necessary to attract everything existing in the world. I shall do all I can to send somebody abroad in the autumn of 1921 to collect plant materials. We need to delegate a collecting mission to Africa where cultivated plants have practically never been studied.[18]

The collecting of variability, in order to be systematic and avoid the "multitudinous chaos of innumerable forms" typical of traditional plant collections, had to fill up "all the gaps in the series" by searching, if necessary, the "entire world" to collect plant materials. Instead of the networks of correspondents established by Vilmorin or Strampelli, Vavilov was proposing state-sponsored surveys to complete the series for every crop in the world. Through this cumulative method, the resource of global genetic variability could be brought in its entirety to St. Petersburg (by then called Leningrad).

The surveys were not to cover every territory of the planet in an arbitrary way. According to Vavilov's theory of the centers of origin of cultivated plants, the area of the greatest variability of a domesticated plant was also its region of origin, where its wild strains should still exist.[19] By 1926, after missions to Iran and Mongolia in the early 1920s to which he added the study of materials from Asia and Africa, Vavilov was in a position to propose five such major centers of variability: "The areas of origin and type-formation of the most important cultivated plants which at the same time are the foci of a wealth of types, belong mainly to the mountain areas of Asia (Himalayas and its system), the mountain systems of northeastern Africa and the mountain areas of southern Europe (the Pyrenees, the Apennines and the Balkans), the Cordilleras and the Southern spurs of the Rocky Mountains."[20]

Although the number and the extent of those centers were to undergo revision in later publications, Vavilov's insights would prove invaluable for every subsequent plant breeding program. After presenting his findings in 1927 at the First International Wheat Congress held in Rome and at the Fifth International Genetic Congress in Berlin, Vavilov became an inspirational figure for breeders all over the world. But let us set Vavilov's international stature aside for now and focus on his relations with the Soviet regime.

According to Vavilov, his world collection would be the motor of the Soviet plant industry, providing peasants with improved seeds produced by VIR geneticists. In 1931 he urged the Soviet government to assume control of the replacement of farmers' landraces with the new strains, rather than waiting for results from the "spontaneous processes that were relied upon in the past." In the same year he boasted that one single newly introduced variety of maize was now grown on 1.5 million hectares, and that in a few years "all the seed material in the Soviet Union would be standardized."[21] Stalin would use similar words when bragging to Winston Churchill that "we have improved the quality of the grain beyond measure. All kinds of grain used to be grown. Now, no one is allowed to grow any sort but the standard Soviet grain from one end of the country to the other."[22] Vavilov's promise was clear: genetic variability was a foundational resource for Stalin's Great Break of 1928–1930, which combined collectivization of the peasantry, crash industrialization, and a planned economy. VIR's seeds, whose acronymic name fittingly recalled the industrial nature of the undertaking, would contribute to a new agriculture in which, Stalin said, "small, backward, and fragmented peasant farms [would be replaced by] consolidated, big, public farms, provided with machines, equipped with the data of science, and capable of producing the greatest quantity of grain for the market."

Owing to Vavilov's ultimate fall from grace and the triumph of Lysenko-ism, Vavilov's proximity to the Soviet leadership in the 1920s and the 1930s has been neglected.[23] But only by placing Vavilov in the context of Stalin's Great Break can we understand the enormous resources available for his scientific enterprise. In 1929 he was made president of the newly founded VASKhNIL, the Lenin Academy of Agricultural Science, created under the banner of "mobilizing science in the service of socialist construction." In 1932–33 the research system controlled by the VASKhNIL included 1,300 institutions employing about 26,000 specialists, and was without equal in the entire world.[24] And 185 of those institutions were engaged in plant breeding. The largest and most important of them was the VIR, directed by Vavilov himself. This concentration of power was characteristic of the Soviet organization of science. A few chosen scientists had direct access to the Bolshevik leadership and thus were able to funnel funds from powerful state agencies into their research empires. In the case of Vavilov, the funds were from the People's Commissariat of Agriculture (Narkomzem).

It was the Narkomzem that funded most of the work of assembling the collection of plants at Leningrad. At least forty expeditions were organized by Vavilov and his aides to collect plant varieties for this first global gene bank. The 250,000 samples of cultivated plants and their wild relatives stored in metal cases in Leningrad at the end of the 1930s also included material from another 140 expeditions across the vast Soviet empire. Following the perverse logic of plant breeders since the introduction of their pedigree selections, as landraces were replaced by modern bred varieties in the farmers' fields, the variability formerly present in the landraces was to be preserved *ex situ* to sustain further breeding. This appropriation of a "resource" (as Vavilov liked to call it) by the Soviet state was intended not only to increase the output of Soviet agriculture but also to enable the colonization of barren tracts of land previously deemed too inhospitable for sustaining human communities. The VIR gene bank was meant to supply genes for cold resistance or higher yields, thus putting flesh on the bones of Soviet dreams of settling northern Arctic regions and vast areas of Siberia.[25]

Vavilov's theories and institutions provided a clear road map for any country willing to participate in the world scramble for plant genetic resources. Major countries sent expeditions to the Himalayas, to Ethiopia, to Turkey, and to Peru to harvest this new precious resource of genetic diversity. National collections like those of the United States and Germany took shape in the interwar years. The expedition craze of the 1920s and the 1930s brings to mind the eighteenth-century expeditions that made exotic animals and plants public patrimony by placing them in museums of natural history and in botanical gardens.[26] If by the end of the eighteenth century no European power could sustain imperial ambitions without a botanical garden, in the pre-World War II years a well-endowed gene bank became a crucial attribute of any country expressing continental ambitions. This was true not only for the Soviet Union and the United States but also for Nazi Germany.

Plant Breeding and Nazi Imperial Ambitions

Food issues are currently a burgeoning topic in the historiography of the Nazi regime. Food is now an indispensable ingredient in any analysis of Nazi interpretations of German defeat in World War I, of the *volkisch* dimensions of the regime embodied in the motto Blut und Boden (Blood and Soil), or of Nazi colonization of Eastern Europe. Also, the apparently minor field of plant breeding has now come to occupy center stage in any competent discussion of science and the Nazis, including close scrutiny of the work done at the institutes of the Kaiser-Wilhelm-Gesellschaft zur Förderung der Wissenschaften (Kaiser Wilhelm Society for the Advancement of Science).[27, 28] It is revealing that the generous 1932 budget of the society—about 5.7 million Reichsmarks—would more than double during the Nazi years, reaching 14.4 million Reichsmarks in 1944. And among the society's many institutes, the Kaiser-Wilhelm-Institut für Züchtungsforschung (Kaiser Wilhelm Institute for Plant Breeding, henceforth abbreviated KWIPB) was most generously endowed, followed by the one for animal breeding.

The early Nazi interest in the KWIPB was justified by achievements such as Reinhold von Sengbusch's breeding of sweet lupine in the early 1930s. In contrast to the bitter common forms of lupine, the sweet ones that came out of the KWIPB's experimental plots were edible by cattle, a property that promised a significant improvement in Germany's food autonomy.[29] Imports of cattle fodder were actually the main stumbling block to German agricultural autarky, and the Nazis repeatedly insisted that Germany's lack of agricultural autarky had been one of the causes of its defeat in World War I. Visions of sweet lupine growing on Germany's extensive tracts of poor sandy soil sustained the promise of freeing the country from the abhorred dependency on supplies from Argentina, the United States, or Canada. More to the point, Erwin Baur, the KWIPB director, noted that thanks to sweet lupine, Germany would be in a much better position in a future European war. (During World War I, the Allied blockade caused some 600,000 deaths by starvation among the German and Austrian populations.)

As in the Soviet Union, the new seeds were inscribed with the mission of rationalizing German fields. The Seed Law of 1934, passed only one year into Nazi rule, promised to eliminate many of Germany's landraces by forcing breeders to get rid of their less valuable varieties and replace them with more productive and pest resistant ones.[30] Of the 454 varieties of wheat, 438 were banned. Of the 1,500 varieties of potatoes cultivated by German farmers in the 1910s, no more than 74 remained in 1937.[31] The process was orchestrated by the all-powerful Reichsnährstand, the Nazi institution that controlled the regime's food policies (wheat prices, the baking industry, seed potatoes, agricultural exhibitions, and so on) and which, with its 17 million members, was the regime's biggest single organization.[32] The intensive intervention in and centralization of agricultural policy by the Nazi state was supposed to serve the double purpose of preparing the country for war and guaranteeing that the peasantry remained the backbone of the German national community. If the *volkisch* rituals organized by the Reichsnährstand that celebrated the peasantry under the banner of Blood and Soil might suggest that the Nazi leadership was actually reneging on its commitment to the industrialization of the country, we shouldn't forget that the Nazi peasantry was itself envisioned as a community wedded to a state-standardized, streamlined style of agriculture on the basis of the very modern products of the KWIPB geneticists.[33]

As close followers of Vavilov's work, German plant geneticists were well aware that any ambitious breeding program would require a well-stocked gene bank for quick access to global crop diversity. As was noted above in reference to Strampelli's wheats, cultivars developed only by selection from landraces were poorly adapted to intensive use of fertilizers, and breeders thus became increasingly dependent on exotic material for their crossings. It is therefore hardly surprising that German expeditions, most of them directed by scientists from the KWIPB, were sent to the Vavilov centers of genetic diversity in Latin America, Turkey, Spain, and Ethiopia. Though Erwin Baur had been touting the importance of cultivated primitive races outside Europe since the 1910s, huge nationally celebrated expeditions on the model of the Soviet example were not staged until the Nazi years.[34] In 1935, the celebrated German Hindu Kush expedition brought plant hunters to India, Afghanistan, and Iran. In 1938–39 the SS sponsored Ernst Schäfer's expedition to Tibet, where wheats and barleys were collected along with archeological objects and anthropological data in search of an Aryan heritage.[35] In the aftermath of these expeditions, both the SS and the Kaiser Wilhelm Society founded new plant breeding institutes whose first task was to store and manage the seed collections and explore their genetic diversity in order to contribute to what Minister of Agriculture Herbert Backe called the "nutritional freedom" of the expanding Reich.

The new institutes were the Kaiser Wilhelm Institut für Kulturpflanzen-forschung (Institute for Research on Cultivated Plants), near Vienna, and the SS's Ahnenerbe Institut für Pflanzengenetik (Institute for Plant Genetics), near Graz.[36] The Institute for Research on Cultivated Plants was conceived as the center of a network of collection stations "from the polar sea to the Mediterranean area, from the Atlantic to the extreme continental region, from the sea coast to the Alps."[37] This network would guarantee German access to the plant genetic resources of the entire European continent. In addition to expeditions like those just mentioned, the institute also used Nazi domination of southeastern Europe to tap the genetic resources of the Balkans, particularly northern Greece and Albania.[38]

This new Kaiser Wilhelm Institute for Research on Cultivated Plants was founded in 1943, even as the tide of war was turning against Germany. Interestingly, the promoters of the institute justified their cause by arguing that the wild forms of cultivated plants had to be saved from the ravages of war. Hans Stubbe, the director, also mentions in his report on the Balkans expedition that the recent Italian occupation of Albania had lasted long enough to replace many of the traditional landraces with the new Italian wheats developed by Strampelli. As in the Vavilov collection, the gene bank served not only to provide raw material for the breeder's work but also to conserve the resource that was vanishing by the very fact of breeding. However, in this case it was not only breeders that were enabled to work with the collected material. The extensive collections of wild and primitive forms of cultivated plants as well as traditional landraces were to be subjected to the simultaneous scrutiny of a battery of disciplines: taxonomy, genetics, physiology, and biochemistry.[39]

The same rhetoric of saving the endangered resource of crop diversity from possible destruction was used by German scientists after the war to justify their looting of the Vavilov Institute's network during the Nazi occupation of the Soviet Union. In this period several trainloads of seeds were sent back to Germany and Austria to augment the collections of the Kaiser Wilhelm Institutes of Plant Breeding and of Cultivated Plants.

The accessions of the looted Soviet collection enabled German breeders to launch their ominous investigations into an ersatz rubber based on the Kok-saghyz, a kind of Russian dandelion. Because Germany lacked access to the world's centers of rubber production, a potential shortage of this crucial raw material was a major concern for the Nazi war machine. In 1943, Heinrich Himmler was nominated Plenipotentiary for All Issues related to Plant Rubber. And while historians of the Holocaust ignore the role of IG Farben's synthetic rubber plant in the expansion of Auschwitz at their peril, the agricultural experiment station that was also part of the camp complex routinely gets short shrift. The main objective of the station, with direct collaboration by KWIPB scientists, was the breeding of Kok-saghyz plants with higher rubber content. In 1943 and 1944 there were vast Kok-saghyz plantations in Poland and in the Ukraine, where women and children were used as slave labor. Few cases so clearly reveal the significance of plant breeding for the Nazi regime and the importance of controlling plant genetic resources for the domination of Europe.

The looting of the Vavilov Institute by Nazi scientists suggests something else about the status of plant genetic resources. It is not only that the Nazis and the Soviets had explored the local commons of diverse landraces as no other power, with the exception of the United States, had done before. The crop diversity amassed in the VIR or the Kaiser Wilhelm Institutes was, in contrast to what we saw in the cases of Strampelli and Vilmorin, no longer part of a breeders' commons. The Nazi plundering clearly shows that we are now dealing with crucial nationalized resources sustaining imperial ambitions. It is of course true that before the war there were exchanges of varieties between the different national collections, evidenced by perennial tales of some interesting plant brought into the country as the result of a visit by a breeder to some other national breeding institution. But in view of the hundreds of thousands of specimens deposited in the VIR collection, it is obvious that such sporadic contacts didn't allow for a systematic exchange of seeds among breeders. The Nazi and Soviet collections were first and foremost national collections, and not until after the war would a new trans-European commons of crop variability emerge. In order for that to happen, an entirely new structure of exchange had to be designed and put in place.

Eurocommons

In 1945 the Kaiser Wilhelm Institute for Research on Cultivated Plants was transferred from Vienna to Gatersleben, a village that would soon find itself in the German Democratic Republic. The institute's director, Hans Stubbe, reported to the Kaiser Wilhelm Society that the collections had been saved in their entirety. They included not only precious material collected during 1941 and 1942 in the southeastern parts of Europe but also the material taken from Soviet institutions.[40] Apparently this did not prevent the Soviet occupying forces from collaborating with Stubbe to convert the Gatersleben institute into one of the major European gene banks of the postwar years.[41]

Soviet control over Eastern Europe stimulated the founding of several national gene banks, modeled on the Vavilov Institute, whose tasks were the collection, preservation, and exploration of genetic resources for use in plant breeding. The Bulgarian gene bank was founded in 1947 as an adjunct to the K. Malkow Institute for Introduction and Plant Resources at Sadovo. Czechoslovakia's scattered collections were brought together in 1951 under the centralized control of the Division of Genetics of the Institute of Crop Production in Prague-Ruzyně. In the Hungarian People's Republic, the Institute of Agrobotany, founded in 1959 at Tápiószele, was designated to conserve the country's genetic resources.[42]

Nevertheless, in all these cases the collections were formed around land-races—older and more recent cultivars native to the respective country. Only the sharing of accessions with the Soviet collections—namely that of the VIR, now called the N. I. Vavilov Institute of Plant Industry—enabled the national gene banks of Eastern Europe to increase the diversity at their disposal. This was facilitated by the overarching framework of the Council for Mutual Economic Assistance (Comecon), within which a Technical and Scientific Council for Genetic Resources was established in 1973 to institutionalize the interchange of material between the different gene banks of the Soviet-bloc countries. This Council for Genetic Resources was headed, as might be expected, by the director of the N. I. Vavilov Institute of Plant Industry. The council would later evolve into the Genetic Resources Network of the Comecon, involving seven different countries. One of its main tasks was to standardize protocols for documenting the accessions of each bank in order to facilitate the circulation of material. In addition, the actual management of each bank, especially as regards storage conditions and rejuvenation protocols, were normalized. It was clear that a moral economy could flourish only on the basis of standardized norms and procedures.[43] This was thus the first step in ensuring access for all Comecon breeders to the crop diversity stored in national gene banks; in other words, it was the first step in embedding access to and use of plant genetic diversity within a transnational cosmopolitan commons.

The postwar development of gene banks in Western Europe lagged behind those in the Eastern Bloc. Oddly, the leading protagonists in the East and the West had both headed competing German plant breeding institutes during the Nazi years and had in fact been fierce rivals. Hans Stubbe, former director of the Kaiser Wilhelm Institute for Cultivated Plants, assumed a major role in the conservation of plant genetic resources in Eastern Europe; Wilhelm Rudorf, head of the Kaiser Wilhelm Institute of Plant Breeding since 1936, did the same in the West. While Stubbe launched a new career with the support of the Soviets, Rudorf had become a protégé of the British occupiers.[44] He held on to the reins of power in the KWI for Plant Breeding—renamed the Max Planck Institute for Breeding Research—as it was transferred to West Germany after the end of the war. His tight connections with Nazi rule were no hindrance to the reestablishment of close ties with fellow West European plant breeders. In 1956, together with Jean Bustarret of the French Institut National de la Recherche Agronomique and J. C. Dorst of Wageningen University in the Netherlands, Rudorf launched the European Plant Breeders Union—EUCARPIA—which today is still the major scientific society for plant breeders in Europe.

It was Wilhelm Rudorf, as head of EUCARPIA from 1959 to 1962, who championed the idea among West European breeders of conserving genetic resources on a continental scale, in line with his previous Nazi experience of thinking in pan-European terms.[45] In 1961 EUCARPIA tried to establish a European Potato Gene Bank at Braunschweig (in the Federal Republic of Germany), with Rudorf as its main supporter. In spite of lengthy discussions with different West European governments, and later with the OECD, it proved impossible to agree on financial support for the gene bank. The problem was resolved, in 1974, only after the government of the Federal Republic, agreeing to bear all expenses pending the voluntary accession of other countries, signed an agreement with the Dutch government to participate in what became the German-Netherlands Potato Gene Bank.

The Potato gene bank betrayed the preference of EUCARPIA breeders for transnational European facilities rather than national gene banks. In 1968 a subcommittee of the breeders' society proposed a European network of gene banks to avoid the expense of exhaustive national collections for each country. This network was to consist of the existing banks of the socialist countries (forming a Central and Eastern Europe node), a Northwestern gene bank (which would become the Braunschweig gene bank), a Southern Europe bank (including the Mediterranean region), and a Scandinavian facility.

The timing coincided with a 1967 conference, held by the UN's Food and Agriculture Organization, that warned of the dangers of the rapid depletion of genetic resources in the wake of the Green Revolution in Mexico, India, and the Philippines.[46] This is not the place to repeat the well-known story of the consequences of the monocrop fields of the Green Revolution cultivated with standard varieties developed mainly by American breeders. Suffice it to say that through EUCARPIA European breeders were associated from the onset with the world gene bank craze of the 1970s and the 1980s. But it does deserve mention that European initiatives for banking crop diversity predated that global movement, for Europe had been experiencing its own green revolution since the first decades of the twentieth century.

The first node of the European network of gene banks was the Germplasm Institute at Bari, set up by the Italian National Research Council in 1970 and devoted to Mediterranean Basin crops. Italian breeders collaborated closely with the Food and Agriculture Organization and with the Consultative Group on International Agricultural Research, which was controlled by the World Bank. The Bari facility would become the center for the Mediterranean Germplasm Program of the IPGRI (International Plant Genetic Resources Institute) and one of the IPGRI's four centers of preservation of wheat germplasm. The same kind of arrangement was established between the IPGRI and the other transnational European gene banks. In the later 1970s, the Braunschweig facility became the main conservation center for the IPGRI of grain legumes, potatoes, and beet roots. In 1979 the Nordic Council of Ministers, representing Denmark, Finland, Iceland, Norway, and Sweden, launched the Nordic Gene Bank at Lund, Sweden. The collection was a merger of the separate national collections; it also established strong ties with the IPGRI, being responsible for the preservation of one of the major world collections of peas.[47]

The willingness of international agencies related to the United Nations or the World Bank to cooperate with the European initiatives followed from the fact that about two-thirds of the world's plant germplasm collection was maintained in Europe.[48] This was due, of course, not to the historical crop diversity present in European fields, but to the history of collecting genetic resources particularly by Nazi Germany and the Soviet Union, which had established the largest collections by far on the continent. It was impossible to imagine a global commons that ignored European collections. On the other hand, a crucial component of any transnational gene bank is standardization of descriptors of the stored material to facilitate proper evaluation and circulation. The close connections between European facilities and both the IPGRI and the FAO were intended to ensure that the European network being formed was part of a unified recording system able to support a global commons.

In 1971 a Gene Bank Committee was officially established by EUCARPIA at its Sixth Congress at Cambridge with a mandate to foster cooperation between the different banks. Six years later, in 1977, this committee was instructed to ensure equal representation for gene banks from Western and Eastern Europe. One of its explicit aims was to foster convergence between Eastern and Western gene banks in their approaches to managing genetic resources.[49] And in order to transform the heterogeneous stored diversity into a commons accessible to every European scientific breeder, it was essential to produce a Thesaurus for the International Standardization of

Gene Bank Documentation that would standardize the descriptors of collected material.[50]

By the end of the 1970s, EUCARPIA's efforts, together with those of the FAO and the IPGRI, bore fruit in the form of the European Cooperative Program for Crop Genetic Resources (ECP/GR), a program dedicated to ensuring a "full and free exchange of available plant genetic resources and related data" in order "to make this material available to all European plant breeders."[51] By 2002 the ECP/GR was proud to announce that it had established 50 central crop databases for cereals, grain legumes, vegetables, fruit, forages, and industrial crops. These databases are managed by the 32 institutes from 19 countries participating in the ECP. They have the dual role of providing users with information on the germplasm maintained in Europe and providing working groups with a tool for management of the collections, e.g., rationalization and backups. In 2003 the ECP/GR made available online the EURISCO, the European Internet Search Catalogue, automatically updated from the different European plant genetic resources inventories.

In the final analysis, much of the work that ultimately led to the building of the Svalbard infrastructure entailed agreements on such standards. But besides establishing uniform database protocols, another critical issue was the conservation procedures. Since 1984, duplicates of the Nordic Gene Bank had been stored in an abandoned coal mine on Svalbard.[52] The seeds were packed in airtight glass ampoules (not in the sophisticated foils of the current facility), which were then placed in wooden boxes and stacked inside metal containers. When in 1989–90 it was first proposed to build the Global Seed Vault in Svalbard, the infrastructure was based on natural cooling through permafrost at an average temperature of –3.5°C, whereas the standard established by FAO/IPGRI for long-term conservation was –18°C. Inability to meet these international Gene Bank Standards was one of the main reasons, aside from the many uncertainties surrounding legal issues of plant variety property rights, that justified the dismissal of that first Svalbard project. To produce a new cosmopolitan commons, there has to be a prior consensus on the technologies that sustain it, whether they be interchangeable databases, conservation temperature, collection procedure, or seed packages.

Access and Commoners

Any serious discussion on commons must engage access policy, which for new plant varieties (intentionally produced) comes down to intellectual property rights and the restrictions on use that they establish.[53] Intellectual property rights for plant varieties date from the 1930s. However, such restrictions did not become widespread in Europe until the 1960s, after the creation of the International Union for the Protection of New Varieties of Plants (UPOV) instituted a new regime in the form of Plant Variety Rights (PVRs). The UPOV was based on the French Certificate of Plant Protection (Certificat d'Obtention Végetale), which supported plant variety protection in the form of plant breeders' rights.[54] This regime—designed by Jean Bustarret, one of the founders of EUCARPIA—reflected his convictions on how innovation should serve European agriculture in the aftermath of World War II.

Bustarret openly praised the Nazi Seed Law, which had established a system of seed certification and had restricted the sale of non-approved seed. This law had informed Bustarret's own practices during the Vichy regime (1940–44). In those years he exploited the corporatist regime of Marechal Pétain to disseminate the public breeders' select varieties—particularly of wheat and potatoes—among French farmers.[55] In Bustarret's vision, France could overcome its wartime (and postwar) shortages only by exclusive recourse to the varieties of the official catalog, such as his potato BF 15, and he actively campaigned for the banning of unreliable traditional landraces. The pure lines and their stable properties were to be the foundation on which European agriculture would be rebuilt, and these would thenceforth be the only officially certifiable varieties. Certification demanded that they fulfill the criteria of distinctiveness, homogeneity, and stability, conditions that the heterogeneous and unstable landraces could never satisfy. These were doomed to disappear from European fields and to survive only in gene banks as plant genetic resources for the breeder.

By no means did these measures amount to some kind of conspiracy against French or European farmers. They were, rather, the outcome of principles articulated in the Common Agriculture Policy (CAP) established under the Treaties of Rome (1957), with Jean Bustarret as an active participant. The CAP, the oldest, most integrated and still one of the most controversial EU policies, aimed first and foremost at European self-sufficiency in food production, a constant food supply, and price stability for consumers.[56] A secondary aim was to keep income levels in the countryside high enough to avoid a rural exodus. In other words, the guiding principles of Nazi and Vichy agricultural experts during the war years—food autarky and a strong peasantry—had also become cornerstones of the largest and best-funded common EU policy.

But generous subsidies were not enough to accomplish such aims. It was also necessary to achieve a drastic improvement in the productivity of

European farmers. Here Bustarret's breeders' varieties re-enter the narrative. These had to be homogeneous enough for mechanized agriculture, responsive to fertilizers, tolerant to pesticides, and stable and distinct enough to support anti-fraud enforcement and property-rights protection. Bustarret's UPOV can thus be seen as the translation of CAP policies into plant variety property rights. This model was now to be implemented in other European countries, and in fact entered into force in 1968 after ratification by the Federal Republic of Germany, the Netherlands, and the United Kingdom. By 1977 the signatories also included Belgium, Denmark, France, Italy, and Switzerland.[57] Membership in the UPOV went from these eight original European members to about 20 by the early 1990s. By 2003 there were 52 members.

In its first incarnation, the UPOV required member states to provide intellectual-property protection for at least 15 years on plant varieties. Nevertheless, this variety certificate differed significantly from a patent. The UPOV certificate had been explicitly designed to allow other breeders to use protected varieties in their breeding programs, the so-called breeders' exemption, thus preserving an open-access commons of genetic resources among breeders. In addition, farmers had a right to sow seed saved from their own crops. Under this regime, the breeder's work *de facto* enlarged the commons of genetic resources that could subsequently be accessed by any other breeder. The seed multipliers and distributors were the ones who had to pay royalties to breeders for the commercialization of their varieties.[58] In other words, the breeders' information commons was kept intact.

In 1966 the European Community created a Common Catalogue of seeds. Each member state was required to maintain a national catalog of officially recognized varieties that could legally be marketed in its territory. All varieties entering the catalog had to be tested for distinctiveness, homogeneity, and stability. The Common Catalogue thus constituted the main vehicle for the standardization and homogenization of European fields in the second half of the twentieth century. In France, for example, only those farmers making use of seeds from the Common Catalogue were eligible for funding from the generous Common Agriculture Policy.[59]

Enlarging the Community of Commoners

Radically separating the spheres of production (farmers) and those of innovation and conservation (breeders) by building a commons to which only breeders have free access has been criticized fiercely since the 1980s. The criticisms were fueled by two diametrically opposed sources of discontent.

At one pole, many farmers resisted the passive role assigned to them as mere consumers of plant materials developed by academic and commercial breeders. At the other pole, starting in the early 1980s, private interests began an assault on public breeding programs, chastising their privileged access to genetic resources.

The privatization of public breeding programs, exemplified by the sale of the Plant Breeding Institute (PBI) in Cambridge to private companies at the end of the 1980s, was a clear sign that the era of open access for breeders had come to an end.[60] The PBI was first sold in 1987 to Unilever, which in turn sold it to Monsanto in 1998, an acquisition in tune with the increasing investments made by huge agrochemical companies in the seed sector. The seed business had become a serious game, with big players taking over from the small companies of the previous decades. The small companies were vulnerable in part because their business model was not organized around amassing intellectual property rights. These small companies depended mostly on the relationships of trust they were able to develop with farmers regarding their selected seeds, but the big companies sought security in patents. Five companies currently control one-third of global seed sales and 38 percent of agricultural biotechnology patents.

The promise that genetic manipulation could introduce genes (especially genes for disease resistance) into plant genomes made biotechnology crucial for chemical giants like Monsanto and Du Pont. With the triumph of molecular biology and its ability to manipulate genes, plant varieties lost their status as the basic unit of plant breeding in favor of genes. This new status of the genetics laboratory in breeding work brought the breeders' commons to an abrupt end. The tools and techniques of molecular-biology research brought a sharp increase in private investment in genetic amelioration and a rush to patent new plant varieties and processes.[61] With the biotechnology boom of the 1980s, the capital costs of developing a new variety skyrocketed, amplifying previous pressures by corporate breeders to protect investments and reviving the patent debate. The famous case of the development of Golden Rice starkly revealed these pressures, with the commercialization of the new variety demanding negotiations with some 40 patent owners. The venerable practices of breeders like Vilmorin or Strampelli freely using other breeders' works to develop their varieties were thus hamstrung by the proliferation of the patented techniques crucial to any current innovation work in plant breeding.[62]

The 1991 revision of the UPOV recognized the new trend, transforming plant variety protections into something very similar to a patent. No longer are there breeders' exemptions or farmers' privileges; now breeders can freely use other breeders' varieties to create new varieties only if they make major changes to the plant's genomes, and if farmers want to reuse seed now they have to pay additional royalties.[63] Of course, the end of breeders' exemptions undermines plant genetic resources as an information commons. According to Bent Skovmand, a renowned expert on the conservation of genetic resources, the only programs that would have easy access to funds in the era of privatized biotechnology would be those of potential interest to private industry. This was the case with gene banks and their emphasis on the conservation of wild materials and landraces, an excessively long-term activity for the fast pace of private companies. Skovmand's opinion cannot be easily cast aside if one considers that he was responsible for the CIMMYT (International Maize and Wheat Improvement Center) gene bank between 1988 and 2003, the year he was hired as director of the Nordic Gene Bank.[64] The role of gene banks was no longer to feed the egalitarian republic of breeders working at national institutes or international centers for development aid, but instead to shower biotech corporations with genetic resources.

The dominant approach in the conservation of plant genetic resources has been to identify endangered crop varieties and to preserve their precarious germplasm in gene banks. This tacitly defines the commons as a static entity, a non-renewable resource. The gene hunter goes out in the field, collects diversity, and conserves it *ex situ* in an infrastructure entirely detached from the *locus classicus* of diversity production—the farmer's fields. Now, in spite of the many efforts described above to standardize the data on seeds stored in a gene bank, many collections are still poorly characterized and are rarely used. Also, even if breeders (commercial or academic) were to make intensive use of the collections, the increase in genetic variability resulting from such cross-breeding would still be extremely low. In point of fact, population geneticists have explored the dangers of placing too much confidence on *ex situ* conservation of genetic resources.[65] Geneticists have shown how, for example, a heterogeneous population of soft wheat cultivated in different environments evolves in different directions, thus increasing genetic diversity in relation to the initial population. The conclusion here is that farmers should be taken seriously as cultivators of genetic resources.

Moreover, farmers are more likely to experiment with the many different varieties present in gene banks than breeders who are constrained by the productivity standards expected from commercial varieties. Thus, if farmers' fields are seriously considered as crucial places for the management of the commons, not only may the common resource be more intensively used, with an increase of influxes from the gene banks to the fields, but the variability itself may increase as plants are tested in a much wider variety of environments, replenishing the gene bank with more variability. The combination of *ex situ* conservation and *in situ* initiatives appears to be a more reasonable system than the one in place, which radically severs farms and gene banks. In fact, many participatory breeding initiatives worldwide, with farmers' plots recognized as crucial elements of the conservation system of genetic resources, have already demonstrated the validity of such an approach. The French Semences Paysannes (Peasant Seeds) network, the Italian regional schemes of agricultural conservators of landraces, and the Spanish Red de Semillas (Seeds Network) are all good European examples.[66] Since the end of the 1980s, Europe has witnessed a multiplication of networks of alternative farmers, amateur horticulturalists, gardeners, and slow-food enthusiasts who have emphasized the importance of conserving and enlarging crop diversity by promoting the interchange of traditional European landraces. In Britain, for example, Garden Organics' Heritage Seed Library currently has about 11,500 members, each of whom receives six or seven packets of traditional landrace seeds ordered from an annual catalog. There are also about 250 volunteer "seed guardians" who adopt a particular variety by cultivating its seeds in their own gardens.[67] In France in 2004, a "Save Farmers' Seeds" petition was signed by about 50,000 people.[68]

It is true that all these movements had their roots in resistance to the definition of crop diversity as a scientific commons and were inspired by a nostalgic vision of a pre-modern commons, a golden era of European landraces produced by farmers and their informal local networks. However, these are modern times, and we are not seeing a return to naive localism. What we are seeing since the first decade of the twenty-first century is a growing appreciation by scientists of European farms or even home gardens as sites for the conservation and production of diversity. This has inspired efforts not only to integrate farmers into the commons but also to offer such local initiatives the means and protocols to scale up and form transregional and transnational networks of seed exchange.[69] In other words, what we are observing is not a comeback of the old local commons, but an elaboration of the cosmopolitan commons of crop diversity.

Gene banks are still important sources for the distribution of gene resources, but they are no longer at the center of the system, having become hubs in a distributed network.[70] The idea is not to eliminate gene banks, but to decenter them and to restore the connections between breeders and farmers. In this vision, gene banks play a significant role in the moral economy of the commons (as repositories and accounting centers) but are no longer the central institution or the exclusive locus of variety.[71] By redefining the commoners and abandoning exclusive access to the common resource by breeders, it has been possible to enlarge the commons, both in terms of commoners and in terms of resources. European fields are radically different places if, on the one hand, gene banks are taken as the only source of genetic diversity and genetic resources defined as an exclusive scientific information commons, or if, on the other hand, farmers are invited to participate in genetic conservation. Different ways of defining and managing the commons are different ways of building Europe.

Notes

1. For a detailed description of the vault and its functions and purposes, see Cary Fowler, *The Svalbard Global Seed Vault: Securing the Future of Agriculture* (Global Crop Diversity Trust, 2008) (available at http://www.croptrust.org).

2. On the problems with the use of this term when referring to crop diversity, see Lindsay Ash, Plants, Patents and Power: Reconceptualizing the Property Environment in Seeds in the 19th and 20th Centuries (master's thesis, University of Vienna and Leipzig Univerity, 2009).

3. Donald L. Plucknett, Nigel J. H. Smith, J. T. Williams, and N. Murthi Anishetty, *Gene Banks and the World's Food* (Princeton University Press, 1987), pp. 8–12.

4. For a recent informed discussion on the challenges facing Europe in regard to the erosion of its traditional agricultural diversity, see M. Veteläinen, V. Negri, and N. Maxted, eds., *European Landraces: On-Farm Conservation, Management and Use* (Bioversity International, 2009).

5. Renée Vellvé, *Saving the Seed: Genetic Diversity and European Agriculture* (GRAIN/Earthscan, 1992), pp. 53–59.

6. Valeria Negri, Nigel Maxted, and Merja Veteläinen, "European Landrace Conservation: An Introduction," in *European Landraces*, ed. Veteläinen et al.

7. Alfred W. Crosby, *The Columbian Exchange; Biological and Cultural Consequences of 1492* (Greenwood, 1973).

8. The literature dealing with botanical gardens is immense. Two significant examples are Richard Drayton, *Nature's Government: Science, Imperial Britain, and the Improvement of the World* (Yale University Press, 2000) and Londa Schiebinger and Claudia Swan, eds., *Colonial Botany: Science, Commerce, and Politics in the Early Modern Period* (University of Pennsylvania Press, 2005).

9. Jean Gayon and D. T. Zallen, "The Role of the Vilmorin Company in Promotion and Diffusion of the Experimental Science of Heredity in France, 1840–1929," *Jour-*

nal of the History of Biology 3 (1998): 241–262; Christophe Bonneuil, "Mendelism, Plant Breeding and Experimental Cultures: Agriculture and the Development of Genetics in France," *Journal of the History of Biology* 39, no. 2 (2006): 281–308.

10. Vellvé, *Saving the Seed*, pp. 37–38.

11. D. Grigg, *The Transformation of Agriculture in the West* (Blackwell, 1992). One US ton per acre is equal to 2.24 metric tons per hectare.

12. Landraces are currently defined as dynamic populations of cultivated plants with historical origins, distinct identities, and lack of formal crop improvement. They are often genetically diverse, locally adapted, and associated with traditional farming systems. See Karl Hammer and Axel Diederichsen, "Evolution, Status, and Perspectives for Landraces in Europe," in *European Landraces*, ed. Veteläinen et al.

13. Bonneuil, "Mendelism," p. 17.

14. On Strampelli see Tiago Saraiva, "Fascist Labscapes: Geneticists, Wheat, and the Landscapes of Fascism in Italy and Portugal," *Historical Studies in the Natural Sciences* 40, no. 4 (2010): 457–498; Sergio Salvi, *Viaggio nella Genetica di Nazareno Strampelli* (Pollenza, 2008).

15. Vellvé, *Saving the Seed*, pp. 30–31.

16. Christian O. Lehmann, "Collecting European Land-races and Development of European gene Banks—Historical Remarks," *Kulturpflanz* 29, no. 1 (1981): 29–40.

17. Igor G. Lokustov, *Vavilov and His Institute: A History of the World Collection of Plant Genetic Resources in Russia* (International Plant Resources Genetic Institute, 1999); Peter Pringle, *The Murder of Nikolai Vavilov: The Story of Stalin's Persecution of One of the Great Scientists of the Twentieth Century* (Simon & Schuster, 2008).

18. Lokustov, *Vavilov*, pp. 82–83.

19. Ibid., pp. 84–90.

20. Ibid., p. 85.

21. Michael Flitner, "Genetic Geographies. A Historical Comparison of Agrarian Modernization and Eugenic Thought in Germany, the Soviet Union and the United States," *Geoforum* 34, no. 2 (2003): 175–185.

22. Quoted in Plucknett et al., *Gene Banks*, p. 16.

23. Arrested and charged with espionage in 1940, he died of malnutrition in prison in 1943. See David Joravsky, *The Lyssenko Affair* (Harvard University Press, 1970).

24. Nils Roll-Hansen, *The Lysenko Effect: The Politics of Science* (Humanity Books, 2005). For a general overview of the Soviet organization of scientific research, see Nikolai Krementsov, *Stalinist Science* (Princeton University Press, 1997).

25. D. R. Harris, "Vavilov's Concept of Centers of Origin of Cultivated Plants: Its Genesis and Its Influence on the Study of Agricultural Origins," *Biological Journal of the Linnean Society* 39, no.1 (1990): 7–16.

26. Antonio Lafuente and Nuria Valverde, "Early Modern Commons," *HoST* 2 (fall 2008): 13–42.

27. Susanne Heim, *Plant Breeding and Agrarian Research in Kaiser-Wilhelm Institutes, 1933–45: Calories, Caoutchouc, Careers* (Springer, 2008); Susanne Heim, ed., *Autarkie und Ostexpansion: Pflanzenzucht und Agrarforschung in Nationalsozialismus* (Wallstein, 2002); Thomas Wieland, *"Wir beherrschen den pflanzlichen Organismus besser, . . ."*: *Wissenschaftliche Pflanzenzüchtung in Deutschland, 1889–1945* (Deutsches Museum, 2004).

28. For a review of literature on science and Nazism divided into three main categories—racial hygiene and biomedical research, autarky, and militarization—see Susanne Heim, Carola Sachse, and Mark Walker, "The Kaiser Wilhelm Society under National Socialism," in *The Kaiser Wilhelm Society under National Socialism*, ed. S. Heim, C. Sachse, and M. Walker (Cambridge University Press, 2009). That volume, which surveys the results of the Max Planck Society's Research Program on the History of the Kaiser Wilhelm Society in the National Socialist Era, has thoroughly revised the existing historiography of science and Nazism. Two earlier influential volumes were *Science, Technology, and National Socialism*, ed. M. Renneberg and M. Walker (Cambridge University Press, 1993), and *Science in the Third Reich*, ed. M. Szöllösi-Janze (Berg, 2001).

29. Wieland, *"Wir beherrschen den pflanzischen Organismus besser"*, pp. 178–186.

30. Jonathan Harwood, "The Fate of Peasant-Friendly Plant Breeding in Nazi Germany," *Historical Studies in the Natural Sciences* 40, no. 4 (2010): 569–603.

31. Tiago Saraiva, *Fascist Pigs: Genetics, Food, and Fascism*, forthcoming.

32. Gustavo Corni and Horst Gies, *Brot, Butter, Kanonen. Die Ernährungswirtschaft in Deutschland unter der Diktatur Hitlers* (Akademie-Verlag, 1997).

33. Saraiva, *Fascist Pigs*.

34. E. Baur, "Die Bedeutung der primitiven Kulturrassen und der wilden Verwandten unserer Kulturpflanzen für die Pflanzenzüchtung," *Jahresbericht Deutsche Landwirtschaft Gesellschaft* 29 (1914): 104–109.

35. Michael Kater, *Das Ahnenerbe der SS: 1935–1945. Ein Beitrag zur Kulturpolitik des Dritten Reiches* (Oldenbourg, 1997).

36. On these institutions, see Ute Deichmann, *Biologists under Hitler* (Harvard University Press, 1996), pp. 214–218 and 258–264; Bernd Gausemeier, "Genetics as a Modernization Program," *Historical Studies in the Natural Sciences* 40, no. 4 (2010): 429–456.

37. Quoted by Susanne Heim in Olga Elina, Susanne Heim, and Nils Roll-Hansen, "Plant Breeding on the Front: Imperialism, War, and Exploitation," *Osiris* 20 (2005): 161–179.

38. H. Knüpffer, "The Balkan Collections 1941–42 of Hans Stubbe in the Gatersleben Gene Bank," *Czech Journal of Genetics and Plant Breeding* 46 (2010): 27–33.

39. Gausemeier, "Genetics as Modernization Program," pp. 447–452.

40. Heim, *Plant Breeding*, p. 178.

41. Lehmann, "Collecting European Land-races," pp. 29–40.

42. Ibid.

43. For more on the importance of standards for databases, see Geoffrey C. Bowker, *Memory Practices in the Sciences* (MIT Press, 2008).

44. Heim, *Plant Breeding*, p. 185.

45. J. G. Hawkes and H. Lamberts, "EUCARPIA's Fifteen Years of Activities in Genetic Resources," *Euphytica* 26, no. 1 (1977): 1–3. On the Nazi trans-European approach to agricultural production, in which W. Rudorf was deeply involved, see Herbert Backe, *Um die Nahrungsfreiheit Europas. Weltwirtschaft oder Großraum*, second edition (Wilhelm Goldmann Verlag, 1943).

46. R. Pistorius, *Scientists, Plants and Politics: A History of the Plant Genetic Resources Movement* (International Plant Genetic Resources Institute, 1997).

47. Lehmann, "Collecting European Land-races," p. 35.

48. L. Maggioni, "The ECP/GR, an Example of Cooperation for Crop Genetic Resources in Europe," in *European Cooperative Programme for Crop Genetic Resources, Report of the Ninth Steering Committee* (IPGRI, 2003).

49. Hawkes and Lamberts, "EUCARPIA's Fifteen Years of Activities in Genetic Resources," pp. 1–3.

50. It is important to stress the significance of producing standards for the management of commons. Geoffrey Bowker, in *Memory Practices in the Sciences*, takes the development of flexible, stable data standards and the generation of protocols for data sharing to be among "the central issues for science and technology in the context of the new knowledge economy." He asserts that "you can only protect through policy interventions that which can be named, that which can be shown to have been important in the past." Bowker is referring here to the standardization efforts developed for biodiversity conservation and the processes of federating databases "in order to create tools for planetary management." See Bowker, *Memory Practices*, pp. 120–136.

51. Maggioni, "The ECP/GR."

52. Marte Qvenild, "Svalbard Global Seed Vault: A 'Noah's Ark' for the World's Seeds," *Development in Practice* 18, no. 1 (2008): 110–116.

53. I follow mainly Derek Byerlee and Harvey Jesse Dubin, "Crop Improvement in the CGIAR as a Global Success Story of Open Access and International Collaboration," *International Journal of the Commons* 4, no.1 (2010): 452–480; Glenn E. Bugos and Daniel J. Kevles, "Plants as Intellectual Property: American Practice, Law, and Policy in World Context," *Osiris* 7 (1992): 75–105.

54. C. Bonneuil, E. Demeulenaere, F. Thomas, P. Joly, G. Allaire, and I. Goldringer, "Innover autrement? La recherche face à l'avènement d'un nouveau régime de production et de régulation des savoirs en génetique végétale," *Dossier de l'environnement de l'INRA* 30 (2006): 29–51.

55. On plant breeding and the Vichy regime, see Christophe Bonneuil and Frédéric Thomas,"Purifying Landscapes: The Vichy Regime and the Genetic Modernization of France," *Historical Studies in the Natural Sciences* 40, no.4 (2010): 532–568.

56. A. C. L. Knudsen, Defining the Policies of the Common Agricultural Policy: A Historical Study (PhD thesis, European University Institute, Florence, 2001); W. Grant, *The Common Agricultural Policy* (Macmillan, 1997).

57. Vellvé, *Saving the Seed*.

58. Bonneuil and Thomas, *Gènes, Pouvoirs et Profits*, pp. 215–218.

59. Guy Kästler, "Europe's Seed Laws: Locking Out Farmers," *Seedling*, July 2005.

60. Paolo Palladino, *Plants, Patients and the Historian: (Re)Membering in the Age of Genetic Engineering* (Manchester University Press, 2002).

61. The turning point here is the US Supreme Court's decision stipulating that for patenting purposes life is no more than chemistry, which led to the granting in 1985 of a utility patent to Keneth Hibberd, a scientist at a subsidiary of Molecular Genetics research, Inc., for a type of genetically engineered corn. See Bugos and Kevles, "Plants as Intellectual Property," p. 102.

62. Jack Ralph Kloppenburg Jr., *First the Seed: The Political Economy of Plant Biotechnology*, second edition (University of Wisconsin Press, 2004), pp. 146–147.

63. Vellvé, *Saving the Seed*.

64. Susan Dworkin, *The Viking in the Wheat Field: A Scientist's Struggle to Preserve the World's Harvest* (Walker, 2009).

65. This point is taken from p. 45 of Bonneuil et al., "Innover autrement?"

66. S. Brush, ed., *Genes in the Field: On-Farm Conservation of Crop Diversity* (IPGRI/IDRC/Lewis Publishers, 2000). See also the websites of these initiatives at http://www.redsemillas.info and http://www.semencespaysannes.org.

67. Bob Sherman, "Seed Saving in the Home Garden: Garden Organic's Heritage Seed Library," in *Crop Genetic Resources in European Home Gardens*, ed. A. Bailey, P. Eyzaguirre, and L. Maggioni (Bioversity International, 2009).

68. Elise Demeulenaere and Christophe Bonneuil, Haricot d'ici et blés d'antan, réinventer la variété dans les collectives "haricot tarbais" et "Réseau Semences Paysannes Blé," unpublished paper (available at http://www.semencespaysannes.org).

69. See *European Landraces*, ed. Veteläinen et al. (Bioversity International, 2009); *Crop Genetic Resources in European Home Gardens*, ed. A. Bailey, P. Eyzaguirre, and L. Maggioni (Bioversity International, 2009).

70. This spatial contrast between the role of gene banks in a delegation system and a participative and distributed one is made on p. 46 of Bonneuil et al., "Innover Autrement?"

71. For a Spanish example, see A. C. Perdomo Molina, F. Varela, M. Ramos, and C. de la Cuadra, "Avance del estudio sobre la disponibilidad del material vegetal presente en los bancos Españoles de conservación de recursos fitogenéticos Españoles," paper presented at VII Congreso Sociedad Española de Agricultura Ecológica, Zaragoza, 2006.

8   Under a Common Acid Sky: Negotiating Transboundary Air Pollution in Europe

Arne Kaijser

As the night shift came off work at the Forsmark nuclear power plant, 100 kilometers north of Stockholm, at 7 a.m. on Monday, April 28, 1986, events quickly took a dramatic turn. Passing through the routine contamination control, the workers all showed enhanced levels of radioactivity on their clothes. Further investigation revealed a thin layer of radioactive dust on the grounds all around the power station, but no evidence of leakage or any other mishap. The plant was evacuated nonetheless. Only a skeleton crew remained to monitor operations. At 10 a.m., the contamination was reported to the Swedish Radiation Protection Agency, the SSI, in Stockholm.[1]

The SSI immediately put its nuclear incident protocol into operation, assembling a team of diverse experts to investigate the situation. Radiation measurements were collected from various places in Sweden. The SSI's sister organizations in the other Nordic countries were consulted, and the Finns reported that they had also observed increased radiation levels. By 1 p.m. the nuclear specialists had reached the conclusion that the radiation stemmed from a reactor, not a bomb test. On the basis of meteorological analysis of wind speeds and directions, it took about another 3 hours to identify four nuclear stations in Eastern Europe as possible sources of the contamination.[2] When these findings were presented to the Swedish Minister for Energy and Environment, Birgitta Dahl, she immediately instructed the Swedish ambassadors in the relevant countries to demand immediate explanations from their hosts. At 6 p.m., the Soviet government reluctantly admitted that about 40 hours earlier an accident had indeed occurred in a nuclear power plant near the town of Chernobyl in the Ukraine.[3]

In the weeks after the accident, large parts of Europe suffered the effects of radioactive fallout.[4] Mass media in Western Europe gave generous coverage to the increased radiation levels, and this caused much anxiety in the affected countries. Many people were afraid to let their children play outside, milk and food were dumped, and the agencies for radiation protection had difficulty reassuring the general public. In short, Chernobyl provided a public demonstration that air pollution does not respect national borders and that all inhabitants of Europe can be affected by toxic emissions anywhere else on the continent.

The spread of radioactive particles after the Chernobyl accident was an unusual form of long-range air pollution. The radioactive dust that spread across Europe was very specific and unusual, which made it possible to identify Chernobyl as the source. In many other cases of air pollution, pollutants are emitted from many similar plants or appliances and are mixed in the atmosphere before falling to Earth. This makes it difficult to identify who is polluting whom, particularly if the pollutants spend a long time in the atmosphere and travel far before coming down. Moreover, if the pollutants cross national borders and the polluter and the victims are thus subject to different jurisdictions, it is hard for the latter to exert any influence on the former.

This chapter deals with the atmosphere as a medium for the transport of pollutants. Pollutants stay up in the air for different lengths of time, and thus we can distinguish among local, regional, and global air pollution. Pollutants that fall to the ground quickly do not travel far and remain local; pollutants (such as sulfur) that stay in the air for a few days can drift hundreds of kilometers from the source of emission, across national borders, and pollutants (such as chlorofluorocarbons and carbon dioxide) that "survive" in the atmosphere for many years can easily spread around the globe. Today, air pollution is an obvious topic for international negotiations. The major focus now is on global air pollution linked to climate change, and the annual UN Climate Change Conferences attract much attention from politicians and from the media. By contrast, there was little media attention when the first negotiations about trans-border air pollution were initiated in Europe in the early 1970s. Those negotiations, which concerned the regional environmental problems of "acid rain" caused by sulfur emissions, are the topic of this chapter.

In view of Europe's high degree of industrialization and the generally small size of European countries, it is not surprising that regional, trans-border air pollution was first discovered there. It is more surprising, in view of the inevitably opposing interests of countries that were net exporters of sulfur emissions and those that were net importers, and in view of the deep political divisions of the Cold War, that the negotiations led to international treaties and protocols that helped bring substantial decreases in emissions of sulfur and a number of other pollutants.

I believe that the comprehensive framework of cosmopolitan commons theory presented in chapter 2 is helpful for analyzing these negotiations and their outcomes. It offers a means of engaging with the complex nature of the problems involved, which encompass geophysical characteristics of the atmosphere above Europe and of soils and waters in different parts of the continent; various technologies causing, abating, and monitoring emissions; and an array of actors: the policy makers, scientists, and technical experts involved in the negotiation processes; the power companies and industries responsible for most of the emissions; and the farmers, foresters, and fishers whose lands and waters were affected by pollution. In the 1970s, environmental scientists and experts began to recognize that the atmosphere above Europe constitutes a shared resource-space with the capacity to transport chemical substances over long distances, threatening ecological systems in many parts of Europe. I will argue that this recognition by scientists and experts that European countries affect one another through air pollution—that they are "communities of fate" in this respect—led to the growth of a cosmopolitan commons for long-range trans-border air pollution. This commons emerged in 1979, when a convention was signed that established a set of rules and an organizational arena for further negotiations on how to handle this pollution. This commons was fairly weak at first, but in the 1980s and the 1990s it was gradually strengthened by protocols on emission cuts. Besides seeking to understand the commons-building process, the chapter strives to analyze the special characteristics of the cosmopolitan commons that developed to control transboundary air pollution. In the concluding section, these characteristics will be discussed in relation to some of the other commons that are analyzed in this volume.

The Era of the Chimney

As a prelude to our exploration of the building of a cosmopolitan commons to cope with transboundary air pollution, it is useful to review how air pollution was understood before the 1970s. In the eighteenth and nineteenth centuries, large quantities of coal were burned in the industrial towns and regions of Europe, and as a result the air became increasingly polluted. Other industrial processes, including the smelting of metal ores, also contributed. Consequently, from the late nineteenth century on, many industrial regions in Europe suffered from chronic air pollution. Surprisingly, the inhabitants became somewhat inured to it. In fact, the extent of the pollution was noticed only on those few rare occasions when it was inhibited for one reason or another.

One such hiatus occurred in January of 1923, when the Ruhr district was occupied by French troops because Germany had failed to pay its war reparations. The German workers went on strike, and industrial production across the entire Ruhr ground to a halt. As a result, the sky over the Ruhr cleared up for the first time in many decades. The harvests that summer were much more bounteous than normal, trees grew faster than ever before, and housewives did not have to scrub their homes twice a day. However, the strikes led to an economic crisis, with hyperinflation and mass poverty—especially among the families of the strikers. Therefore, it was a relief for the population in the Ruhr district when the strikes were called off in September. When production resumed, the "normal" smoke came back, and it was seen as a blessing.[5]

The unexpected benefits that stemmed from the clearing of the sky during the long strike had drawn attention to the severity of the air pollution, and in 1924 an official committee was set up to investigate its effects. One of its main conclusions was that it was not the visible smoke and ash but rather the invisible sulfuric acids that were the most damaging pollutants. However, the committee could not identify any feasible measures for reducing these emissions and concluded "that the battle against air pollution by large industries seems to have little chance of success." Instead, it proposed the planting of species of deciduous trees that were resistant to sulfuric acids.[6]

The rather laconic attitude of this committee was typical of the way air pollution was seen in many heavily polluted industrial areas of Europe and elsewhere until the middle of the twentieth century. Air pollution was seen as an almost inevitable consequence of industrial production that gave employment and prosperity to the very people who were affected most by it. This is not to say that there was no resistance at all. For example, in Manchester, in England's industrial heartland, an elite group of urban reformers were pursuing investigations of air pollution as early as the second half of the nineteenth century. However, the group was not able to muster enough political support to actually get laws enacted that would curb emissions. Among this group of reformers was a chemist, Robert Angus Smith, who systematically investigated the content of rainwater in and around the city of Manchester in the 1850s and the 1860s. He emphasized the detrimental effects of sulfur emissions, and the term "acid rain" was introduced in his 1872 book *Air and Rain: The Beginnings of a Chemical Climatology.*[7]

The only feasible strategy for coming to grips with the most dangerous health effects of air pollution seemed to be the building of higher and higher smokestacks. With sufficiently high stacks, the gases and particles produced by industrial processes could be emitted high into the atmosphere, where they were dispersed to concentrations that were deemed harmless. (It was not yet realized that these ever-higher smokestacks would ultimately transform air pollution into an international phenomenon.) However, under certain meteorological conditions this strategy of dispersal failed miserably, even as a local solution. For example, in December of 1952, London was hit by a terrible "smog." A combination of cold weather (and thus much combustion) and windless conditions that lasted for a week led to very high concentrations of a variety of pollutants in the city, stemming from both household furnaces and industrial plants. As a result, 4,000 people lost their lives. The debate that ensued about preventing such smog in the future eventuated in the so-called Clean Air Act of 1956. This act imposed severe restrictions on the burning of coal in major urban areas. As a result, air pollution levels decreased significantly. Such measures had become technically and economically feasible in the 1950s because electricity, gas, and oil were providing new alternatives to coal combustion.[8]

The United States and Japan, having experienced similar smog in a number of their own large cities, soon followed the British example and introduced legislation to reduce local emissions of certain effluents. Furthermore, new environmental protection agencies were established to enforce these laws. By the 1960s, similar kinds of national measures had been taken in most of the industrialized countries of the West. The focus was primarily on mitigating the deleterious public health effects of air pollution.

Until the late 1960s, air pollution was thus seen as a local issue that could be handled by national authorities. However, it was appreciated that large industrial plants located near national borders could easily be a source of serious trans-border pollution. In the 1930s and the 1940s there had been a spectacular conflict of this kind in the western part of North America. A huge lead and zinc smelting plant located in the little British Columbia town of Trail, a few miles from the US border, was identified as the source of heavy pollution experienced across the border by farmers in the state of Washington. The American farmers protested vigorously and succeeded in gaining the support of President Franklin D. Roosevelt, who wrote a letter of protest to the prime minister of Canada. The two governments set up a special arbitration tribunal. In its final decision, in March 1941, the tribunal articulated a general principle that has since become a pillar of international environmental jurisprudence: "No state has the right to use or permit the use of its territory in such a manner as to cause injury by fumes in or to the territory of another or the properties of the persons therein, when the case is of serious consequences and the injury is established by clear and convincing evidence."[9] As a result of the arbitration, the Canadian government took on the responsibility of ensuring that the smelting company in Trail reduced its effluent burden and financially compensated the American farmers who had been affected by the pollution. The Trail Smelter Convention and its general principle have become famous in international law.[10] It is important to emphasize, however, that the Trail Smelter case did not challenge the common view that air pollution was a local problem.[11]

The (Re)Discovery of "Acid Rain"

On October 24, 1967, Sweden's largest newspaper, *Dagens Nyheter*, published an article on "The Acidification of Precipitation" in its culture section. Though the title was rather boring, the message was new and radical: serious air pollution was taking place on a large scale, and "there is a more or less permanent cupola of polluted air over Europe." The author—Svante Odén, an associate professor of soil science—claimed that sulfur, emitted in large quantities by power plants and industries all over Europe, was transported very long distances through the air and was transformed into sulfuric acid by chemical reactions in the atmosphere. When the sulfuric acid reached the ground as a constituent of precipitation, it lowered the pH levels of streams and lakes. He asserted that altered pH levels had reduced fish populations in many lakes in Sweden and Norway and that increasing acidity could be expected to lower the biological productivity of forests and cultivated land—especially in the Nordic countries, whose soils were unusually sensitive to acidification.

Odén was not the first to discover acid rain; as was noted above, Robert Angus Smith had done so a century before. Yet Odén was the first to claim that acid rain was not merely a local or regional phenomenon, but an affliction suffered over large parts of Europe. His article was based on measurements collected since the mid 1950s by members of an international research collaboration called the European Air Chemistry Network under the aegis of the International Meteorological Institute at Stockholm University. The main purpose of this collaboration was to measure the amount of nutrients available to agriculture through the air. Odén had been given the task of assembling and analyzing the data, and while doing so he saw a link between the increasing acidity of rain and recent observations of drastically declining fish populations in lakes in Western Sweden. This led him to propose a connection between decreases in pH levels due to acid rain and declining fish populations.[12]

Thanks to its publication in the leading Swedish newspaper, Odén's alarming message became an instant sensation. In particular, "acidification of precipitation" caught the attention of officials at Sweden's new

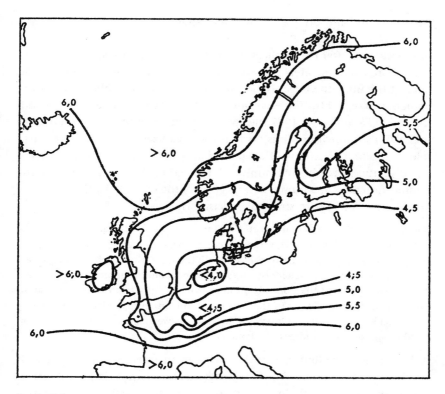

Figure 8.1
This map published in Odén's article in *Dagens Nyheter* had the following caption:
"A map of the pH of precipitation over Europe in 1962. Note the acidic finger that
extends up over the Baltic."

Environmental Protection Agency, established only a few months earlier. It
was also a welcome new issue for the Social Democratic party. Environmental issues had become important, especially among young voters, and 1968
was an election year. But many in Sweden were skeptical of Odén's message.
Predictably, industry spokesmen questioned his findings, fearing that they
might encourage the introduction of taxes on sulfur.[13] The most important
political implication of Odén's article was the international character of
the "new" air pollution—the fact that emissions in one country could have
major effects in others. Purely Swedish countermeasures would therefore
not suffice; instead, the new air pollution had to be dealt with on an international political level. But how could the Swedish government hope to get
the issue onto the international agenda when no one saw air pollution as
an international problem?

Putting "Acid Rain" on the International Agenda

In the months after the publication of Odén's article, the Swedish government searched desperately for arenas in which to launch the issue of acid rain. At the time, the Organisation for Economic Co-operation and Development was one of the few environmentally active intergovernmental bodies, and in December of 1967 the Swedish Minister of Industry proposed to the OECD's Council of Ministers that the organization take a more active role in tackling questions of air pollution. However, when the representative of the Swedish Environmental Protection Agency introduced the acidification issue at a subsequent meeting of the OECD's Air Quality Committee, he was met with strong skepticism. He recalls:

I submitted a written and oral report describing how we saw the acidification problem from the Swedish side. The results were received with great skepticism by the group of air quality experts. This was, after all, the time when high chimneys were seen as the solution to problems of emissions from combustion plants. One by one, the delegates declared that, outside an inner zone of only a few kilometers, sulfur emissions were no problem. In their opinions, the Swedish results seemed highly questionable. There was absolutely no question, as we claimed, that the pollution could be spread over hundreds of kilometers.[14]

Now that air pollution is widely seen as an international problem, it may be hard to understand how controversial the claim of large-scale transborder air pollution was in the late 1960s. As the Swedish representative noted, building high smokestacks was seen as an easy and effective way to come to grips with air pollution. It was based on a perception of the atmosphere as a virtually limitless sink.[15] Were this conception to be abandoned, new abatement technologies would clearly be needed, and this might lead to high costs for many power companies and industrial firms. In spite of its skepticism, the committee appointed an Expert Group to take a closer look at the Swedish data.[16]

Swedish government representatives also presented the acid rain findings at meetings with Nordic neighbor countries, which, according to Odén's research, were also strongly affected by imported acid rain. Not surprisingly, the message got a much warmer reception from those countries than at the OECD, and the Nordic Research Council decided to initiate a joint research program to investigate the acidification issue further. This Nordic research team contacted the Expert Group appointed by the OECD, and together they made plans for a more ambitious survey. In April of 1972, the OECD decided to initiate a so-called Co-operative Technical Program to measure the long-range transport of air pollutants. Austria, Belgium,

Denmark, Finland, France, the Netherlands, Norway, Sweden, Switzerland, the United Kingdom, and the Federal Republic of Germany agreed to participate.[17]

The Swedish government continued looking for additional international arenas in which to address the issue of acidification. A very attractive option appeared in December of 1968, when the General Assembly of the United Nations voted to convene a Conference on the Human Environment. Sweden offered to host the conference, which was to be held in June of 1972, and was thus able to influence the agenda. The Swedish government understandably prioritized the issue of acid rain and appointed an expert group to prepare a study on the subject. The group was chaired by Bert Bolin, Sweden's leading meteorologist, and included Svante Odén as a member. In 1971 it presented its report, *Air Pollution Across National Boundaries: The Impact on the Environment of Sulfur in Air and Precipitation*, which was based on research conducted in the previous year. The report introduced the concept of *transboundary air pollution* and demonstrated that sulfur dispersed over Europe had a tendency to move toward the east and the northeast because of the prevailing winds. The report emphasized that the transboundary character of air pollution had political implications, asserting that "adjacent countries intervene in each other's economies through the effects of atmospheric pollutants."[18]

However, the report's reception at the conference was somewhat of a disappointment to the authors and the Swedish government. Though the report and its findings were strongly supported by Norway and the other Nordic countries, which had in the meantime also discovered acidification in their environments, a large majority of the participating European countries saw acidification as mainly a Nordic problem: the Nordic countries' lime-deficient soils were incapable of buffering the acids. The non-Nordic countries did not want to impose expensive restrictions on their power plants and industries to protect fish and forests in the Nordic countries.[19] Another disappointment for the conference's organizers was occasioned by the long shadow of the Cold War. The East European countries boycotted the conference in solidarity with the German Democratic Republic, whose delegates had been refused visas by the Swedish government.

However, the conference was quite successful in other respects. It led to the establishment of the United Nations Environment Program and to the adoption of a Declaration of Principles. Particularly important was Principle 21: "States have, in accordance with the Charter of the United Nations and the principles of international law, the sovereign right to exploit their own resources pursuant to their own environmental policies, and the responsibility to ensure that activities within their jurisdiction or control do not cause damage to the environment of other States or of areas beyond the limits of national jurisdiction."[20] In the fall of 1972, this principle, with its roots firmly planted in the Trail Smelter arbitration, was also adopted by the Eastern European countries in a session at UN headquarters in New York.[21]

Joint Research Projects

A number of international research projects to monitor long-range air pollution were carried out in the 1970s. In 1973, the OECD survey mentioned above was launched, with twelve countries participating under the leadership of Sweden and Norway. Data on the emission, the dispersion, and the deposition of pollutants (primarily sulfur and sulfur acids) were collected from the whole of Western Europe. By 1977, the OECD was able to publish a final report that confirmed the scope and severity of transboundary air pollution.[22] Parallel to this study, other organizations, particularly the United Nations Economic Commission for Europe (UNECE), also initiated projects to monitor and map air pollution, which now included scientists from Central and Eastern European countries. Gradually these various efforts coalesced, and UNECE took the lead in establishing an ambitious Cooperative Program for Monitoring and Evaluation of the Long Range Transmission of Air Pollutants in Europe (also known as the European Monitoring and Evaluation Program, or EMEP).[23]

These research projects brought together researchers from many countries and from numerous fields, including meteorology, chemistry, limnology, forestry, geology, computer science, and mathematical modeling. The set of scientific specialties was indicative of the complex and comprehensive nature of the pollution processes involved, yet gradually a common view of transboundary air pollution and its environmental effects developed. The researchers also built up mutual trust; to use a concept coined by Peter Haas, they became an "epistemic community."[24] A particularly important aspect of these scientific efforts was the development of sophisticated mathematical models of the geographical flows of pollutants and the chemical transformations to which they were subject. In the 1950s and the 1960s, access to ever more powerful computers had enabled meteorologists to develop mathematical models of atmospheric systems for weather forecasts (see Paul Edwards' chapter in this volume). These models could also be applied to pollution studies. The Meteorological Institutes in Stockholm and Oslo had been leading centers of such research, and it is therefore not surprising that they came to play major roles in the EMEP and OECD programs.[25]

Another institute involved in the EMEP program was the Hydro-meteorological Institute in Moscow, but its scientists lacked a computer powerful enough to calculate the transport models. Their Western colleagues within the EMEP tried hard to help them in the late 1970s. Thanks to "some quiet and some not so quiet diplomacy" aimed at securing funding in convertible currency for such a computer and at overcoming US export restrictions, a modern computer was delivered to the Hydro-meteorological Institute.[26] This anecdote illustrates both the importance of mathematical modeling and the degree of mutual understanding and trust that emerged within the epistemic community of air pollution researchers.

Research in the 1970s confirmed the long-suspected asymmetry of the acidification problem. Long-range air pollution did not spread evenly in all directions, but rather followed the prevailing wind patterns, and thus moved primarily toward the east and the northeast (owing to the rotation of the Earth). While some countries, notably the Nordic countries, were net importers of pollutants, others were net exporters. This asymmetry was aggravated by the fact that some of the net importing countries had soils with low lime content, which made them more vulnerable to acidification.

Even if scientists from different European countries were gradually able to reach a consensus about the extent and the nature of the acidification, its asymmetry made it an extremely difficult political problem.[27] The net importing countries called for the introduction of efficient measures to diminish the emissions, while the exporting countries had few incentives to impose extra costs on their polluting industries. What complicated the matter even more was that the exporters (among them Germany, Great Britain, and France) were big and powerful nations, whereas some of the main importers, the Nordic countries, were fairly small, at least in terms of population and political power. Poland, Czechoslovakia, and Bulgaria, major exporters situated on the other side of the Iron Curtain, seemed, from a Nordic perspective, to be even less susceptible to influence. Deadlock seemed almost inevitable. However, political developments at a high level, far removed from environmental politics, intervened to prevent an impasse.

Pollution and Cold War Politics

In the mid 1960s, after the Cuban missile crisis brought the world to the brink of nuclear war, the two superpowers embarked on the road to a lasting détente. The major symbol of this détente process was the Conference for Security and Co-operation in Europe (CSCE), which was in session from 1973 to 1975, with the United States, Canada, and 33 European countries, including the Soviet Union, participating. The end result was the Helsinki Act, signed in 1975, which entailed an agreement to cooperate in three areas, or "baskets": armaments control, human rights, and economic affairs. The third basket had a sub-basket for cooperation on environmental issues.

It turned out to be difficult to achieve any progress in the three main areas. Both superpowers were reserved about arms control, the Eastern Bloc was not inclined to promote human rights, and economic cooperation proved difficult to achieve. But both sides were still eager to have something come out of the Helsinki agreement. When the Soviet Union proposed cooperation in the environmental arena, the Western countries responded positively. Both sides expected negotiations in this area to lead to rather undemanding obligations, and the United Nations Economic Commission for Europe (UNECE) was chosen as a convenient arena.[28]

Soviet policy makers and experts looked for suitable environmental issues and identified acid rain as an interesting candidate. They knew that it was of great concern to the Scandinavian countries. Moreover, a preliminary study by Soviet scientists had revealed that the Soviet Union was also a major net importer of air pollution. The study indicated that acid rain caused annual losses of more than 150 million US dollars to Soviet agriculture.[29] At the beginning of 1978, Norway's Minister of Environmental Protection, Gro Harlem Brundtland, was invited to Moscow. Upon arrival she argued for an international convention on the reduction of sulfur emissions. To her surprise, she met with unexpected sympathy from her hosts. This led to an alliance between Norway and the Soviet Union aimed at organizing such a convention within the framework of the UNECE. Subsequent negotiations in Geneva were thus championed by what must have seemed, from a Cold War perspective, like an odd or even an unholy alliance between Norway and the Soviet Union. The Soviet Union promised to "take care of the position of other socialist countries during the negotiations," as Valentin Sokolovsky, the chief Soviet negotiator, put it, while Norway agreed to agitate for a convention among Western European countries.[30] Norway, in close cooperation with its Nordic neighbors, was able to exert a kind of moral pressure on the Western European countries. In 1977 Norway's case was considerably strengthened by the OECD's final report, which confirmed what the Scandinavians had been arguing for a decade: that large-scale trans-border air pollution was indeed taking place in Europe and that it was causing severe environmental damage in many parts of Europe.

The Nordic countries wanted a convention with binding commitments for reducing sulfur. Many West European countries believed that they did not have any domestic acidification problem, however, and were thus not prepared to take on the costs of reducing their emissions. Opposition collapsed when the Nordics dropped the demand for reduction commitments. Britain and France agreed to a weaker convention of this nature, and the most stubborn holdout, the Federal Republic of Germany, finally succumbed to pressure from all the other countries. The Soviet Union kept its promise about convincing its allies, and in November of 1979 the Geneva Convention on Long-range Transboundary Air Pollution (LRTAP Convention), was agreed upon and signed by representatives from 33 countries. Most of the European countries signed the treaty, as did the United States and Canada.[31]

Even if the LRTAP Convention did not impose any binding obligations on the signatory countries, its adoption can still be seen as the creation of a cosmopolitan commons for handling long-range transboundary air pollution. The convention was based on a moral economy rooted in a straightforwardly expressed collective commitment. In the preamble it referred to Principle 21 of the 1972 UN Stockholm Conference Declaration and also to the Helsinki Act of 1975. The guiding principle of the convention, formulated in Article 2, was that "the Contracting Parties, taking due account of the facts and problems involved, are determined to protect man and his environment against air pollution and shall endeavor to limit and, as far as possible, gradually reduce and prevent air pollution including long-range transboundary air pollution."[32]

Most importantly, the LRTAP Convention established an international regime for further negotiations: it formulated a set of rules and created an organizational structure with a permanent secretariat and an executive body composed of government officials from the signatory countries meeting once a year. In addition, a number of *ad hoc* working groups, research programs, and task forces were set up, and they provided the growing epistemic community of researchers and experts on air pollution with a stable transnational platform. Finally, the convention secured the continuation of EMEP, the monitoring and evaluation program that had been established in the mid 1970s. All these elements formed a cosmopolitan commons for LRTAP, though still a fairly weak one.

Waldsterben and "Tote-Board Diplomacy"

In the spring of 1982, ten years after the UN conference on the environment, Sweden hosted a conference on Acidification of the Environment. The German biochemist Bernhard Ulrich presented a study arguing that a million hectares of forests in Central Europe were at risk of acid deposition, and that 100,000 hectares were already dying. The phenomenon of *Waldsterben* (death of forests) was particularly pronounced in the Black Forest, according to Ulrich.[33] These findings had an enormous impact in the Federal Republic of Germany, where the Green movement was growing fast and gaining political momentum. In October of 1982, the Christian Democrat Helmut Kohl came to power as chancellor of the Federal Republic of Germany. Under pressure from the media and the growing Green movement, the new government took a new stance toward long-range air pollution and began to support the Nordic countries in their struggle for concrete actions to reduce emissions. The Christian Democrats were far less in thrall to the coal industry and its trade unions than the Social Democrats had been.[34]

It was not only concern for the health of German forests that led the new government to change its tune, but also the growing feasibility of new technology for flue gas desulfurization (FGD) at large power plants. In the 1970s, Japanese industry had made great advances in this technology in order to satisfy the Japanese government's strict environmental policies. In the early 1980s, a few German companies were quite successful in further advancing FGD technology to the point that reductions of SO_2 emissions of up to 95 percent could be achieved at power plants fired by coal or fuel oil. In a familiar pattern, the cost of the new technology decreased as production volumes increased, and by 1983, it became feasible for the German government to enact a law that made FGD compulsory for large power plants.[35]

With the Federal Republic of Germany as a new ally and with affordable and efficient technological options at hand for decreasing emissions, there seemed to be few obstacles to a decisive advance in the war against transboundary air pollution. At a meeting of the Executive Body of the LRTAP Convention in 1983, representatives from the Nordic countries proposed an amendment to the convention in the form of a protocol aiming at a 30 percent reduction of SO_2 emissions by 1993, using 1980 emission levels as a basis. They received support for their proposal from Germany, Switzerland, and Austria—which all had domestic Green oppositions concerned about *Waldsterben*—and from Canada and the Soviet Union. Actually, the Soviet Union argued for a 30 percent reduction of *transboundary fluxes* rather than of *total emissions*, which meant that the Soviet Union would not have to reduce emissions in parts of the country that did not affect other European countries. The nine countries mentioned above, soon known as the "30 Percent Club," were a minority of the convention's members, and in particular the US and the UK argued that there was no scientific justification for the 30 percent target. The Executive Body was thus unable to agree on a joint protocol.[36]

However, the "30 Percent Club" continued to exert pressure on the other countries, in particular by organizing two high-level conferences in 1984, first in Munich and then in Ottawa, presenting new data on environmental degradation caused by acidification. As a result, more and more countries joined the club. When the Executive Body met in 1984, 20 of the LTRAP Convention's 33 members were in favor of a binding protocol on reduction of sulfur emissions. According to Article 12 in the LRTAP Convention, amendments had to be adopted by consensus, and 13 countries still refused to reduce their emissions. A deadlock seemed inevitable. However, an elegant diplomatic solution was found and agreed upon: those countries that were willing to reduce emissions should be allowed to sign a protocol that would be binding only for its signatories, while the other members of the convention would not restricted by it. A Working Group was set up to draft a protocol, and this Sulfur Protocol was signed by 21 countries at the next Executive Body meeting, held in Helsinki in 1985 to commemorate the tenth anniversary of the Helsinki Act. The signing of this protocol was a significant step forward for the "30 Percent Club," even though it meant that "free riding" by some commoners was officially accepted: a number of big net exporters of sulfur, including Czechoslovakia, the German Democratic Republic, Greece, Poland, Spain, and the United Kingdom, none of which had signed the protocol, could continue to emit sulfur as usual.[37]

However, the members of the "30 Percent Club" persisted in what can be called moral-economy diplomacy toward the free riders. Marc Levy has used the metaphor of a tote board to describe this. A tote board is a device often used in the US for charity fund raising. It shows how much various parties have donated. Making the size of the contributions visible to all creates incentives for increasing donations. The Sulfur Protocol worked in a similar fashion. It clearly showed which countries were willing to cut emissions and which countries were not. Governments that did not sign were free to allow their citizens and industries to discharge effluents without limit, but in doing so they ran afoul of their own domestic Green movements and suffered a kind of ostracism from the community of "compliant" countries. The tote-board mechanism also put pressure on countries that had signed the protocol but failed to comply, inasmuch as the monitoring of emissions made their failure public. Thus, even though the Sulfur Protocol had no material sanctions for non-compliance, it could still exert considerable moral pressure on members and on non-members alike.[38] An important tool for this tote-board diplomacy was the impartial monitoring of emissions and subsequent public reporting under the aegis of EMEP. In 1984, the signatories to the LRTAP Convention voted an annual budget to secure the long-term financing of EMEP's monitoring activities, which suggests that the EMEP was seen as "the backbone" of the convention.[39]

Figure 8.2
The meteorological station in Oulanka, located in northern Finland very close to the Russian border, is one of the several hundred stations all over Europe that form the EMEP network. All these stations continuously analyze the chemical composition of the local precipitation and air. Source: http://gaw.empa.ch. Reproduced with permission.

Through the monitoring activities of EMEP, scientists had been able to demonstrate that a number of other substances besides SO_2 caused severe transboundary pollution. Nitrogen oxides contributed to acidification, and so-called volatile organic compounds produced "summer smog," not least in Germany. In the ensuing years, additional protocols for other pollutants were prepared and ratified: one for nitrogen oxides (NO_x) in 1988, one for volatile organic compounds in 1991, a second for sulfur in 1994, one for heavy metals in 1998, and a "multi-effect" protocol in 1999. As with the Sulfur Protocol, signing these additional protocols was not mandatory for members of the LRTAP Convention. But partly as a result of the tote-board mechanism, more and more countries signed the various protocols. With all these protocols, and with more and more countries signing them, the cosmopolitan commons for LRTAP had become considerably stronger than at the outset in 1979.

"Best Available Technology" versus "Critical Loads"

The first Sulfur Protocol was a short and fairly simple document, consisting of only four pages and bereft of appendixes. The central sentence in the document stated that "the Parties shall . . . develop without undue delay national programmes, policies and strategies which shall serve as a means of reducing sulfur emissions or their transboundary fluxes, by at least 30 percent as soon as possible and at the latest by 1993."[40] There were no specifications as to *how* these reductions should be achieved; that was left to the signatory governments. The later protocols have gradually become more and more complex, complemented with appendixes aimed both at stipulating technical means for reducing emissions and at reducing the ecological impacts of emissions. Since 1994 the protocols have also mandated an Implementation Committee (formally established in 1997) to oversee implementation of the measures and compliance by the signatories.

In the 1990s a controversy emerged within the epistemic community between advocates of a technology-oriented approach based on the concept of "best available technology" and an effect-oriented approach based on the concept of "critical loads."[41] The technology-oriented approach was developed first. It was rooted in a long-standing UNECE tradition, dating back to the mid 1960s, of "working parties" in which technical experts exchanged views and information on technologies for abating air pollution. In the 1960s, the agenda was still limited to local air pollution. In the mid 1980s this tradition was revived within the LRTAP Convention, in the form of a Task Force on Abatement Technologies that was established in preparation for the 1988 NO_x Protocol. In 1991, the Task Force was upgraded to become a permanent Working Group on Abatement Techniques, and a number of expert groups were set up to review abatement techniques of many different kinds—with a particular emphasis on operational experiences and costs. The work of all these groups was essential for most of the protocols that were developed in the late 1980s and later.

The technology-oriented approach had a strong influence on the NO_x Protocol signed in 1988, which was considerably more complex than the Sulfur Protocol signed three years earlier. The protocol itself is eight pages long and has a sixteen-page Technical Annex, said to be "an integral part of the Protocol." The signatories took on a twofold obligation: to ensure that their total emissions in 1994 would not exceed their 1987 emissions and to apply emission standards based on the "best available technologies which are economically feasible." The Technical Annex specified what the "best available technologies" were, and was to be regularly updated to keep pace with technological developments.

Though most of the sulfur effluents originated from large power plants and factories where flue gas desulfurization could be applied, the NO_x emissions tended to originate from cars. This propelled catalytic converters to the forefront of negotiations about the NO_x Protocol. Catalytic converter technology had been developed in the United States for the purpose of reducing NO_x emissions from cars in response to increasingly stringent standards for automobile emissions. Some technical experts saw catalytic converters as the "best available technology" for reducing NO_x emissions from cars; others thought they were too expensive and thus not "economically feasible." The British, French, and Italian car manufacturers lobbied their governments to avoid emission standards based on converters; the German and Swedish car manufacturers did the opposite. The latter had an important stake in the US market, and it was in their interest to conform to US regulations that stipulated catalytic converters. However, they did not carry the day. The new NO_x Protocol did not mandate the introduction of converters, and could therefore aim only at a freeze on emissions. But parallel to the signing of the protocol, twelve countries publicly pledged a 30 percent reduction in nitrogen oxides by 1998, a feat that could be achieved only by introducing catalytic converters on a large scale.[42]

The controversies about catalytic converters and other abatement technologies led some countries to question a purely technology-oriented approach. They argued that this approach conferred undue benefits on producers of clean technologies and ignored not only different capacities for implementation but also the differential ecological effects of emission reductions. Therefore they advocated an effect-oriented approach based on the "critical loads" concept as a complement or alternative to the technology approach.[43] Nordic scientists had developed the concept of "critical loads" in the 1970s and the 1980s. It refers to the vulnerability of a specific ecosystem to pollutants. The critical load is the maximum quantity of pollutants that an ecosystem can tolerate without being irrevocably damaged. Proponents of the "critical loads" approach argued that the most cost-effective way of reducing air pollution in Europe was to replace the ongoing pursuit of uniform reductions in all countries with differentiated reductions based on the dispersion patterns of different kinds of emissions and the differential sensitivity of ecosystems in various parts of Europe.[44] This approach thus addressed the asymmetry of long-range air pollution.

To become relevant for policy purposes, the "critical loads" concept had to be buttressed with complex computer models capable of producing an integrated assessment of the deposition of pollutants over different parts of Europe on the basis of different scenarios for future emissions. Such a model, called RAINS (Regional Acidification Information and Simulation), was developed in the late 1980s at a scientific institution that is of interest on its own account, the International Institute for Applied Systems Analysis (IIASA).[45]

The establishment of the IIASA was the outcome of a "bridge-building" initiative intended to reduce East-West tensions by creating a common research institution in the expanding field of systems analysis. A non-governmental organization, it was established in 1973 by scientific institutions in twelve countries, on both sides of the Iron Curtain. It was located within a neutral country—in Laxenburg, Austria. The IIASA's research focused on making global models in different domains—for example, energy, agriculture, and forestry. An IIASA project on transboundary air pollution initiated the early 1980s culminated in the RAINS model. The model was designed to estimate the long-term environmental impacts of different emission-control policies for all of Europe at a resolution of 150 × 150 kilometers. The model had six modules to be studied for each pollutant: emissions and costs of reducing them, atmospheric transport and deposition, soil acidification, lake acidification, groundwater sensitivity to acidification, and forest sensitivity to acidification. With this comprehensive model, color maps could be produced illustrating how different emissions scenarios would affect ecological systems in different parts of Europe. The IIASA scientists later also developed a user-friendly version of the RAINS model that allowed policy makers to test the environmental effects of various emission-reduction strategies.[46]

The RAINS model thus provided a sophisticated, scientifically based tool for developing efficient reduction policies, and in the late 1980s an effect-oriented approach became institutionalized within the LRTAP regime.[47] A special Task Force on Integrated Assessment Modeling was set up, and the negotiations for the Second Sulfur Protocol, adopted in 1994, were primarily based on an effect-oriented approach complemented by a technical approach. The basic obligation for the signatories was now to ensure "that the depositions of oxidized sulfur compounds in the long term do not exceed critical loads." In addition, it was stated that the signatories "shall make use of the most effective measures for the reduction of sulfur emissions."[48] This protocol has five appendixes, the first two addressing environmental effects and the latter three addressing abatement technologies. When negotiating this protocol, the negotiators discovered that the effect-oriented approach allowed for more political maneuvering. Countries that had been reluctant to sign earlier protocols requiring uniform reductions, fearing they would be too costly, could now be offered more lenient terms with lower reduction

Average accumulated exceedance (Lagrangian model)
of acidity critical loads (in eq/ha/yr)

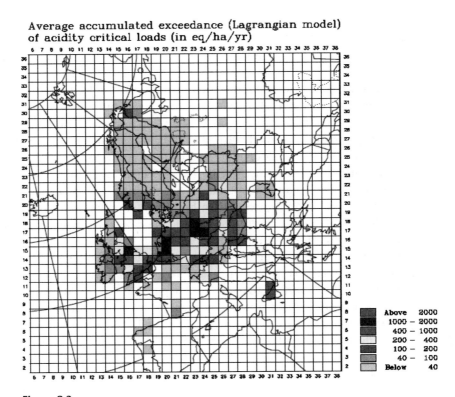

Figure 8.3
An example of an estimate of the critical load of acidity in 1996 based on the RAINS
model. Reproduced, with permission but not in color, from *Transboundary Acidifica-*
tion, Eutrophication and Ground Level Ozone in Europe, EMEP Summary Report 2001,
Norwegian Meteorological Institute.

targets or longer time limits.[49] This seems to be one reason why a growing
number of countries signed the protocols in the 1990s.

The environmental-effect approach was even more prominent in the
negotiations in the second half of the 1990s, leading to the so-called
Gothenburg Protocol that was signed in 1999. This was a new kind of protocol, based on a "multi-pollutant/multi-effect" strategy. The aim was to
"reduce emissions of sulfur, nitrogen oxides, ammonia and volatile organic
compounds that are caused by anthropogenic activities and are likely
to cause adverse effects on human health, natural ecosystems, materials
and crops—due to acidification, eutrophication or ground-level ozone."[50]
The protocol was thus a complex and multifaceted agreement based on

simulations made with the RAINS model. However, for some countries the emission-reduction requirements were rather modest, which suggests that some political maneuvering took place in order to get as many as countries as possible to sign the protocol.

The Effectiveness of the LRTAP Cosmopolitan Commons

Under the auspices of the LRTAP Convention, delegates from the signatory countries were able to negotiate and agree on a series of agreements, or protocols, in which the signatories agreed to reduce their emissions and to introduce different kinds of measures and policies. Furthermore, the number of countries that signed these protocols increased. It is also clear that the emissions of many pollutants has decreased substantially since 1979, when the convention was established. For example, from 1980 to 1989 the total emission of sulfur from all the European countries that had signed the LRTAP Convention decreased by 23 percent. Furthermore, there was a marked difference between the signatories and the non-signatories of the Sulfur Protocol. The former cut back their emissions by an average 29 percent; the latter achieved no more than an 8 percent decrease.[51] The period between 1980 and 2000 witnessed an overall reduction of sulfur emissions in Europe by nearly 70 percent. In the years 1990–2000, emissions of nitrogen oxides and volatile organic compounds were reduced by 25–30 percent, lead by 60–70 percent, and mercury by 50 percent.[52] Since 1980, according to assessments made with the RAINS model, these reductions have also resulted in wider areas' receiving less than the critical load of different pollutants.[53]

Does this mean that the cosmopolitan commons was effective in coping with transboundary air pollution in Europe? This is a hard question to answer.[54] On the one hand, this commons is often looked upon as a very successful example of international environmental cooperation. It clearly raised public awareness of air pollution in most European countries, and thereby fostered a readiness to enact and conform to legal emission standards. It also facilitated the emergence of an epistemic community of scientists and technical experts, who achieved a thorough scientific understanding of pollution processes. Many countries also introduced specific legislation to enforce emission cutbacks, for example by taxing emissions or making the use of abatement technologies compulsory, such as FGD or catalytic converters. Furthermore, by the mid 1990s the dynamism of the LRTAP Convention had infected the EU, which in 1997 adopted an Acidification Strategy based on the IIASA's RAINS model. The European

Union had at its disposal more effective tools with which to enforce compliance with its strategy than the LRTAP Convention had.[55] Overall, there are grounds for arguing that the commons has been an effective form of organization of the struggle to decrease emissions.

On the other hand, it can be argued that many of the emission cuts were outcomes of processes that had little to do with environmental policies and would have taken place even without the LRTAP Convention. The energy sector was the largest emitter of sulfur, and, although it succeeded in lowering its emissions substantially, its cutbacks were not motivated primarily by environmental concerns, but rather by the adoption of more efficient and profitable process technologies. One such change was the shift to North Sea oil, which had become the source of choice for Western Europe in the 1980s, not least for security reasons. This oil had much lower sulfur content than that from the main Arab suppliers, so use of it decreased emissions. Another change was a switch from coal and oil to other energy sources. The 1970s and the 1980s saw a rapid increase in the consumption of natural gas in most European countries. Since natural gas contains almost no sulfur, this also led to substantial reductions of sulfur emissions. Norway and the Soviet Union became the two largest suppliers of natural gas. Both countries cherished ambitions to expand their sales of natural gas to the European continent in the late 1970s, and it is not unlikely that this shared interest motivated their unholy alliance against acid rain in the late 1970s.

Many European countries built nuclear power plants in the 1970s and the early 1980s. The decisions to build these plants had been made long before the LRTAP Convention came into force. But when the nuclear plants were commissioned, they often replaced fossil-fuel plants, and thus reduced emissions. Furthermore, the sharp hike in oil prices that occurred in the mid 1970s provided strong incentives for introducing more energy-efficient technologies, and in many European countries energy consumption increased less than expected, which also kept emissions in check. Finally, in Eastern Europe, the dramatic decline in industrial production subsequent to the political transition of 1989 resulted in major reductions of pollutant emissions. This was another development hardly driven by environmental policies.

Thus, in a number of cases it can be argued that reductions of emissions had little to do with efforts made by the LRTAP commons. But such a view is too simplistic. The commons contributed to a new mindset among many European policy makers and politicians regarding environmental issues. Though few politicians were prepared to accept high costs for reducing emissions, they were prepared to exploit windows of opportunity for legislating emission cutbacks made possible by developments in other fields. In this way the LRTAP commons also contributed to emission cutbacks that seemed to have little to do with environmental policies.

Concluding Discussion

In the introduction I stated that the ambitions of this chapter were to understand the emergence and gradual strengthening of the LRTAP commons and to analyze the special characteristics of this commons. In this concluding section I will first discuss the special characteristics of the LRTAP cosmopolitan commons by contrasting them to those of the other cosmopolitan commons that are analyzed in this volume. I will then summarize what I see as the main factors in the emergence and further development of the LRTAP commons.

One feature that the LRTAP commons shares with the others is that the dilemma it seeks to overcome was induced by technology. It was the rapidly increasing combustion of coal and oil in Europe after World War II— for generating power, fueling industrial processes, heating buildings, and powering cars—that led to much higher emission levels than previously, levels that produced tangible environmental effects over considerable distances. In addition, the use of high smokestacks, intended to decrease local emissions, also contributed to the increase in long-range air pollution.

In one respect, the LRTAP commons differs from almost all the others: in contrast to airplane crashes, interfering radio emissions, and some types of river pollution, which were tangible and obvious also to laymen, *the negative consequences of long-range air pollution were hard to notice*. It also differed from local air pollution, which was visible, smelled, and at times palpably affected human health. By contrast, long-range air pollution was invisible, did not smell, and harmed the environment in indirect and unforeseen ways. In addition, it was very difficult to trace and verify a link between environmental damage to lakes, soils, plants, and animals in one location and emissions from power plants hundreds of kilometers away. Such a link could be established only on the basis of careful measurements and complicated modeling carried out by scientists.

Like many of the other commons, the character of the LRTAP commons was shaped by certain *geophysical traits and topologies of Europe*. The dispersion of air pollution was not as isotropic as radio waves and not as linear as water pollution in a river, but somewhere in between. In the short term, patterns of dispersal can be highly unpredictable, as was demonstrated in the weeks after the Chernobyl accident when changing winds spread radioactive pollutants in many different directions. But in the long run, dispersal follows the prevailing wind patterns and thus moves primarily in eastern and northeastern directions. This implied an asymmetry: some countries, notably the Nordic countries and the Soviet Union, were net importers of pollutants; other countries were net exporters. This asymmetry was aggravated by the fact that some of the net importing countries had soils with low lime content, which made them more vulnerable to acidification, whereas some of the exporting countries had environments that seemed to be fairly insensitive to acidification (or so it seemed in the 1970s). Many of the exporting countries, furthermore, had abundant coal deposits, which literally fueled coal-based power production. Thus, different kinds of geophysical traits and political topologies influenced the character of the LRTAP commons and not least the relations among the stakeholders or "commoners."

Another characteristic that the LRTAP commons shares with many of the others is that it was and still is very *complex in terms of actors*. Scientists played a critical role in demonstrating the extent of long-range pollution. The early 1970s witnessed the launch of a number of ambitious international research projects aimed at monitoring, mapping, and modeling emissions and their effects. Through these projects, researchers in different countries developed common models, theories, methods, and standards; in the process, they developed trust in one another's findings and increasingly even in one another's motives. Gradually an epistemic community of scientists and experts emerged, founded on a consensus both about the nature of the transboundary pollution and about the necessity of international cooperation to come to grips with it. In short, they developed a moral economy—a shared system of cooperation, transparency, and trust. The members of this epistemic community played an important role as advisors in the negotiations leading to the LRTAP Convention and its subsequent protocols.

However, these scientists and experts were not commoners in the sense that they were directly involved as stakeholders, either as victims of pollution or causers of it. Of course, in the broadest sense all Europeans (and thus also scientists) can be seen as victims of pollution. But the primary victims were owners or users of environments that were affected by pollution, particularly farmers, foresters, and fishers in areas with sensitive soils (without lime). And the major perpetrators were fossil-fuel-burning power companies and industrial plants. The former wanted international policies to reduce emissions; the latter opposed such policies. However, these groups did not participate directly in the negotiations on how the regime for the commons should be designed. The negotiators—representatives of governments in the various countries—performed as the "acting" commoners trying to represent the most influential interest groups in their countries.

Let me now turn to the emergence and gradual strengthening of the LRTAP commons. This process was strongly influenced by the geophysical traits of long-range air pollution, or rather by the way these traits were perceived among the acting commoners. There was a marked polarization during the 1970s, when Nordic governments, representing many victims, were early proponents of emission cuts while governments of countries with large polluters resisted them. However, thanks to participation in joint research projects, and with an assist from Cold War politics, the two camps eventually came to agreement on core guiding principles, recognizing LRTAP and agreeing to ratify the 1979 LRTAP Convention. This marked the emergence of a cosmopolitan commons (though a fairly weak one, insofar as the negotiators were unable to achieve a consensus either on the severity and the effects of the pollution or on the need for concrete actions).

During the 1980s, the balance of opinion gradually shifted. It started with the discovery of *Waldsterben* and the emergence of a strong political opposition to acid rain in Germany. Other countries also came around to a more negative assessment of long-range air pollution as they discovered the effects of acidification in their own environments. These changing perceptions motivated a growing number of countries to join the "30 Percent Club," with its agenda of pursuing obligatory emission cutbacks. The first Sulfur Protocol of 1985 ushered in a second phase of the LRTAP commons. This was marked, as we have seen, by a rather unorthodox commons regime under which some commoners imposed restrictions on themselves by promising to reduce their own emissions while allowing others to be free riders, rejecting all constraints.

However, the former persisted in efforts to convince the latter, through further research, monitoring, and negotiations, that their interests too would be served by a universal cutback in emissions. At the same time, they paved the way for new commitments by developing more flexible protocols under which a country could set more ambitious or less ambitious targets for emissions reduction. As a result, in the 1990s more and more countries opted to join the protocols and to reduce their emissions accordingly. By the end of that decade nearly all of the European countries had signed at least some of the protocols and we can characterize this development as a third phase of the LRTAP commons, a phase marked by institutional momentum and a kind of irreversibility.

How can we understand this gradual change and the growing willingness to undertake emission cutbacks? There were three main factors. One was the increasing knowledge of the environmental consequences of LRTAP thanks to the continued monitoring and modeling activities within EMEP. On the basis of these findings, many net exporters discovered that their own environments were also affected, and that emission cutbacks would benefit them too. A second factor was the development of relatively inexpensive "technological fixes" for end-of-pipe emission control, along with a large-scale introduction of substitute power-plant technologies that eliminated sulfur emissions, notably nuclear power and natural gas. The third factor has to do with politics. In the 1970s, Cold War politics facilitated an unholy alliance between the Soviet Union and the Nordic countries in pursuit of a convention for LRTAP. In the ensuing decades, the close economic and political relations among the countries involved played an important role. Even countries lacking any intrinsic environmental motive to cut back on their toxic emissions could be swayed to do the "right thing" by the evident frustration among other countries that saw themselves as victims of these emissions. In the long run, frustrations in one sphere of relations between countries will also affect other spheres. Until the fall of the Iron Curtain, the East European countries were not very receptive to the frustrations of Western European countries. But by the mid 1990s, a growing number of East European countries seeking entry to the EU were compelled to agree to the European Union's much stricter environmental policies as a condition for membership.

The story told in this chapter may be a source of comfort to those who are frustrated about the pace and difficulty of ongoing international negotiations on climate change. It shows that it is indeed possible to achieve a successful environmental commons among heterogeneous countries and actors pursuing very different interests. However, achieving such a global cosmopolitan commons will clearly demand not only commitment and purposefulness, but also patience and an understanding of and a respect for differences.

Acknowledgments

I would like to thank Anna Kaijser, Lars Lundgren, Måns Lönnroth, and two referees for valuable comments on the manuscript. I also want to thank the other authors of this volume for fruitful discussions and Nil Disco and Eda Kranakis for extraordinary efforts to help improve both the content and the form of my text. A grant from the Bank of Sweden's Tercenary Foundation and a fellowship at the Netherlands Institute of Advanced Studies enabled my research.

Notes

1. Swedish government, *Efter Tjernobyl*, DsI 1986:11, 38–39.

2. Ibid. and interview with the former head of SSI, Gunnar Bengtsson, May 7, 2010.

3. Interview with Birgitta Dahl, January 18, 2010.

4. On the much more severe local effects around Chernobyl, see Paul Josephson, *Red Atom: Russia's Nuclear Power Program from Stalin to Today* (University of Pittsburgh Press, 2005).

5. Franz-Josef Brüggemeier and Thomas Rommelspacher, *Blauer Himmel über der Ruhr. Geschichte der Umwelt im Ruhrgebiet 1840–1990* (Klartext Verlag, 1992), pp. 50–52.

6. Ibid., p. 52 (translation by Arne Kaijser).

7. Peter Brimblecombe, *The Big Smoke: A History of Air Pollution in London since Medieval Times* (Methuen, 1987), pp. 138–141.

8. Ibid., pp. 161–178.

9. Keith A. Murray, "The Trail Smelter Case: International Air Pollution in the Columbia Valley," *British Columbia Studies* 15 (1972), p. 84.

10. Philippe Sands, *Principles of International Environmental Law I* (Manchester University Press, 1995), pp. 243–244.

11. Murray, "The Trail Smelter Case," pp. 68–85. See also Rebecca Bratspies and Russel Miller (eds.), *Transboundary Harm in International Law: Lessons from the Trail Smelter Arbitration* (Cambridge University Press, 2006).

12. Lars Lundgren, *Acid Rain on the Agenda: A Picture of a Chain of Events in Sweden, 1966–1968* (Lund University Press, 1998). Odén was the first to unify "knowledge about acid precipitation in the fields of limnology, agriculture, and atmospheric chemistry," according to Elis B. Cowling ("Acid Precipitation in Historical Perspective," *Environmental Science and Technology* 16, no.2,1982, p. 114A).

13. Lundgren, *Acid Rain on the Agenda*, pp. 152–155.

14. Göran Persson, ". . . och än faller regnet," in *Miljö för miljoner '85* (SNV, 1986), pp. 9–10 (translation by Arne Kaijser).

15. A similar perception was also applied to water. Rivers, large lakes, and seas were seen as sinks that could make emissions harmless. See Joel Tarr, *The Search for the Ultimate Sink: Urban Pollution in Historical Perspective* (University of Akron Press, 1996), chapter 1.

16. Göran Persson, "The Acid Rain Story," in *International Environmental Negotiations: Process, Issues and Contexts*, ed. G. Sjöstedt, U. Svedin, and B. Hägerhäll Aniansson (Utrikespolitiska Institutet and Forskningsrådsnämnden, 1993), p. 107.

17. Ibid., pp. 107–108.

18. Bert Bolin et al., *Air Pollution across National Boundaries: The Impact on the Environment of Sulfur in Air and Precipitation* (Royal Ministry for Foreign Affairs, 1971), p. 12. Martin Letell emphasizes the significance of the concept "transboundary air pollution" that was introduced in this report; see Letell, Governable Air: Studies on the Science and Politics of Air Pollution in Europe, STS Research Report 12, Göteborg University, 2006.

19. Lundgren, *Acid Rain on the Agenda*, p. 288.

20. The declaration can be downloaded from the website of the United Nations Environment Program.

21. Philippe Sands, *Principles of International Environmental Law I* (Manchester University Press, 1995), p. 37.

22. Persson, "The Acid Rain Story," p. 108.

23. Toni Schneider and Jürgen Schneider, "EMEP—Backbone of the Convention," in *Clearing the Air: 25 Years of the Convention on Long-Range Transboundary Air Pollution*, ed. J. Sliggers and W. Kakebeeke (United Nations, 2004).

24. Peter Haas, "Introduction. Epistemic Communities and International Policy Coordination," *International Organization* 46, no. 1 (1992): 1–35.

25. Schneider and Schneider, "EMEP—Backbone of the Convention."

26. Ibid., p. 39.

27. Gunnar Sjöstedt, "Special and Typical Attributes of International and Environmental Negotiations," in *International Environmental Negotiations: Process, Issues and Contexts*, ed. G. Sjöstedt, U. Svedin, and B. Hägerhäll Aniansson (Utrikespolitiska Institutet and Forskningsrådsnämnden, 1993), p. 32.

28. Thomas Gehring, *Dynamic International Regimes: Institutions for International Environmental Governance* (Peter Lang, 1994), chapter 2.

29. Valentin Sokolovsky, "Fruits of Cold War," in *Clearing the Air*, ed. Sliggers and Kakebeeke, pp. 7–17.

30. Ibid., p. 10.

31. The US and Canada have been members of the UNECE since its establishment in 1947.

32. The LRTAP Convention and all its subsequent protocols can be downloaded from the UNECE website.

33. Marc Levy, "European Acid Rain: The Power of Tote-Board Diplomacy," in *Institutions for the Earth: Sources of Effective International Environmental Protection*, ed. P. Haas, R. Keohane, and M. Levy (MIT Press, 1993), pp. 92–93.

34. Dieter Jost, "Waldsterben a Breakthrough," in *Clearing the Air*, ed. Sliggers and Kakebeeke, pp. 15–17.

35. Lars Lindau, Andrzej Jagusiewicz, and Endre Kovacs, "Software and Hardware, No Protocols without Technologies," in *Clearing the Air*, ed. Sliggers and Kakebeeke, p. 50.

36. Levy, "European Acid Rain," pp. 93–94; Gehring, *Dynamic International Regimes*, pp. 145–148.

37. Gehring, *Dynamic International Regimes*, pp. 155–157.

38. Levy, "European Acid Rain," pp. 77, 93–94.

39. EMEP is commonly characterized as "the backbone" in the literature on LRTAP.

40. Article 6. The protocol can be downloaded from the UNECE website.

41. Karin Bäckstrand, *What Can Nature Withstand? Science, Politics and Discourse in Transboundary Air Pollution Diplomacy*, Lund Political Studies 116, Lund University, 2001, pp. 176–179.

42. Levy, "European Acid Rain," pp. 95–99.

43. Ibid., p. 129.

44. Jørgen Wettestad, "The ECE Convention on Long-Range Transboundary Air Pollution: From Common Cuts to Critical Loads," in *Science and Politics in International Environmental Regimes: Between Integrity and Involvement*, ed. S. Andresen et al. (Manchester University Press, 2000).

45. Harvey Brooks and Alan MacDonald, "The International Institute for Applied Systems Analysis, the TAP Project, and the RAINS Model," in *Systems Expert, and Computers: The Systems Approach in Management and Engineering, World War II and After*, ed. A. Hughes and T. Hughes (MIT Press, 2000).

46. Ibid.

47. The model became an "obligatory point of passage" for the whole regime, according to Rolf Lidskog and Göran Sundqvist ("The Role of Science in Environmental Regimes: The Case of LRTAP," *European Journal of International Relations* 8, no. 1, 2002, p. 92).

48. Article 2. The protocol can be downloaded from the UNECE website.

49. Bäckstrand, *What Can Nature Withstand?* pp. 212–220.

50. Article 2, UNECE protocol.

51. Levy, "European Acid Rain," p. 114.

52. Lars Lindau, Andrzej Jagusiewicz, and Endre Kovacs, "Software and Hardware," in *Clearing the Air*, ed. Sliggers and Kakebeeke, p. 46.

53. Jean-Paul Hettelingh et al., "Air Pollution Effects Drive Abatement Strategies," in *Clearing the Air*, ed. Sliggers and Kakebeeke, pp. 77–78.

54. The discussion below builds on Jørgen Wettestad, "Acid Lessons? LRTAP Implementation and Effectiveness," *Global Environmental Change* 7, no. 3 (1997), and on Levy, "European Acid Rain."

55. Jørgen Wettestad, *Clearing the Air: European Advances in Tackling Acid Rain and Atmospheric Pollution* (Ashgate, 2002), chapter 5.

III   Temporal Layering and Interlinking of Cosmopolitan Commons in Nature's Spaces

9 Changing Technology, Changing Commons: Freight, Fish, and Oil in the North Sea

Håkon With Andersen

The sea is a strange and surprising place. Even though it occupies more than two-thirds of the surface of the Earth, it is not subject to the traditional tesselation into countries and nation-states that we see on land. On the contrary, the sea was regarded as a kind of free space, not subject to territorial claims and national sovereignty; or so it seemed at least up to the middle of the twentieth century. This two-thirds of the Earth's surface that is open for all to enter and to use would seem to be one of the largest undivided shared resource-spaces known to man. At the same time, the opposite is the case: it is unusual to view the sea as a commons in the sense put forward in this book. We require a more "civilized" sea. In this chapter, I propose to confine the idea of the sea to a particular historic sea that can be analyzed as a commons in its proper historical setting—a setting in which technological changes have encouraged new types of commons among a number of nation-states. The North Sea is such a place.

"The North Sea" usually refers to the rather shallow sea bordered by Britain (and its islands, the Orkneys and the Shetlands), Belgium, the Netherlands, Germany, Denmark, and Norway. This is the narrow definition of "the North Sea"—or, if you like, the core of the concept. Commercially "the North Sea" has a wider scope, including the west coast of Britain, the Celtic Sea, and the Irish Sea. Northward it includes the Norwegian continental shelf up to the Barents Sea.[1] Even the strait known as the Skagerrak is often included. In any case, what these areas have in common are shallow waters and large sedimentary basins. The North Sea's sea bed is the continental shelf of Europe, generally no more than between 100 and 200 meters below the surface. Usually, the continental shelf is regarded as a continuation of the land masses out into the sea before the sea floor falls to the really great depths of the open ocean. In the North Sea there is one important exception to this: the Norwegian Trench, which is very close to the Norwegian coast and is more than 400 meters deep.

This sea has for centuries been a cornucopia of resources open for all to use, and used it they have. The North Sea is one of the busiest seas of all. It is rich in resources, including fish, oil, and gas, and for hundreds of years it has been a primary hub in the international sea traffic system. Around the periphery of the North Sea we find some of the world's busiest harbors: Hamburg, London, Antwerp, Rotterdam.

As available or novel technologies have enabled humans to make better or more efficient use of their resources, the North Sea has been put to many uses. It is the main thesis of this chapter that the commons has changed in character as new and different technologies have been introduced and have challenged established governance regimes, at least to the extent that the North Sea has been governed, managed, and controlled through agreements across borders and institutions. Not only has technology shaped governance; it has also played a role in turning the geographical area of the North Sea into three rather distinct commons superimposed in space.

Far from being tragedies, the uneven history of these resources are neverending stories of negotiations, conferences, treaties, and unilateral actions, all increasingly embedded in moral economies. The resulting transnational commons regimes controlled risks and prevented degradation and depletion of the resources. Though never perfect and always at the mercy of actors trying to bend the rules or to renege on agreements, they nevertheless worked as regulated commons, with different rules for access and use depending on different uses of the resource-space and the technologies available.

The sea has always been a contested space, and the North Sea is no exception. What makes the North Sea particularly interesting is that the countries that surround it are not only coastal nations but also maritime powers.[2] Only the Mediterranean exhibits similar features. The combinations of coastal and maritime powers make the development more interesting since the coastal interests and the maritime interests may be rather contradictory. To clarify this, we have to look at the transformation of the North Sea, the common space, in relation to these different interests and in relation to the dramatic changes that occurred after World War II, when a large group of newly established states joined the international negotiation system as coastal powers. The international focus shifted from the North Sea to the seas and oceans in general and from the powerful cosmopolitan maritime powers to the more local coastal powers simply because of the latter's numerical preponderance.

In what follows, I will focus on the transformations of the North Sea— that is, the dynamic changes related to the use of resources and the types of fortunes and risks that were at stake. I will examine the role of technology in this changing process and the dynamics over a long time period, from the middle of the nineteenth century until today.

In pursuing these topics, one must pay special attention to the coastal zones, inasmuch as these are the borderlands of the common space: the unclear zones and the "gray areas" where the cosmopolitan commons meet national jurisdiction and sovereignty. This was where the big maritime powers, with their interest in a *mare liberum* (free sea), met the coastal nations, with their focus on resources in and under the water closer to shore.[3]

The coastal zones were an area not only of political conflicts but also of physical disasters. Risk and safety are of particular importance when we study such a capricious place as the North Sea, and the most dangerous places are the difficult waters in the coastal zones. They are often the most physically challenging to ships and their crews. They are also the areas where we see the confrontation between different control strategies—between open access and private or state ownership, and between transnational governance and national or regional authority.

The dangers and risks of sea travel are exactly why national sovereignty is so difficult to uphold on the high seas. But the border zones—the harbors, the near coastal waters, the offshore sea bed—are a different story. There is a surprising history of negotiations and agreements, not always signed, but nonetheless silently accepted, often only after some time and after violent protests. Might this define a particular European style in managing and supervising commons as technologies shift? Even after devastating wars, earlier agreements silently resurfaced to rule peacetime relations. Although agreements were slow to be accepted, weakened by quibbling and by liberal interpretations of the rules, in the end they commanded some sort of mutual understanding and respect.

There are many resources in the common space, but we can rather crudely divide them into three categories on the basis of their physical attributes. Traditionally the surface of the sea is the realm of travel and sea power—where ships navigate and where sea wars are fought. This is the domain of the maritime powers: the liberal and non-governmental *mare liberum*. As a shared resource, the sea as a transport corridor is akin to a public good—like air or sunlight—to which access cannot easily be restricted and which is resistant to depletion—i.e., *prima facie* non-subtractable. The second group of resources is found not on but in the sea: all sorts of creatures that can be harvested (from whales to fish and shellfish). Here the main interest of each coastal power is in profiting as much as possible from the resources. Being finite and vulnerable, these resources are subject to depletion and pollution. The resources in the third and last group are below the sea, in the sea bed. Typical examples are oil and gas, but there are many as yet unexplored resources. The last point reminds us that we are not at the end of history and that the future will write new chapters in the book of the North Sea commons. The sea bed is annexed by coastal nations in their ceaseless endeavor to enrich themselves. But in contrast with fisheries, excavation on the sea bed is a highly capital-intensive activity. It is a challenging but promising technological frontier for "big actors" such as states and multinational corporations. Sea bed resource extraction is thus a rather different enterprise than the fisheries. The sea bed represents high modernity and sophisticated technology.

These three basic physical locations—the surface, the bulk of the sea, and the sea bed—are also linked with different border zone infrastructures: the first with harbors and lighthouses, the second with fishing vessels, their gear, and fish-processing plants, and the third with oil and gas wells, pipelines, and processing terminals. The multiple commons also give rise to a continuous discussion about jurisdiction over the sea and over what is in it and under it. How far from the coast should the jurisdiction of the nation-state reach, and what kind of sovereignty should it involve? Or should the entire sea remain open for all, as the maritime powers had always claimed? How wise is it for nations to exercise unilateral sovereignty over their coastal zones? There were also questions as to who "owned" the fish and, more recently, who owned the oil and gas under the North Sea. Were these resources to be "common goods?" National attempts to control all these spaces and resources have been augmented by international and transnational negotiations—sometimes they have succeeded, sometimes not. However, in the longer run the results have been silently accepted, even if the negotiations are now about seas and oceans in general and take place under the aegis of the United Nations and its subcommittees or of the International Court of Law instead of conferences for the North Sea.

I will stick to these three main groups of resources based on physical location: ship traffic—that is, risks, regulations and harbors on the surface, fishing and fish processing in the body of the sea, and oil and gas with pipelines and terminals on the sea bed and on shore. These three perspectives on the North Sea reveal how dynamic cosmopolitan commons can be, showing transitions between open-access public goods, commons governed by regulated access provisions, and commons "undone" (i.e., reverting to national sovereignty). Changing technology shaped these choices and transitions. Nationalization has governed the development of the oil and gas resources. Fishing is somewhere between the open access of navigation and total national control, depending critically on agreements and accepted quotas.

The Surface: Shipping and Trading

That the surface of the sea is a shared resource is not as obvious as one might think. In 1608, the Dutch statesman and legal scholar Hugo Grotius advanced a clever and historically robust argument for the freedom of the oceans for all travelers. This argument was made in the context of the ongoing conflict of the Dutch and the Portuguese over free access to the sea. Grotius' *mare liberum*, or freedom of the sea, would prevent anyone exercising a monopoly over the sea trade. However, Grotius' book (titled *Mare Liberum*) hardly settled the matter. In the next century, England challenged the Dutch and their liberal concept of the *mare liberum* by adopting the Navigation Ordinance (1651). The Navigation Ordinance and its successors prohibited third-party countries from importing goods to England from other European countries (and allowed only English ships to import goods from the rest of the world). The dispute over the freedom of the sea was resolved in practice a hundred years later by restricting national territorial claims to the distance a cannonball could be fired from shore, the origin of what in time came to be known as the outer extent of territorial waters: a line three nautical miles out from shore.[4] With the exception of these universally recognized national zones, the greater expanse of the sea's surface was to be open for all.

How was the nautical commons used? How was it strengthened? What threatened its common character and use? One might expect the most obvious and spectacular threat to be piracy. However, in the usual sense of the term there was no piracy on the North Sea in the eighteenth and nineteenth centuries. But there was something almost similar. During the Napoleonic wars, privateers or corsairs received Letters of Marque and Reprisal that allowed them to take prizes—that is, to seize enemy merchant ships and bring their crews and goods to their home harbors. As Denmark/Norway was in alliance with France against Britain, privateering in the North Sea was common, and many a seaman ended up imprisoned in the enemy country while fortunes were made on the conquered goods. It was only with the Treaty of Paris, which ended the Crimean War in 1856, that international maritime law outlawed privateering and corsairs in war.[5] Neutral ships were sunk during both World War I and World War II, but that was often done in an attempt to blockade harbors and countries (a practice that was not abolished in 1856). There were, on the other hand, no legal constraints on unlimited submarine warfare.

Another threat, eliminated even earlier, was the scavenging of wrecks, sometimes linked with deliberate efforts to trick ships into stranding in order to plunder them. With the consolidation of nation-states in the eighteenth century, this activity disappeared.

Nation-states vigorously pursued the exploitation of the nautical commons, particularly in the nineteenth century. A prerequisite was the establishment of lighthouses and beacons along the maritime borders, in the coastal zones. Sailing vessels were at particular risk along coastlines. They had to navigate with the wind, which put them at a great risk in the neighborhood of reefs and rocks. The maneuverability of sailing ships was limited because the wind's direction restricted the course they could follow. In addition, sea traffic increased substantially during the second half of the nineteenth century as economies developed, grew larger, and became more international. More crowded shipping lanes increased the risk of accidents.

Another significant event for the North Sea as a practical nautical commons was the repeal of the Navigation Act by the British Parliament in 1849. This opened Britain up to third-party countries providing carrier services. This change turned out to be an important stimulus for the expansion of merchant fleets across the North Sea and in the Baltic.[6] The abolition symbolized the removal of one of the largest obstacles to an economically liberal seafaring regime. With this shift, a true *mare liberum* could at last be realized. The North Sea became the hub of a global, open and liberal seafaring regime. This was possible only because the building of lighthouses, harbors, and ports, the control of piracy, and the revision of laws allowed the North Sea to function as an effective and less precarious open-access transport commons.

The commons had gotten rid of pirates and national favoritism, but there remained another threat, a real threat to life and goods. It was the weather—more accurately, the relationship between weather and the degree to which ships and crews could deal with it. Once more we face the *mare liberum* and the liberal regime. Freedom for ship owners implied that nation-states could not force them to comply with standards of maintenance or construction. The ships were beyond national jurisdiction when sailing in the common space; on the open sea they neither had, nor recognized, any master but the captain and the shipping company. To be sure, traditionally the flag state's laws applied to the seamen on board the ship, but otherwise vessels were considered to be in international waters and under no jurisdiction. Hence it would be difficult for flag states to maintain strict control over their own fleets, since in many cases the ships would not be in their waters. It could be argued that, since it would probably always be in ship owners' and ship builders' interest to keep their ships sailing, a system of private insurance or underwriters could help to discipline the seafaring captains and their ships in cases where the flag state was not able to do so.

The establishment of this liberal regime in the second quarter of the nineteenth century was formative for the identity of the powers along the North Sea. It defined them as first-class maritime powers, and it turned them into staunch defenders of a *mare liberum* and an economically liberal international maritime policy. But throughout the nineteenth century that liberal policy resulted in a terrible toll in human lives and goods. Accidents at sea were frequent, as was loss of life. In the 1860s, in a ten-mile strip along the English coast alone, an average of more than 800 seamen were lost per year. That amounts to about 8,000 lives lost in the 1860s alone, most of them along the North Sea coast between Newcastle and London.[7] Around the turn of the twentieth century, the rather antiquated Norwegian fleet lost about 500 seamen each year. Even that paled in comparison with the mortality rates among fishermen in northern Norway in the second half of the nineteenth century.[8]

Some found these losses highly disturbing and felt compelled to take action, the most renowned among them being the Englishman Samuel Plimsoll. His campaign against what he called "coffin ships" came to have significant international consequences in the twentieth century. Plimsoll's "coffin ships" were overinsured and overloaded old ships that were of more value to their owners if sunk than if they reached their destination. Under the liberal shipping regime, Parliament had no say over these coffin ships, and sailors could be imprisoned for three months if they refused to honor a contract to serve on them.[9] Plimsoll's campaign against these ships was long resisted by Parliament, but in 1876 they finally mandated the use of load lines ("Plimsoll marks") for all ships in British harbors. These marks indicated the maximum depth to which ships could be loaded and thus enabled the authorities to prevent them going to sea with insufficient freeboard. Fifty-six years later, in 1930, an international agreement made load lines obligatory for all ships in international waters.[10]

Another controversial figure who tried to rectify the terrible situation was Magnus Andersen, later to be the first director of the Public Administration for Maritime Affairs in Norway. He was as upset by the loss of seamen's lives as Samuel Plimsoll. However, he sought the solution in improved lifesaving equipment and lifeboats. In 1886, to prove his point that even a 20-foot boat could save lives, he and a friend attempted to cross the North Atlantic in a small open fishing boat, and nearly succeeded.[11]

The forces working against Plimsoll and Andersen were effectively the same in both countries: ship owners were as well represented in the British

Parliament as they were in the Norwegian government. They had their own ideas about reducing the loss of life; they preferred a system whereby an insurance company or underwriters could realistically assess what exactly they were insuring and what kinds of risks were involved. We shall look at this system for underwriting traffic in the nautical commons, but first it is important to clearly differentiate the two approaches (legally mandated Plimsoll lines versus insurance and underwriting): one emphasized the responsibility of the flag state to intervene in the liberal shipping regime to secure safety and fairness; the other advocated a system of voluntary insurance and classification to achieve the same result.

Let us now look more closely at the ship owners and their alternative. Since the late eighteenth century, British insurance underwriters had relied on certified statements by a third party (in this case Lloyd's) as their common agent for insuring ships. Before the Napoleonic wars, the London underwriters had organized this as a collective service in the form of Lloyd's Register of Shipping (1764), which became an independent entity in 1834. Before 1851, however, its coverage was limited to the British fleet.[12] Lloyd's assessed the quality of each ship, gave it a class rating, and provided the necessary certificates. The statement of class was important for the underwriters who insured the hull. The certificates were important for the clients whose goods were to be shipped, inasmuch as the latter also had to be insured in foreign ports. Samuel Plimsoll attacked the system of underwriting that prevailed in London, arguing that it jeopardized safety on the North Sea. Underwriters were men of means who typically agreed to guarantee only a small part of the total cost of insurance (usually a few hundred pounds) at a rate that was deemed reasonable. Hence a ship might need 50 to 100 underwriters to be properly insured. Plimsoll's argument was simply that, for such a small amount of money for each underwriter, no one would really look into the risk they were taking on. Thus, he argued, it was an easy matter to overinsure the "coffin ships." To remedy this situation, the position of Lloyd's as a third-party classification company had to be reinforced, which would force the underwriters to calculate their actual risk and hence to demand a fair price for the insurance. However, class and classification remained voluntary, so it was still possible for a "coffin ship" to sail without class.

For Norway and other new maritime nations, the situation was a bit different. There were no men of means to act as underwriters so mutual insurance clubs grew up along every fjord along the south coast of Norway. Mutual insurance provided much better protection against fraud than the system of underwriters. It was in many ways a pre-modern institution, however. Mutual clubs were like literary clubs in the sense that the members knew each other and admitted new members by ballot. It goes without saying that only trustworthy "pillars of society" (to quote Ibsen) were inducted as members.[13] The system worked well for the insurance of the hulls, but there was no way to obtain certificates that would be valid in, for instance, British harbors or the Baltic, so that cargo could also be insured. Certificates had to be bought at great cost from either Lloyd's or that firm's Belgian-French competitor, Bureau Veritas. The Norwegian solution was to unite all the mutual clubs in a new society able to provide inexpensive classification and, most important, certificates for the ship that could convince clients to hire the ship to transport their goods. In 1864 such an organization was formed. Since the name Norwegian Lloyd's had already been taken by an insurance company, they chose to call it Det Norske Veritas (The Norwegian Veritas). A long struggle ensued to get the certificates of Det Norske Veritas accepted in as many foreign harbors as was possible. Later, when insurance companies took over the work of the mutual clubs, Veritas became a guarantor of quality and international certification, much like the British Lloyd's. In Germany, Germanischer Lloyd similarly covered a large part of the German fleet. Bureau Veritas was the only classification company that continued to sell classification and certificates commercially.[14]

Toward the end of the nineteenth century, the classification companies and the insurance system became the system of choice for ship owners. However, that liberal system was under persistent attack from national authorities. In Britain the Board of Trade increasingly imposed unilateral requirements for ships entering British ports, and in Norway and other countries the public demanded that national authorities should have the last word on a ship's seaworthiness. Different solutions were pursued, but the most typical was the pragmatic solution whereby national authorities accepted as seaworthy all cargo ships that had been classed as such by any of the large recognized classification companies or societies. For passenger ships, the authorities most often would do part of the survey themselves, inasmuch as passengers were considered to be innocent third parties needing extra protection.[15]

The public was particularly aware of accidents with passenger ships. The Titanic disaster in 1912 was the signal for an international initiative to augment safety. An international conference held in late 1913 and early 1914 resulted in the signing of a Convention for the Safety of Life at Sea, yet ratification stalled with the onset of World War I. A further conference held in London in 1929 resulted in the signing of a new treaty (with the same name) that addressed the matter of safety equipment on passenger ships.[16]

As we move further into the twentieth century, we find fewer conflicts about sea traffic centering specifically on the North Sea. Instead, the latter became subsumed under global initiatives for international safety at sea. After World War II, new actors took over this work and carried it forward. The UN Maritime Organization (known first as the Inter-Governmental Maritime Consultative Organization and later as the International Maritime Organization) became the main actor in this sphere. International seas need international agreements and organizations.[17] In addition, the large classification societies established a global organization, the International Association of Classification Companies.[18]

The Bulk of the Sea: Fisheries

Curiously, the three different commons we are discussing became politically salient at successive moments in time. The fact that they became pivotal at different times is evidence of the temporal specificity of technological change. Commons "go critical" when technology and organizational change establish new orders of problems and open up new possibilities. For the nautical commons, we saw that shipping in the North Sea faced great challenges with regard to safety and quality from the middle of the nineteenth century up to World War I, but that ways of handling these problems were also developed. After the Great War, sailing ships disappeared and international shipping—utilizing steel-hulled motor ships—expanded enormously and became more capital intensive. The North Sea became merely a province of international shipping. It was the era of the great maritime powers. Internationally competitive and liberal, they fostered safety through private solutions such as classification societies serving as overseers of technical standards and ship quality, and even made this compulsory in their respective national laws.

The North Sea as a fishing commons is similar to the nautical commons in the sense that changes in technology brought forth new opportunities and challenges for the participants. The fishing commons emerged later within the same geographical area and with the participation of the same nations that were already involved in shipping, though now as coastal rather than maritime, powers. As with shipping, new technologies were applied and efficiency and capital intensity increased, though not always at the same time. Moreover, the timing differed according to the fish species. Fishing for herring was industrialized much earlier than fishing for cod, for example.

The start of the twentieth century was a time of great change. Small engines increasingly replaced oars and sails on fishing craft, and fishing boats began to be fitted with decks. An important consequence was the increased range of the boats, which now could go further offshore. Bigger and faster boats also accelerated a transition from coastal-community-based local fishing, often without the benefit of harbors, to a more mobile practice that made use of dedicated fishing-boat harbors. From then on both engine power and boat size increased, as did the fleet's range, the size of catches, and the need for more capital to fund fishing operations. The coming of trawlers after World War II and the advent of full-fledged factory ships with on-board freezing equipment in the 1960s lifted the last of the restraints on the range of operation and the size of catches. However, local coastal fisheries persisted. With this divergence in fishing technologies, the contradictory interests of high-seas fishing fleets and local fleets became much more visible. When high-seas fishing became dominant, the North Sea became merely one of many regional marine management systems, leaving little room for the local perspective. This development culminated in the integration of the North Sea into the North-East Atlantic Fisheries Commission.[19]

The North Sea has always been home to plentiful stocks of fish, which provided employment and income to a large fishing population along its shores. Initially, and indeed until quite recently, one man's fishing did not influence another's very much. However, as the scale of fishing increased, and the underlying technologies changed, the threat to the common resource became ever more manifest. Herring, for instance, were notorious for disappearing from places where they had been plentiful the year before, and no one could say where they had gone or even if they had simply disappeared. Other species were much more dependable, but still it became increasingly desirable to ensure more predictable fishing conditions.

As early as the 1870s, there were initiatives to draw up an international convention to regulate the fisheries in the North Sea. In 1882, an international convention that established a common framework for fisheries management was signed by Belgium, Denmark, France, Germany, and Great Britain. The purpose was to prevent conflicts, particularly in the southern portion of the North Sea, where distances between different nations were small and common fishing grounds within easy reach. That early agreement and the protocol that was added in 1887 were followed by several other agreements relating to fisheries and fishing grounds in the North Sea. In 1908, a convention concerning the limits of territorial waters was ratified by the coastal nations. Another convention signed in 1946 regulated the size of the mesh in fishing nets. There were also many failed attempts to reach agreements. Indeed, with the exception of defining the territorial waters (a three-mile limit from the coast), comparatively little progress was achieved by formal negotiation. Nonetheless, the failed negotiations were occasions for discussion and learning, and their value should not be underestimated. The various national actors became familiar with each other's standpoints on many questions, and each was consequently able to think strategically about the different positions and either accept a situation or find ways to alter it or to get around it.

After World War II, the volume of fishing and the capacity of fishing fleets in the North Sea expanded dramatically. This was due primarily to the development of large-scale fishing vessels, which often incorporated processing plants and new cooling and freezing technologies. No longer were fishing nations confined to their own coasts. Fishing became a truly international activity. But at the same time, coastal states became increasingly aware of "nearby" resources in but also under the sea much further offshore than the nearby territorial waters. The UN tried to resolve these contradictory trends at two international conferences in 1958 and 1960, without achieving much success aside from bringing the problem to light and convincing coastal states that a consensus was not in sight. That led many nations to undertake unilateral actions. By the late 1960s these tensions moved the UN to convene a Third Law of the Sea Conference, which lasted for ten years, from 1973 to 1982. In the end, that conference arrived at definitions and principles pertaining to economic zones, continental shelves, and the development of the deep sea floor. Even if the mountains of work that went into all these conferences did not result in ratified general agreements, the conferences did provide guidelines for national regulations and national strategies.

The political scientist Arild Underdal has described three fundamental "requirements" for fisheries commons, emphasizing their respective conflicts. Viewing these three requirements in a historical framework, he has described a logic governing their emergence through time. The three requirements are defined by the following questions: Who has the right to fish? How should fishermen behave when fishing? How should fishing stocks be preserved for future fisheries?

Let us start with the discussion of fishing rights. If not everybody is to have access, who is to be excluded and on what grounds? Here the approach typical of coastal states dominated: fishing rights should be limited to a coastal state's inhabitants as far offshore as the state could claim sovereignty. This was the first limitation of the freedom of the seas, and it marked the introduction of fishing limits nationally and regionally. As we have seen, states around the North Sea generally considered three nautical miles to be the appropriate limits to territorial waters and national fishery jurisdiction. The exception was Norway, which claimed four miles.[20] In the first half of the twentieth century, as noted, fishing technology had already made it possible to fish further out with larger boats. As a result, fishing vessels from many nations were operating just beyond the three- or four-mile territorial limit, harvesting what coastal nations deemed to be national resources. The international laws, the management of the resources in the commons, and the technologies being applied were obviously not in balance.

Between 1958 and 1962 the Nordic countries unilaterally extended their territorial waters to twelve nautical miles from shore. The United Kingdom did not respond unilaterally and invited most European states to a fishing conference in 1963. This conference was not to end in an agreement, although it came close. Nonetheless, it introduced a principle that would have great implications in the years to come. The conference established a uniform six-mile limit to territorial waters, but in addition it ceded jurisdiction to the coastal state over waters between six and twelve miles from shore on the condition that rules and laws did not discriminate among nations that had historically fished there, and on the condition that all signatories had the right to continue fishing in this zone if they had been doing so for the last ten years. However, the nation-state set the total quotas (or negotiated quotas) for catches and allocated these quotas to the fleets of the various "guest" nations according to each nation's historical fishing tradition and the negotiated agreements. Further expansion of national sovereignty came in the middle of the 1970s with the establishment of the 200-mile Exclusive Economic Zone (EEZ) as the limit to coastal jurisdiction. Thus, the zone of national sovereignty was extended further out into the North Sea, and a further zone combining national sovereignty with a system of limited international usufruct was established.

The second "requirement" in fisheries related to rules of conduct. How are people to behave in the commons? This problem was initially addressed by the pathbreaking North Sea Fisheries Convention of 1882. That convention included agreements about markings on fishing boats and gear, appropriate behavior on the fishing grounds, and types and use of signals.[21] Fighting on the fishing grounds and destruction of fishing gear were not unknown, and no doubt an agreement was useful. It took until 1967 before newer rules were agreed upon, but there had been several bilateral agreements on damage claims and related matters between countries in the intervening years.

The third "requirement" that Underdal discusses is related to conservation. Before the twentieth century, the supply of fish usually had seemed endless, and nothing humans could do seemed to affect that. Under those circumstances, the fish in the North Sea represented a public good (like air or sunlight) and the only challenge was to catch them. It was not until the late 1930s that an international agreement between ten European countries was reached on the minimum sizes of the mesh in fishing nets and the minimum size for some species. This first regulation was put in place to safeguard stocks by making sure that young fish could escape from the nets or would be thrown back into the sea. After World War II it became increasingly evident that further measures would be needed to prevent depletion of several species of fish. The highly efficient boats and fishing gear now threatened the sustainability of the fishing commons, and the situation called for more rigorous measures. Fish now began to be treated as a limited, shared (common-pool) resource, and no longer as a public good that could be harvested without limit. Several conferences were convened, without much success.

The 1882 conference, which had focused on the North Sea, had been the first conference of its kind anywhere. After World War I, however, the North Atlantic became the contested area. Disputes erupted between Norway and the Soviet Union in the early 1920s and between Iceland and the UK in 1958 and again in 1972-3. After 1950, the UN played an increasingly prominent role in these disputes, and fishery issues became generalized to all the world's oceans. This was the time of the new coastal powers in the so-called Third World. Extending this narrative of commons upscaling, we also see the North Sea becoming a kind of local European Union sea at the start of the twenty-first century, and as such mainly governed from Brussels. However, we should first look at developments in the 1960s and the early 1970s.

It took four years (1966–1970) to hammer out an agreement on temporary bans on herring fisheries in the North Sea. The most important innovation in these years was the establishment of a permanent international fisheries commission, mandated to provide ongoing recommendations for fisheries management to the flag states along the coast of the North East Atlantic, including the North Sea. The idea was that this would obviate the need for a new conference every time new issues came up. Ironically, the commission was itself the product of a conference that was turned into a more permanent commission: the North East Atlantic Fisheries Commission (NEAFC), which first met in 1959 (although the international convention that formally established it was not ratified until 1963). Despite this convention, it appeared that a "tragedy of the commons" would prevail. As the pressure on several fishing stocks increased, and inasmuch as the commission could only make recommendations, states began to make bilateral and regional agreements.

The forces that were globalizing fishing and fishing treaties continued to work in the other direction. The United Nations and its Food and

Agriculture Organization got heavily involved, helping to establish a system based on what were called Regional Fisheries Management Organisations. The NEAFC became just one of a dozen similar fishing organizations that managed fisheries on the high seas around the world. The slow but steady progress of these regional management organizations gradually improved matters, more in some places than in others. In any case, they began to set quotas, to establish fishing seasons, to improve monitoring of catches, and to get flag states and coastal states to inspect and control catches.[22] A crucial aspect of this new system for the commons was the introduction of the 200-mile Exclusive Economic Zones. Though these zones are not formally national territory, the coastal states are obliged to police them and regulate them in non-discriminatory fashion. Coastal states therefore had to establish coast guards to police activities in their zones—notably to inspect fishing ships' catches and their gear (e.g., the mesh sizes of their nets) and to make sure officially required reports and logbooks were in order.

Still, these measures are not as effective as one might wish. Fishermen are still able to assume the roles of pirates and free riders in the commons, and those who do so have now acquired their own place in the world of UN and OECD acronyms: their activities are denoted as Illegal, Unreported, and Unregulated (IUU) fishing. Note that illegal, unreported, and unregulated are all considered undesirable behaviors that should be punished. By default, the IUU category implies that all fisheries should be regulated, reported, and "legal"—that is, all fisherman should be registered and hold permits to fish, their catches should be reported, their mesh sizes should be correct, and they should catch fish only in their designated areas and at designated times. This is a long way from the open-access regime that prevailed a century earlier. However, even if everything is regulated, the sea is still an international commons beyond national control (even if the 200-mile EEZ gives coastal states privileges and policing authority). It remains to be seen how well this enhanced international system will work in the future.[23]

An important aspect of fishing is its gambling aspect. Fishing could create small fortunes for fishermen when they were lucky (the herring industry is a good case in point) or misery when the sea seemed empty and bleak. Fishing was also a kind of game in which the fishermen's very lives were at stake. All this served to increase the level of superstition and the aura of gambling that pervaded fishing culture. In this gamblers' community, fishermen themselves developed pragmatic rules of moral conduct, and national or regional authorities later codified these rules. Prohibitions against interfering with another's nets or lines must have been commonplace. From the Lofoten Archipelago we know that it was a sacred duty to help rescue endangered men and equipment. It was also considered obligatory to inform the clergy if life was lost at sea and the seamen came from another part of the country (as was common in the seasonal fisheries at Lofoten).[24] That fish needed protection and conservation seemed, however, to be beyond imagining. The North Sea seemed to be so big that no amount of fishing could ever affect the size of the fishing stock. This may be one of the really big differences between the sea as a commons and common pastures: knowledge of the sea's limits was almost entirely lacking. What the fishermen could see and experience was the surface of the sea and the often threatening weather above it. They knew storms and waves and good fishing grounds by experience. However, fish, their reproduction, and what went on under the surface were only anecdotally understood. Oceanography, fish biology, and fish ecology were products of the second half of the nineteenth century, when these research areas were first institutionalized and began to develop.

Studying fish and their migration scientifically became ever more important for the coastal states along the North Sea because such knowledge could have substantial economic implications. However, inasmuch as the pelagic and demersal fish of the sea were quite indifferent to borders and geographical limitations, it proved easier to study fish populations in fjords, rivers, and lakes.[25] In the 1880s and the 1890s, several biological stations were opened in Norway to study fish in the fjords (with the further aim of learning whether they could be cultivated). At the same time, several expeditions were mounted to study fish and other animal life at greater depths in the sea.[26] Marine research and oceanography became important tools in the effort to uncover the secrets of the hidden world under water, the common resource on which the coastal population lived. The parallels with seismic research a hundred years later are remarkable.

By the end of the twentieth century, regional commissions were routinely basing their recommendations on science and scientific advice regarding the size of quotas for fish brought on shore. Fish biology and fish ecology today define the terms for the management of the resources. They have established the criteria for managing fishing commons from the beginnings in the second half of the nineteenth century until the present. During that time, fishing equipment became so effective that the ocean could easily have been emptied—and the North Sea could have become utterly dead—had not restrictions been put in place. So, despite all the inefficient congresses, tiresome commissions, and partial agreements, the fishing commons has worked so far, though it has never been perfect. Illegal fishing operations continue, though they are not as large in scope as they otherwise would have been.

Since the 1970s, however, a very different form of fishing industry has developed in the North Sea, and it bears a close resemblance to the commons as pasture. This is the fish-farming industry, dominated by salmon farming and mostly confined to territorial waters. Both the United Kingdom and Norway have developed large-scale salmon farming industries within their own nationally controlled zones of the North Sea. The scale of this industry is approaching that of fishing on the high seas.[27] However, fish farming is still heavily dependent on fishing on the high seas inasmuch as the fodder for the farmed fish consists in part of fish oils, which can be obtained only from other marine species. So once again it is not easy to avoid the contradictions of shared resources, even if the area for harvesting the fodder has increased and the areas for production have withdrawn to zones under national jurisdiction.

Under the Sea Bed: Oil and Gas

Whereas the lives of the fish in the high seas were gradually studied and revealed, another layer of the North Sea commons continued to be neglected: the seabed and the layers beneath it. For many years people wondered what sorts of treasures it might harbor, especially whether there might be any oil and gas. Yet before the 1960s the dream of petroleum resources under the sea bed seemed exactly that: a dream.

Oil and gas are the oldest resources in the North Sea Basin (having formed in the Jurassic and Cretaceous eras), but they are the newest of the shared resources in this space. They also offer a striking example of how a common resource comes into being thanks only to the availability of specific technologies. Without platforms and drilling technologies, the resource cannot be said to be a resource at all, because it is not accessible. But thanks to new technologies, the sea bed of the North Sea became a shared international space of great promise, possibly harboring hidden treasures. On the other hand, no one had as yet any definite idea of what wealth actually lay under the sea bed between Britain, Germany, Denmark, and Norway. Very few believed there could be any oil down there.

How this resource-space came to be regulated and "shared" among neighboring countries and successively parceled out to diverse national companies is an interesting story of an entirely new commons emerging among rather stable and advanced economies. In the global drama of gas and oil, commons play a small part. The usual story is that oil and gas are found and developed in fairly new and still unstable countries that easily succumb to pressures from large multinational corporations by simply letting them have their way under very liberal economic regimes. The exploration and exploitation of North Sea oil and gas reserves depart from this scenario in two distinct ways. First and foremost, it was anticipated that the resources would be found on the continental shelf beyond the national borders of the neighboring states. Second, these states were rather mature and resourceful states with stable democratic regimes and with considerable experience in dealing with large multinationals. They were able to stand up to the companies in a way not usually seen at that time among developing countries.

Another characteristic feature of the emergent commons was the basic uncertainty reflected in the fundamental questions "Where are the oil and the gas, if anywhere?" and "How much is there, and is it possible to bring it up in a safe and economical way?" In addition to all the other uncertainties, exploiting oil and gas under the sea bed was a big economic gamble. And in any case it was clear from the start that exploiting this common resource would require huge investments. Thus, in contrast with fishing and sailing, undersea oil and gas demanded "big actors"—large multinational corporations and nation-states with the capacity to take risks and absorb losses.

The rules were also different when it came to villains and crooks. Here there were no pirates, privateers, wreck plunderers, illegal catches, or illegal fishing. The "bad guys" now were "organized capital" trying to obtain the best possible conditions from the nation states as soon as the latter had solved the question of sovereignty.

So who was to "own" the sea bed and its resources, when they extended beyond a reasonable "economic zone"? Before the prospect of oil and gas under the sea floor arose, the question of jurisdiction at sea was first and foremost directed toward the regulation of fisheries. But oil was something very different. Toward the end of the 1950s it began to seem at least possible (but not very likely) that oil and gas could be found under the North Sea's floor. With experience in the Gulf of Mexico providing models for technologies to get it out safely, the question of access and ownership became more real and pressing. In 1958 a temporary agreement was reached in Geneva between some of the North Sea countries; this agreement split up the continental shelf according to the midline principle. (In essence, the midline principle partitions the sea bed equidistantly from the coasts of two countries that border it and have a claim on it, although practical application of the principle involves a number of complexities.) However, this principle was applied only to waters down to a depth of 200 meters. The agreement was, moreover, troublesome for Germany and for Norway.

Germany's claim was effectively squeezed out by application of the midline principle to the Dutch and the Danish claims, which left no obvious way of making a German zone. The International Court of Justice solved the question of a German zone in 1969–1970.[28] The solution combined the midline principle with the sector principle (explained below), which was used later in the polar regions.

For Norway the problem was the Norwegian Trench, a channel between 300 and 700 meters deep running along most of Norway's North Sea coast. On either side of the trench, the depth is about 100 meters. The continental shelf was considered to be exactly that, a shelf. Strictly speaking, the shelf ended where the Norwegian Trench began. Beyond the trench, however, the shelf arguably continued, since the waters here were again well within the 200-meter limit. Thus, if the international agreements were to be strictly interpreted, Norway would have to abandon its claim to much of the continental shelf adjacent to its territory, because the trench was situated rather close to the shore (between two and ten nautical miles). Like Germany, Norway did not sign the temporary agreement but tacitly interpreted the trench as being inside its economic zone, implying that the real shelf started beyond the trench. Hence, it did not present an obstacle to employing the midline principle in a straightforward way. So, dusting off an old scenario, the Norwegian government unilaterally declared that all natural resources on the sea floor were Norwegian as far out as available technologies would allow them to be harvested, stopping only at the midline to neighboring states.[29] And, equally important, the rights to these underwater resources were to belong to the state. This unilateral claim made one-third of the North Sea's floor Norwegian, but it was nonetheless silently accepted by the other countries because it was based on the same protocol as their own claims according to the 1958 declaration. Besides, at that time no one believed there was much of value to be found. The Norwegian Trench was simply forgotten by all. And that was just as well. Had it been remembered, the unreasonable consequence would have been to validate a British claim to almost all of the North Sea floor, leaving only a tiny coastal strip to Norway. That would not have worked; the midline principle was the only workable solution.

Later another principle was employed, particularly in the Barents Sea. This so-called sector principle also provided a solution for the German part of the continental shelf in the North Sea. The sector principle entailed tracing lines from the North Pole (or in the German case from a constructed point) to some common point, most often where the terrestrial international border abuts the sea.[30] The application of the sector principle instigated a big conflict between Norway and the Soviet Union (later Russia), which was unresolved for more than 40 years until an agreement was reached in 2011 by drawing the border along a line equidistant between what would have been a midline solution and a sector solution. When there was no need for a solution (because no one would dare to drill for oil or anything else in those icy waters), no solution was found. The more feasible the actual drilling became (that is, the more reliable the surface and especially the subsurface technology was judged to be), the more pressure there was on the two states to resolve the disagreement, with the result that a compromise was reached even before exploratory drilling started.

The petroleum adventure in the North Sea began in response to the discovery of a large gas field near Groningen in 1959. By mid 1960, the first minor oilfields had been discovered in the North Sea. However, in the beginning things moved slowly and the area was hardly considered promising. The big breakthrough came in December of 1969, after 37 drilled holes came up empty: the Ekofisk field was found by Philips Petroleum in the Norwegian sector. A bit later, the Brent field was discovered on the British side. By then most of the tools for working in the sea bed were in place and operative. As soon as the common space was allocated to the different nation-states, different regimes for handling access to the sea bed emerged. Restrictions were often severe. The cost of drilling a hole was extremely high, and only the prospect of a large reward could justify the investment. Finding ways to reduce uncertainty and enhance the likelihood of gain became important strategies for the oil companies and, at the same time, hard facts of commercial life that the various governments had to take into account.

Two strategies became important: developing good seismic technologies and gaining a monopolistic position. Both were pursued in the North Sea in the 1960s. Better knowledge of the sea floor and of its sedimentary layers went hand in hand with an intense research effort to develop seismic technologies and high-quality three-dimensional seismic maps. Research and development were the keys to reducing uncertainty. Both acquired a boost as new firms specializing in seismic mapping applied themselves to the development of innovative methods of three-dimensional seismic surveying and registration.[31]

The political strategy of the North Sea states focused on getting the best deal from oil companies working in their zones, while oil companies sought to exclude competitors as much as possible. Upon the division of the continental shelf into different national sectors, most countries passed legislation qualifying sea bed resources as state property. Even prospecting for resources like gas or oil therefore required a state concession. The solutions for dealing with the attendant risks and uncertainties varied from country to country. On the Danish shelf, one conglomerate (DUC, a group of companies led by A. P. Möller) was granted a 50-year monopoly on prospecting and drilling for oil.[32] However, the conglomerate was obliged to explore the shelf according to some systematic plan. The solution chosen in the British and the Norwegian sector was to divide the sectors into "small" rectangular blocks. Seismic surveys carried out by the state identified some of these blocks as more promising than others, and the states would put these blocks up for auction among international oil companies. However, the states had to attract bidders, so they had to make their "promise" look convincing.

One could say that the sea floor of the North Sea had undergone two transformations. First, an openly accessible transnational space was transformed into parcels of new territory added to the different nation-states—a kind of decommonizing process ruled by international agreement. Second, at least some of the nation-states re-commonized the area again, but this time as a special kind of territory within the national borders that could be auctioned off as sections of exploration rights to the highest bidders. The land was not "sold," just the right to explore for, and later to extract, oil and gas. Most surprisingly, this affected neither the common nature of the sea column above (that is, the fisheries) nor the principle of *mare liberum* (the commons of sea traffic).

A closer look at the Norwegian sector is helpful for understanding these dynamics of the block-survey and auction system. Early in the 1960s, before any findings and before any of the larger oil companies had shown interest, a smaller oil company (Phillips Petroleum) contacted the Norwegian government and offered to search for oil on the Norwegian shelf on condition that it would be granted a monopoly of the whole shelf (a "Danish" solution except for the fact that the Norwegian shelf was many times larger). As has been noted, there was a great deal of doubt as to whether there was any oil, so the deal did not look bad. However, the request was denied. Instead, the Norwegian government opted for the block-survey system and opened several promising blocks up for bidding on rather strict terms. Phillips, which had not lost interest, took the bait, and started drilling in 1965, and the drilling culminated in the big Ekofisk discovery of December 1969. With proof of prospective oil profits in hand, the Norwegian government could now tighten the noose, increase taxes, and subject the oil companies to more exacting demands. A rather successful regime was instituted, with international oil companies funding research in Norway, training Norwegians, taking Norwegian companies on as partners, and quickly transforming a country with no experience in oil extraction into an international leader in the field.[33]

By mitigating some of the risks and inefficiencies of earlier technologies, new technologies have also enhanced the resource-extraction possibilities. Underwater production units (on the sea floor) have taken over from the surface platforms. Gone are the days when production rigs filled the North Sea. Today, remote-controlled underwater installations with underwater piping fill the sea floor. "Horizontal" drilling (actually, drilling in all directions) has also increased the efficiency of the fields, and many fields have turned out to produce much higher yields than was foreseen when they were discovered. In addition, technologies to increase well pressure have almost doubled extraction rates from the fields. Smaller underwater production units and piping allow even marginal fields to be processed profitably and safely. Nevertheless, a number of accidents in offshore drilling—the Deepwater Horizon accident in the Gulf of Mexico is only the most recent of many—confirm that there is no such thing as fail-safe technology.

It is interesting to see how a gradual learning process transformed the original knowledge and extraction system established in the Gulf of Mexico into a very specific high-tech, state-governed, North Sea regime for oil extraction. This was not by chance but by design. When the North Sea's oil wealth became manifest in the 1970s, a regime to regulate access to, use of, and development of the former sea bed commons was in place. The commons had been nationalized, and hunting for oil had become a knowledge-intensive activity. The level of precariousness had been reduced significantly, and there was very little opportunity for any sort of trespassing or piracy. Horizontal drilling and the existence of trans-border oil reservoirs might make it enticing to think about extracting oil resources from neighboring countries, but most threats to do so have been neutralized by negotiations and agreements. The originally transnational common resource-space had been transformed into a patchwork of national commons, access to which was strictly policed by the individual nation-states.

Conclusion

Looking back at the multiple uses of the North Sea and its resources, we see two basic strategies at work: reducing uncertainty or precariousness and regulating access to the commons. In the domains of fishing and shipping, these strategies ultimately produced a transnational North Sea commons. Shipping was rendered less precarious through improvements in maps, lighthouses, anchors, and anchor chains—and by ensuring the seaworthiness of ships through classification, load lines, and better hulls. Politically, early efforts to restrict access took the form of Navigation Acts and other strategies of national prioritization, but those were not successful in the long run. Instead, the North Sea as a transport commons became stronger and more valuable as a more liberal access regime emerged, and as life and property were better protected through systems of transparent regulation that insured that ships were properly qualified and had the necessary certificates.

Things were much the same in the fishing commons. Uncertainty and precariousness were reduced by means of better equipment, larger boats, and more adequate biological knowledge. After World War II, sonar (ASDIC) became another important tool in fishing. However, efficiency increased so much that a new form of precariousness emerged: fishing resources seemed on the verge of exhaustion. Again, the political answer was to reduce access and to regulate how much could be harvested each year. But to enforce this, states subjected expanding zones of the North Sea to their authority–ultimately through the designation of Exclusive Economic Zones, although the latter also comprise regulated usufruct rights for fishing fleets from other states. Again, international agreements were signed, opening these zones for limited fishing by participating nations. It was not always easy to adhere to the agreements. In contrast to the oil business, fishing involved many actors and was difficult to monitor. It remained quite possible to violate fishing limits, to overfish, or to catch the "wrong" species or size of fish. Still, the important point was not the loopholes, but rather the fact that the countries were able to reach agreements at all, or at least to agree tacitly to some rules. Establishment and enforcement of more formal rules could be deferred.

The character of the North Sea as a common space has undergone substantial changes. Old problems have been surmounted, and governability and access have been unrecognizably transformed. A brief glimpse into the future of our three old commons might be useful. In shipping, large international organizations—the IMO, the EU, the IACS—are still fighting over safety issues. But currently the main area of concern is the "greening of shipping," e.g., through a shift to propulsion systems that emit fewer greenhouse gases. Both sail and natural gas are under study.[34]

Aquaculture inside territorial waters is now at the cutting edge of fishing technology. The amount of farmed salmon in Norway is now greater than the total fisheries of the UK (both at sea and farmed).[35] However, fishing on the high seas, though increasingly controversial, will continue to be necessary, if only to sustain aquaculture.

In the domain of oil and gas, the North Sea is now subject to a set of strict national commons regimes. However, it remains technologically dynamic. For example, thanks to all sorts of new technology, the Ekofisk field is still producing—in fact it is producing twice as much oil as it did when it was discovered, four decades ago. Underwater production plants are regarded as so safe now that they are slated for use in the icy waters of the Barents Sea, though it remains to be seen if that technology can deliver on its promises. It is telling—and perhaps suggestive of a special European dynamic—that similar partitioning of sub–sea bed commons in other areas of the world (for example, the South China Sea) has fomented new conflicts. Meanwhile, the national demarcations established in the North Sea now seem to have been there forever.

It is tempting to say that a permanent and stable commons is an illusion, at least at sea. Not illusory, however, are the transformations, the innovations, or the conflicts and negotiations over access, use, resources, and modes of extraction. In the face of climate change, even shipping might soon find itself confronted with limited environmental resources. Unless we find a cost-effective way to reduce carbon emissions, it too will have to be constrained by common consent. The transformation of the sea (70 percent of the Earth's surface) as a shifting commons may be a never-ending story.

Notes

1. David B. Keto, *Law and Offshore Oil Development: The North Sea Experience* (Praeger, 1978), p. 17.

2. I borrow the concepts of coastal and maritime powers from *Marine Policy and the Coastal Community: The Impact of the Law of the Sea*, ed. D. Johnston (Croom Helm, 1976).

3. Hugo Grotius, *The Freedom of the Seas; or, The Right which Belongs to the Dutch to Take Part in the East Indian Trade* (Oxford University Press, 1916). Original edition: *Mare Liberum* (Leiden, 1609).

4. Ibid.

5. Hugh Chisholm, ed., "Declaration of Paris," in *Encyclopædia Britannica* (Cambridge University Press, 1911).

6. See, for instance, Richard Price, *British Society 1680–1880* (Cambridge University Press, 1999), pp. 117–118.

7. Samuel Plimsoll, *Our Seamen: An Appeal* (Virtue & Co, 1873), p. 30.

8. Håkon With Andersen and John Peter Collett, *Anchor and Balance: Det norske Veritas 1864–1989* (Cappelen, 1989), p. 55. These numbers include sailors lost in all waters. For fishermen see Eilert Sundt, *På havet* (Pax, 1976).

9. Plimsoll, *Our Seamen*, p. 20.

10. *International overenskomst om Lastelinjer* (International Loadline convention) (Sjøfartskontoret, 1933).

11. Magnus Andersen, *70 års tilbakeblikk* (M. Andersen, 1932), pp. 43–69.

12. Blake, George, *Lloyd's Register of Shipping 1760–1960*. (Lloyd's, n.d.).

13. Henrik Ibsen's play *Samfundets støtter* (*The Pillars of Society*), first published in 1877, deals with the seaworthiness of ships and the moral dilemmas of ship owners and shipyards in having to choose between safety and owners' profits.

14. Andersen and Collett, *Anchor and Balance*, chapter 1.

15. Ibid., pp. 77–81; Magnus Andersen, *Norges første Sjøfartsdirektør. Ansættelse—Virksomhet—Entledigelse* (Feilberg & Landmark, 1912).

16. Available at http://www.mss-int.

17. Available at http://www.imo.org.

18. Available at http://www.iacs.org.

19. Available at http://www.neafc.org.

20. Arild Underdal, *The Politics of International Fisheries Management* (Oslo: Universitetsforlaget, 1980), p. 45.

21. Ibid., p. 48.

22. *Why Fish Piracy Persists. The Economics of Illegal, Unreported and Unregulated Fishing* (OECD, 2005).

23. Ibid.

24. Sundt, *På havet;* Narve Fullsås, *Havet, døden og været: kulturell modernisering i kyst-Noreg 1850–1950* (Samlaget, 2003).

25. Pelagic fish live in the water column; demersal fish are bottom feeders and live on the sea bed.

26. Håkon With Andersen et al., *Æmula Lauri. The Royal Norwegian Society of Sciences and Letters, 1760–2010* (Science History Publications, 2009). See also Vera Schwach, *Havet, fisket og vitenskapen. Fra fiskeriundersøkelser til havforskningsinstitutt 1860–2000* (Bergen: Havforskningsinstituttet, 2000).

27. Available at ftp://ftp.fao.org.

28. International Court of Justice, *North Sea Continental Shelf Cases (Federal Republic of Germany/Denmark; Federal Republic of Germany/Netherlands). Judgment of 20 February, 1969*.

29. Tore Jørgen Hanisch and Gunnar Nerheim, *Fra vantro til overmot?* (=Norsk Petroleumsforening, 1992), pp. 18–21.

30. International Court of Justice, *North Sea Continental Shelf Cases.*

31. Andersen and Collett, *Anchor and Balance*, pp. 395–400.

32. Keto, *Law and Offshore Oil Development*, p. 41.

33. Helge Ryggvik, *Til siste drape. Om oljens politiske økonomi* (Aschehoug, 2009). See also Hanisch and Nerheim, *Fra vanntro til overmot.*

34. *Technology outlook 2020* (Det Norske Veritas, 2011).

35. World Fisheries Production, by Capture and Aquaculture, 2008 (available at ftp://ftp.fao.org.

10 "One Touch of Nature Makes the Whole World Kin": Ships, Fish, Phenol, and the Rhine, 1815–2000

Nil Disco

This chapter examines three cosmopolitan commons that were created to govern resources on and in the Rhine River: a "navigational" commons (which dates back to 1815), a "salmon" commons (which emerged in the 1870s), and a "clean water" commons (which began to take shape after 1950). Despite their heterogeneity and their separation in time, the three commons are historically, institutionally, and legally intertwined, not least thanks to the salmon's intolerance for polluted water. In what follows, I will show how, for the Rhine, the salmon—this "one touch of nature," to quote Shakespeare—has been instrumental in "making the whole world kin"—that is, in shaping interlocking cosmopolitan commons on the river.

The navigational commons was rooted in the Congress of Vienna, held in 1815 to reorganize Europe in the wake of Napoleon's defeat. This commons resembled the navigational commons on the North Sea and other "valorizing" commons discussed in this book—airspace, etherspace, and weather forecasts—in that it was developed to optimize the valorization of a natural resource (the Rhine as a waterway) rather than to protect some already threatened resource or natural value. Both the salmon commons and the clean-water commons are examples of the latter. This chapter's long time frame enables us to examine in some detail how these commons were intertwined, how successive cosmopolitan commons facilitated the exploitation of resources embedded in the Rhine, and how they were also mobilized to combat some of the "negative externalities" that in time revealed themselves as the environmental price of that exploitation.

Nature, especially in the form of gravity, had a strong voice in these developments. Water's compulsion to run downhill is the great maker of rivers. Everything that happens on a river happens against the background of this fundamental energetic fact.[1] It creates a range of asymmetries. Locations on a river are positioned not only absolutely in relation to one another according to conventional notions of distance and direction, but also fluvially by dint of their being upstream or downstream of other locations. This means that rivers generate a geography of their own that is more coherent than that of the surrounding countryside.[2] The river reorganizes standard geography by connecting its locations through the medium of flows of water, and also through objects and substances on and in the water. The Rhine, in particular, was a great instigator of port cities connected by waterborne transport—and later of derivative mining, metallurgical plants, and chemical plants.[3]

On a river, space, time, and economics are compressed and extended in many ways. The water (and added ingredients) that flows by location A today will pass by location B tomorrow and will appear at location C the day after tomorrow. Anything moving upstream (fish, ships) will take longer to travel a fixed distance than it would if moving downstream. But the very facts of fish swimming and ships navigating reveal how the river also supports unique waterborne flows that would be unimaginable in the adjoining countryside. Indeed, we shall see that the cosmopolitan commons that were created both to abet the industrialization of the Rhine and to mitigate the effects of that industrialization are deeply embedded in these flows and in the physical, economic, and political asymmetries they express. This positions my story somewhere in between Kristiina Korjonen-Kuusipuro's story of the Vuoksi (where everything seems to flow in one direction from higher to lower elevations) and Arne Kaijser's account of medium-range transboundary air pollution (where the prevailing direction is only that: prevailing).

The two "conserving" commons stories I discuss in this chapter—fish and phenol for short—emerged in the long shadow of a navigational commons that took shape in the course of the nineteenth century. In other words, by the time these later commons emerged, there was already a community of interest, there was already a moral economy, and there already were structured channels of communication to facilitate the common use of the river for navigation. Second, inasmuch as the navigational commons had facilitated the transformation of the unruly natural river into a streamlined navigational conduit and thus paved the way for the "carboniferous" industrialization and urbanization of the river valley, it also co-produced the very conditions that moved actors to establish the later commons.[4] Thanks to the success of the navigational commons and all that went with it, both the river's natural fauna and flora (including its salmon) and the quality of its waters took decided turns for the worse. The national and transnational struggles to turn this around, to prevent the degeneration of former public goods into precarious "common-pool" or "common-flow" resources, ultimately eventuated in the creation of the two later cosmopolitan commons.

A special place is reserved in this history for the Rhine salmon. As the Rhine turned "carboniferous" and became an artery of modern commerce, the river—once full of this strong, beautiful, eminently edible and commercially valuable fish—became such an impoverished habitat that the salmon, and the salmon fisheries, ultimately died out, despite a conservationist salmon treaty enacted in 1885. Though commercially speaking the salmon didn't actually disappear until the 1930s, the signs of decline in fish stocks were visible as early as the 1850s and they in fact provided the first indication that the triumphant march of commerce and industry were a mixed blessing. In the 1980s, when the last examples of the Rhine salmon were already fading from living memory, the fish and the old salmon treaty were resurrected to breathe new life into the stalemated negotiations on a water-quality commons. The return of salmon to the Rhine was now framed as an indicator of an ecologically restored and therefore clean Rhine. The "extinct" fish, in becoming the symbol of the Rhine's ecological renaissance, also became the medium of its own resurrection, but now as a leitmotif of ecological restoration rather than as a commercial resource.

The Navigational Commons and What It Replaced.

A River of Tolls and Privileges

Since Roman times, transportation on the Rhine had suffered a gantlet of tolls based on the premise that the river "belonged" to the rulers of the territory through which it flowed, and that this warranted a share in the profits of river transport. This tacit assumption was explicated in the early seventeenth century by the Dutch legal scholar and statesman Grotius, who stated that "a River . . . is the property of the people through whose lands it flows, or of him under whose jurisdiction that people is; and they may, if they please, make sluices, and appropriate to themselves whatever that river produces."[5] Grotius derives this ownership explicitly from sovereignty over the banks: "It is sufficient for us, that the larger part of the water . . . is shut up in our banks, and that the River, in respect of our land, is itself small and insignificant."[6] That this was not the case for seas and oceans was the basis of his famous claim to freedom of the seas (*mare liberum*).[7]

The Roman practice of levying tolls on Rhine shipping was continued by their successors on the Rhine, the Franks, and then by the Holy Roman Empire (962–1806).[8] Under the Empire, local potentates along the Rhine or riparian free cities were entitled by the emperor and his vassals (often high churchmen) to levy tolls on passing ships. This central "management" of the collective spoils—which was of course primarily intended to cement relations of fealty—also served to prevent an endless proliferation of local tolls that would have throttled the life out of river shipping altogether. In 1250 there were accordingly "only" twelve toll stations on the Rhine between Mainz and Cologne.[9]

However, as is well known, central power was at best a precarious affair in the Holy Roman Empire. During the interregnum of 1250–1273, when it proved impossible to achieve consensus on a new emperor, low-ranking local nobles along the Rhine seized the opportunity to levy their own tolls.[10] These "robber barons" built mighty castles overlooking the river, using them as strongholds from which to extort tribute from passing ships. This made river transport more expensive; it also probably reduced the take of the legitimate tolls. In 1254, despairing of an imperial solution, a powerful coalition of urban merchants joined with archbishops and nobles officially entitled to collect tolls in the so-called Rhine League. The league lasted only three years, but in that time it succeeded in besieging and shutting down ten "outlaw" castles belonging to five different robber barons. In its way, it was a cosmopolitan navigational commons *avant la lettre*— a moral-economy regime armed to the teeth and dedicated to preserving rules of access to a shared and precarious resource. Nonetheless, even after legitimate imperial order was restored, tolls continued to oppress Rhine navigation until well into the nineteenth century, though their benefits increasingly accrued to states and municipalities rather than to individual nobles. It has been estimated that during the eighteenth century the tolls on the forty-mile stretch between Bingen and Coblenz amounted to one-third of the value of a ship's cargo.[11] Toll stations came and went, and their number varied greatly through time. Though it is hard to be exact, the scholarly consensus is that on the eve of the French Revolution there were 32 active tolls between Strasbourg and the Dutch border.[12]

Charters granted to certain cities and their shippers' guilds by secular and churchly rulers, some of which dated back to the thirteenth and fourteenth centuries, constituted a second "medieval" obstacle to the expansion of navigation on the Rhine. These charters granted the "right of staple" (*Stapelrecht*) and the "right of compulsory transfer" (*Umschlagrecht*). The former stipulated that cargo arriving at the privileged city be unloaded and offered for sale at that city's public market for a period of three days. The latter stipulated that all cargo bound further upstream be transferred to ships owned by members of the local boatsmen's guild. The most important cities so privileged were Cologne, Mainz, and Strasbourg. Thanks to their privileges, each of these cities became the effective "head of navigation" for its own leg of the river, profiting from a role as a central market thanks to the right of staple and from a role as a center of shipping thanks to the right of compulsory transfer. Absent engineering projects that would allow the capacious ships operating in the river's roomy lower reaches to penetrate far upstream, this segmented system of river transport had some justification: every segment could be navigated as economically as was possible with ships adapted to, and of maximum size for, local circumstances. Nonetheless, the local privileges were a heavy burden on cosmopolitan shipping, cutting profits and wasting time, and were widely vilified.

A New Order

The French Revolution transformed this fragmented political and economic landscape forever. By 1797, French revolutionary troops led by General Napoleon Bonaparte were in control of the Rhine's left bank all the way from Alsace to the Dutch border. By 1806, Napoleon, now Emperor and victor of Austerlitz, had been able to consolidate the fragmented political map of the German right bank into a small number of larger states united under French protection in the so-called Confederation of the Rhine. The result was that the river now "belonged" to a limited number of powerful players. This set the stage for reorganizing the Rhine as an international navigational thoroughfare. But first it had to be liberated from its feudal integuments.

A precursor to the Confederation of the Rhine, a joint Franco-German arrangement called the Octroi of the Rhine, had come into force in 1804. This bilateral arrangement reduced the number of tolls to twelve and fixed the toll rates according to three classes of goods, thus eliminating arbitrary and unpredictable fees. The Octroi's crowning glory, however, was the institution of a jointly managed bureaucracy to supervise toll collection on the entire river and to inspect towpaths, harbors and other facilities. In the process, this agency collected not only tolls but also information on the natural condition of the river.[13] Again, as in the medieval Rhine League, we see how interested parties created a moral-economy regime that foreshadowed the cosmopolitan navigational commons.

With the decisive defeat of France in 1813–14, the Octroi formally ceased to exist. However, the Congress of Vienna, convened in 1815 to hammer out a new post-Napoleonic European order, saw the wisdom of Napoleon's vision for the Rhine. Prussia, eager to cash in on the economic promise of its vast new riparian territories in Westphalia and in the Rhineland, pushed vigorously for a liberal economic order on the river—for a Rhine accessible to the flags of all nations and devoid of tolls and privileges—despite the opposition of its own shippers' guilds at Cologne and Mainz. Its work was made easier by Napoleon's earlier consolidation of the political landscape along the Upper Rhine. Instead of having to deal with a motley array of aristocrats wedded to the notion of exacting rents from their possessions, Prussia could henceforth act as *primus inter pares* of a confederation of states with similar modernizing ambitions and interests in free navigation of the Rhine—including, of course, the Netherlands and France.

The great powers that convened at Vienna, in cooperation with the smaller Rhine-riparian states, thus laid the groundwork for a new international regime for Rhine navigation. One of the 32 articles of the final protocol called for the establishment of a Central Commission for the Navigation of the Rhine (CCNR) in which all Rhine-riparian states would be equally represented and which would vote by consensus—that is, each state would have the power of veto. The CCNR was charged with four tasks: to facilitate communication among member states regarding shipping on the Rhine, to establish common rules for Rhine navigation and to ensure their enforcement, to issue interim instructions for tolls on the river, and to produce a comprehensive, long-term regulatory agreement for navigation on the Rhine.[14]

Dutch reluctance to give up profitable sea-harbor taxes long delayed the signing of the "comprehensive agreement." But incessant Prussian pressure and the coming of steam-powered navigation to the Rhine combined to erode Dutch resistance. The 1831 Treaty of Mainz cleared some of the remaining underbrush of medieval tolls and privilege. This process was consolidated by the Treaty of Mannheim, signed in 1868. In modified form, the Treaty of Mannheim is still in force today.

Under the terms of the Treaty of Mannheim, the Rhine-riparian nations effectively surrendered claims of unlimited national sovereignty over their territorial segments of the Rhine. The treaty imposed the following duties:

to eliminate any kind of physical or economic restrictions on shipping
to abide by internationally agreed-on rules for marking channels, reporting river stages, and policing river traffic
to submit plans for dealing with calamities
to organize the licensing of ships' masters and vessels according to international rules
to enforce common regulations on dangerous and environmentally harmful substances
to desist from making changes to the river bedding and from building bridges without prior international approval.

The Rhine navigational commons has had its ups and downs, chiefly as a consequence of the series of Franco-German wars starting in 1870. But, like Rhine commerce itself, it has emerged from each down more vigorous than before. The Rhine navigational commons has obviously become a keystone of Rhine shipping, and indeed of the prosperity of all Rhine-riparian nations, despite chronic international animosity.

The effects of this cosmopolitan navigational commons have been profound, and they did not escape the attention of contemporary observers. The very industrialization and urbanization of the Rhine valley would have faltered but for cheap bulk transport of coal, steel, and coal-tar chemicals,[15] which was possible largely because of the navigational commons described above. But as the Rhine became "carboniferous," farsighted contemporaries also noted that an enormous environmental price was being paid. The incessant coming and going of steamships disturbed fish and wildlife and polluted the waters. Effluents from industries and from rapidly growing cities gradually transformed the Rhine's waters into an exotic chemical soup. Drainage water from coal and potash mines increased the river's salinity greatly. Engineering projects to improve the shipping channel caused erosion and increased sediment transport. The railways that ran through the Rhine gorge forever destroyed the tranquility of the riverscape.

To be sure, shipping was not singled out as the only culprit. Major reclamation and flood-control projects, such as the program carried out by the new Grand Duchy of Baden on the Upper Rhine between 1817 and 1874, were equally seen to be destructive of natural and cultural habitats.[16] The costs of a hydrodynamically "congenial" river included displacement of populations, destruction of such sources of livelihood as fishing and panning for gold, wholesale destruction of breeding grounds for salmon and trout, and uncontrolled bottom erosion coupled with downstream shoaling. Huge projects to capitalize on the Rhine's hydroelectric potential by building weirs and power plants in the river not only interfered with shipping—which made it a big issue in the Central Commission—but also prevented salmon and other anadromous (migratory) fish from making their way past the weirs to their upriver breeding grounds in small tributary brooks and streams. The Rhine's first hydroelectric plant, at Rheinfelden above Basel on the Swiss-German border, commenced operation in 1898. It provided 27 megawatts of power. It also prevented salmon from reaching the Aare River system, which drains nearly half of Switzerland and whose many swift and gravel-bottomed tributaries had been home to breeding salmon.

But although the Rhine navigational commons contributed to the wholesale destruction of the Rhine's fresh-water ecology, it provided an example of the benefits of international cooperation. It also provided an organization, the Central Commission, that would one day serve as a platform for forcing the "side effects" of economic prosperity onto the international political agenda.

Fish

In the case of the Rhine, the fish are bred in a country 300 miles from the sea, and caught in Holland, where the people only invent the most destructive nets that have ever been conceived of—800 yards in length, and worked by steamboats all the year round. . . . If such a river belonged to one State, or to one owner, it is not possible to conceive the extent to which it might be improved by cultivation.
—Thomas Ashworth, *Essay on the Cultivation of a Salmon Fishery*, 1866[17]

Salmon Fishing—Downstream Becomes Upstream

The Rhine's abundance of fish was, no doubt, one of its perennial attractions. And among the Rhine's fish, the Atlantic salmon (*Salmo salar*) was highly prized for its size, its beauty, and its delicate taste. Atlantic salmon could be caught all the way from the Rhine's mouth to far upstream on the headwaters of its many tributaries, even into Switzerland at elevations of 2,000 meters or more above sea level. The salmon's migratory life cycle and its economic and dietary importance shaped a complex web of riverine interdependencies that culminated in the mid 1880s in the establishment of a cosmopolitan salmon commons on the Rhine. However, the protective measures embodied in the commons were too little too late. By the time the Salmon Treaty of 1885 was enacted, Rhine salmon populations were already succumbing to ills that the curbing of overfishing could do little to assuage. In this section I will recapitulate the history of this cosmopolitan commons as a response to a "valorizing" drama of technologies, markets, and property rights and as a "conserving" drama of distributive equity and sustainable fishing—both against the tacit backdrop of the industrial spoliation of the salmon habitat on the Rhine. The moral is that a successful commons has to embody enough perspicacity to address all the forms of precariousness to which a resource may be prone. In this case, blind spots—up to and including disingenuous denial—prevented the taking of measures that might have avoided disaster.

We now know that salmon are anadromous. Born in fresh water, they swim to the sea as juveniles, then return to fresh water as fertile adults to spawn. Their spawning takes place in shallow gravel-bottomed stretches of the headwaters of rivers and their tributaries. This was not always common knowledge, and not until the 1860s did this conception of the salmon's life cycle begin to gain currency, though whether all salmon were anadromous and whether they invariably returned to the precise rivers where they had hatched long remained bones of contention. In any case, everybody knew that fisheries near the mouth of the river had the first shot at the fat and tasty salmon that came from the sea and migrated upriver. Every salmon caught in the river's lower reaches was one less in the nets of upriver fishers and one less that could reproduce. Hence, though Rhine salmon were clearly a "subtractable" common-pool resource (like herring in the North Sea), they were not subtractable "in parallel," only "in series," because they were a flow rather than a static pool. In the case of salmon, the downriver fishers could always subtract first, and the rest had to take the leavings. In addition, at least according to the new salmon science, all the Rhine fishers together determined the total size of future salmon populations by the number of salmon they allowed to pass upstream in order to reproduce.[18]

Up to about 1840, salmon populations and upriver fisheries prospered thanks to the limited technical potential of downriver fisheries. At that time, the traditional fishing techniques in use on the Rhine's lower reaches (in the Netherlands) still allowed plenty of fish through for German, French, and Swiss fishers and for reproduction. A rosy picture of this period was sketched in an 1856 petition to the second chamber of the Dutch parliament by a group of fishermen from Tiel and neighboring towns:

In the past, when salmon fishing was generally done with drift-nets. . . . There was an abundance of salmon; one doesn't have to go back half a century to find that even with a poor market, fishing concessions were sold at high prices.[19]

In this period, the expansion of salmon fisheries was limited by the modest distances over which the highly perishable fish could be brought to market. With pre-industrial means of transport and conservation, markets for salmon were local or at best regional, and catching more would only glut the market and drive prices down. That began to change when industrialization brought steam-powered ice factories and new techniques for shipping salmon by rail. Numerous contemporary commentators pointed this out. One salmon entrepreneur wrote in 1870: "The rapid means of communication, the more efficient manner of transport and packaging, the change in supply and demand that resulted, see here the grounds for higher prices."[20] In the Netherlands in 1840, the wholesale price of a kilogram of salmon was about 30 guilder cents; by 1868 this had trebled to nearly a guilder.[21]

By the beginning of the 1840s a new "industrial" salmon economy was becoming visible, and this encouraged the development of more intensive fishing techniques. These techniques first came together at the Snackert, a venerable salmon fishery located at Ammerstol on the Lek River. (See figure 10.1.) They entailed upsizing the traditional drift net to span the entire river. The nets were called *seines* (*zegen* in Dutch, *segen* in German). Those used at the Snackert were 365 meters long and 10.5 meters deep. A seine was kept upright in the water by floats attached along the top and weights along the bottom. Seine fishing proceeded by holding one end of the net on shore and using a rowboat to pull the other end nearly to the opposite shore. The net was subsequently allowed to drift downstream for a kilometer or so by walking one end along the shore and allowing the boat to drift with the other end. At the downstream terminus of the fishery, the loose end attached to the boat was hauled or winched in to shore by means of a long rope attached to that end of the net. The fishers hoped that the pocket that was formed near the shore when the line was winched in would contain one or more salmon that had swum into the net while heading upstream. This basic process was refined and further mechanized in subsequent years. Now as many as four nets were rolled out in rapid succession. Small steamships were employed to run the loose end of the nets across the river and tow the nets back upstream again. The shore end of the net was attached to a trolley running on rails along the shore. Horses and turnstile winches were used to haul the nets in. (Later, steam and petroleum engines were used, not always successfully.) By 1900, the Snackert employed 52 men under the direction of two bosses. They worked in two twelve-hour shifts, resting while the tide came in. Three seine nets were employed.[22]

The success of the new industrial mode of fishing lured new investors. The Snackert's lessor, the aristocratic Nassau la Lecq family, was not disinclined to grant concessions to other interested parties. This occasioned a gradual downriver displacement of seine fishing on the Lek toward Rotterdam, inasmuch as each new enterprise sought an advantageous position downstream of its predecessor (which it then proceeded to ruin). By 1848 this leapfrog progress had come to a temporary rest at a seine fishery called the Merode, just upstream of Rotterdam at the Kralingseveer. In the absence of regulation, this fishery alone was able to deprive the rest of the Lek of an economically viable salmon population.[23] In an 1851 petition to the Dutch parliament, nine salmon fishers from three towns along the Lek complained that "the once so flourishing salmon fishery has been utterly ruined in the municipalities where the undersigned reside."[24]

Intensive fishing techniques had less of an impact on upriver salmon migration on the other Dutch tributary of the Rhine, the broad and powerful Waal. Owing to its formidable width, its strong currents, and intensive shipping, seine fishing on the Waal was too technically challenging to be

Figure 10.1
Dutch Rhine distributaries. Triangles indicate the sites of mechanized seine fisheries, with the dates of their establishment. Over time a general downriver trend may be observed in the positioning of successive seine fisheries. By the 1870s the distributaries around Rotterdam had become one large salmon trap, leaving only the Nieuwe Merwede and the Waal as a route to the upstream Rhine. Map by author.

profitable. Up to 1870, salmon headed for the Waal did have to run the gantlet of the *steek* (fixed-hoop-net) fisheries in the Biesbosch, a maze of tidal kills and mudflats that separated the river proper from its broad estuaries.[25] But despite the best efforts of the *steek* fishers, enough salmon managed to run this gantlet to support drift-net fisheries further upstream along the Waal and even several large seine fisheries in Prussia between the Dutch border and the Ruhr.[26] The latter were inspired by the big Dutch seine fisheries, though smaller in scale and entirely hand-operated.[27] In addition, even further upstream, salmon were regularly caught with drift nets (especially around Bacharach and St. Goar) and with umbrella nets at countless sites on the headwaters. Nonetheless, from the mid 1840s, declining catches all up and down the river were invariably blamed—not always justifiably—on the rapacious new fishing technologies employed by the Dutch.

Salmon Fishing in Germany
In what became the Prussian Rhine provinces (roughly the territory of present-day North Rhine–Westphalia), fishing rights on non-navigable rivers traditionally accrued to those owning the riverbanks. These non-navigable rivers were generally fast-running, gravel-bottomed Rhine tributaries

that sported trout and an abundance of salmon spawning grounds. The attachment of fishing rights to land ownership created a simulacrum of Garrett Hardin's shared village pasture: everybody did their best to extract as many fish from the common pool as they could, heading the trout and the spawning and juvenile salmon straight for a Hardinesque "tragedy of the commons." Under French rule, in 1814, these riparian fishing rights were replaced by state concessions, but twenty years later the landowners' privileges were restored. By the middle of the century, fish stocks had declined to alarming levels and only ambitious programs of artificial breeding and restocking kept the trout and salmon fisheries solvent. An expert report commissioned by the Prussian government in 1860 recommended a return to the French system of state concessions.[28] Otto Beck, a high-placed government official at Trier, replied that in his opinion a concession system and restocking would hardly be worth the effort, in view of what he called "thievish" Dutch fishing practices. In fact, he invoked the moral economy of the existing navigation commons based on the Treaty of Mainz, which stipulated that the Rhine, as an international river, could not be unilaterally co-opted by the subjects of any nation. In his view, local Prussian measures could be effective only if "the (legally *'jusqu'a la mer'*) free Rhine could be liberated from the thievish (*rauberische*) Dutch salmon traps." He continued:

We hope that the illegal blockade of the Rhine as it has been practiced by the Dutch salmon fishers for a number of years now, will finally be eliminated by the International Rhine Navigation Commission or on the occasion of the closure of a new trade agreement with the Kingdom of the Netherlands. Why should it not be possible to draw other countries like France, who are seriously committed to artificial breeding, into this cause?[29]

This call for what would amount to an international salmon treaty was remarkable in that it built its case for equitable access to the Rhine's salmon on the existing navigational commons, which prohibited any state from unilaterally impeding the free circulation of ships and commodities on the river. This was probably stretching things a bit, but since the Mainz convention (the legal backbone of the navigational commons) was the only binding international agreement on the Rhine at the time, it made sense to try to address the Dutch "salmon blockade" within its framework. Nearly a century later, Dutch efforts to curb Rhine pollution would similarly seek a legal basis in the then-still-valid salmon convention of 1885.

Though it would take another 17 years to achieve consensus on a Salmon Convention, there is no doubt that as of 1868, Beck's diagnosis was widely shared on the Rhine upstream of the Dutch border, even in upstream

Baden, Hesse, France, and Switzerland. Those states suffered their own particular salmon drama as a result of the ongoing destruction of salmon spawning habitats and fishing grounds thanks to the huge Upper Rhine rectification project undertaken between 1817 and 1876 by the Baden government according to plans originally drawn up by its French-trained chief engineer, Johan Tulla. The purposes of that project were flood control, land reclamation, and the suppression of malaria by means of training the river into a single channel, eliminating islands, and cutting off meanders.[30] Figure 10.2, showing a typical before-and-after situation at Plittersdorf, refers to sites called Salmenkopf and Salmengrund, which must have been favored salmon habitats (and favored fishing sites) before they were left high and dry by the rectification. As a result of this and similar reconstructions, salmon fisheries on the Upper Rhine ran into deep trouble, and the reproduction and survival of salmon on the Rhine were further jeopardized. Reflecting on the damage done to the reproductive potential of salmon by nineteenth-century river engineering, the German Fisheries Association had this to say in 1906:

Dams, parallel works and the dumping of dredged sand in the spaces between wing dams . . . contribute to the destruction of secondary channels and backwaters that have hung on here and there, so that finally from a slowly meandering river with its countless channels and bends full of contemplative stillness it becomes a fast rushing gully with straight sides and a much greater sediment-carrying capacity, in which fish can find no place of refuge to rest or to deposit their eggs and hatch their young.[31]

In 1841, in efforts to turn the tide and safeguard the salmon population in the region, France, Baden, and several North Swiss cantons signed a treaty prohibiting the capture of juvenile salmon, a regional delicacy strongly resembling trout. And in 1852, in an effort to augment the salmon's decreasing opportunities for natural reproduction in the region, the French set up an artificial breeding station in the Alsatian town of Huningue with the aim of restocking the Rhine and the Moselle. After the Prussian takeover of Alsace in 1870, this station was promoted to the status of an Imperial Fish Breeding Institute (Kaiserliche Fischzuchtanstalt). By that time, the institute—along with other similar stations along the river—already contributed significantly to the reproduction of Rhine salmon. This compensated not only for the Rhine correction, but also for other "carboniferous" assaults on Rhine salmon. Artificial breeding and restocking of the Rhine system later became so important in maintaining native salmon populations that the required supplies of roe and milt (carried by "ripe" salmon

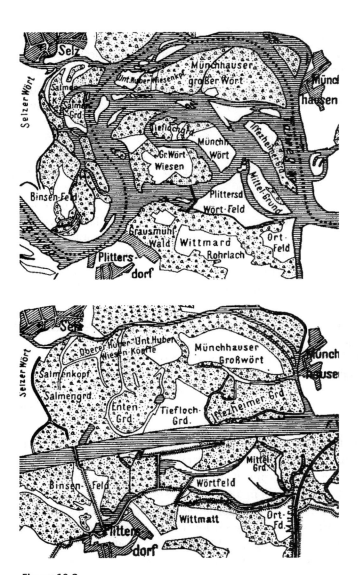

Figure 10.2
The Rhine near the town of Plittersdorf before and after the Tulla regulation in 1818.
Instead of a braided river with shallows and backwaters ideal for salmon spawning,
the Rhine here has been reduced to a single straight shipping channel of uniform
depth. The former spawning grounds are partly drained and cut off from the river.
Source: Friedrich Metz, *Die Oberrheinlande* (F. Hirt, 1925), p. 167.

that survived the journey to the Upper Rhine) became a major legitimization for international agreements to curb intensive forms of salmon fishing in the river's lower reaches.

Toward a National Salmon Commons in the Netherlands

Salmon fisheries became an explicit object of government policy in the Netherlands after 1848, in the wake of the "bourgeois revolution" that fettered the monarchy within constitutional constraints. The series of liberal governments that now held sway over the king were committed to developing national resources and stimulating the economy. Fisheries, and salmon fisheries in particular, were seen as an important national resource: as a source of income for investors and the state, and as a welcome source of protein for the national diet.

The Dutch Law on Hunting and Fisheries of 1852 had settled responsibility for defining a national salmon fisheries policy squarely on the shoulders of the four Dutch Rhine provinces. However, on account of their divergent interests rooted in different locations along the river, the negotiations came to nothing. Meanwhile, Dutch drift-net fishing communities along the Lek and Waal were feeling the dire effects of the persistent lack of regulation. In the following sequel to the rosy vision of the salmon past as expressed in the 1856 petition by the Tiel fishers cited above, we see how the poor catches are blamed on lowered rates of reproduction as well as on unfair competition for available fish:

[N]ow that the blockade has been put into practice, thanks to which the salmon is caught as soon as it comes in from the sea and before it has had the merest chance to deposit its eggs, reproduction is thwarted . . . and as a consequence the salmon catch has been reduced to nearly nothing.

Fishing with unacceptable fishing gear; the blockade of rivers, whether with turning-nets or by means of the use of a number of seines in succession or so disposed as to occupy the entire river and thus to deprive the fish of the possibility of swimming further upstream; all these are unacceptable impositions on the rights of third parties, which should be prohibited by law.[32]

In 1857 this kind of pressure forced the government itself to assume responsibility for drafting a salmon law "after due consultation with the provinces." The outcome was the royal decree of October 10, 1858, "regulating the manner of exercising the salmon fishery."[33] It targeted the seine fisheries, specifying minimum maze sizes for nets and forbidding the use of more than two seines at a time. When the tide was coming in, active seine fishing was prohibited, and nets could not be left hanging in the water in an effort to "turn" the salmon for later capture. There were no specific provisions for operating *steek* fisheries, except that no new *steken* could be set without explicit permission from the Ministry of the Interior. This put a brake on the uncontrolled proliferation of the *steek* fisheries.

The royal decree was a response to the downstream march of the seine fisheries in the 1830s and the 1840s in their incessant pursuit of more advantageous (i.e., more downstream) fishing sites. This "race" eventually forced them into the tidal sea reaches of the rivers. Now the most advantageously situated seine fisheries were faced twice a day with incoming tides that caused the river to flow in the "wrong" direction for a certain period of time. It was nearly impossible to deploy the seines under these conditions, not least because the salmon were disoriented by the shifting currents and swam every which way until the normal seaward river current reasserted itself. However, some ruses were tried. The most popular, called *keren* (meaning turning), entailed leaving a seine floating across the river during periods of slack water or incoming tides in order to prevent salmon that were "waiting" for the river's flow to become dominant again from heading upstream out of the clutches of the fishers before fishing could be resumed. Neither the (by then defunct) upstream seine fisheries on the Lek nor the drift-net fishers on the Waal had to contend with a current that changed directions twice daily; weather and river stages permitting, they were able to fish non-stop.

By keeping the front-row tidal seine fisheries from either fishing or "turning" salmon on the incoming tide, the new law sought to redress the balance among Dutch salmon fisheries. The idea was that the enforced idleness of the downstream fisheries (the more downstream, the longer the tidal pause) would allow enough salmon to swim on to the upstream fisheries to make them solvent again. In fact, the new regulations so increased the "porosity" of the downstream tidal fisheries that soon after passage of the law the upstream non-tidal Snackert could again resume fishing with its large seines after more than ten years of inactivity. The fact that the Snackert could carry on around the clock gave it a decided advantage over the now legally constrained downstream fisheries.

This new arrangement, despite the serious limitations that would soon become manifest, certainly provided the groundwork for a national salmon commons based on guaranteeing equitable access to salmon regardless of the location on the river. It sought to redress the inequities of the market and concession-auction system by imposing artificial restrictions and limits on the privileged tidal fisheries to the benefit of the upstream seine and drift-net fishers. Instead of allowing market forces and capital to rule the acquisition of fishing rights and fish, the new regime utilized the police power of the state to redress at least the most serious inequities in the chances for salmon takes. However, in no way did it directly address the threat posed to the sustainability of the salmon population on the river as a whole by overfishing (or by anything else). Despite increasing evidence that overfishing of salmon, especially in the spawning season, reduced the salmon population and hence reduced everybody's catches, the Dutch salmon regime after 1858 still proceeded on the assumption that, in principle, all the salmon that swam in Dutch rivers belonged to Dutch fishers and in Dutch fish markets.

The Failed Transnational Salmon Commons of 1869

In 1869, the government of the Grand Duchy of Baden invited the other Rhine-riparian states to Mannheim for the purpose of drafting an international convention to regulate salmon fisheries on the Rhine. This came on the heels of the drafting of the Mannheim convention, which did away with the remaining feudal encumbrances on Rhine navigation. In calling for a convention on the regulation of salmon fisheries, Baden sought to valorize its investments in restocking the river with salmon fry and to cash in on its 1841 agreement with France and Switzerland to desist from fishing for young salmon. In the view of the government of Baden, neither of these sacrifices made sense as long as the Dutch continued to decimate the population of fertile salmon making their way upriver. In an internal memorandum proposing the convention, the Baden government argued as follows:

With favorable water levels, only a few salmon, driven by their reproductive instincts to head for the Upper Rhine, have escaped the diligence of the Dutch fishing establishments; so that here, for example, in the last few years it has not once been possible, for the few breeding facilities we have, to acquire the necessary salmon eggs—even against suitable payment. The Dutch fishers are therefore in a position, if they persist in their current manner of fishing, to make salmon in the Rhine a rarity.[34]

The governments of Prussia, France, the Netherlands, Bavaria, and Hesse recognized that they all profited from upstream efforts to increase the salmon population. But failing downstream sacrifices, upstream measures would be futile. The Dutch government, aware that only strict measures could silence international criticism of its fisheries but also faced with complex domestic conflicts of interest, hired the explorer and biologist Francois Pollen to prepare a report on salmon biology and to provide advice during the negotiations. As a member of the emerging international "epistemic community" of salmon experts, Pollen argued that a sustainable salmon population demanded specific and well-timed restraints on fishing along the entire length of the river. The Dutch government, convinced by the success of similar fishing regimes on British rivers, adopted a cooperative stance at the Baden negotiations and ultimately signed a convention that boded ill for the big Dutch seine fisheries.

The stated purpose of the salmon convention was "conservation and the multiplication of the most valuable fish species in the Rhine." Its main provisions were the following:

1. The barriers used in hoop-net fisheries (*steken*) can extend only halfway across the river. Mesh sizes and distance between barriers are subject to minimum dimensions.
2. Minimum mesh sizes for different types of nets, including seines, are specified.
3. No more than two seines can be in use simultaneously; seines may not be fixed across the river (i.e., no "turning"); seine fishing is allowed only at slack water or outgoing tide (current downriver).
4. All seine fisheries must desist from fishing at least 8 hours per day, even where incoming tides last less than 4 hours or are absent altogether.
5. No salmon smaller than 35 centimeters may be sold or offered for sale.
6. Seine fisheries must be closed annually from September 1 to January 1. In this period "ripe" salmon may not be transported or sold, except for purposes of artificial breeding. Fishing with gear other than seines is prohibited for six weeks of this period.
7. A prohibition against discharging "factory wastes" or other substances that are harmful to fish into fishing waters. Local authorities can, however, grant exemptions in the "interests of agriculture or industry."
8. Riparian states containing suitable spawning grounds and nursery habitats for salmon fry are obliged to engage in artificial breeding and to ensure access of "ripe" salmon by constructing fish ladders at dams and other obstacles.
9. Signatories agree to pass appropriate national legislation to enact the convention, establish fines and punishments for infractions, and to provide sufficient surveillance and police personnel.
10. Signatories will each appoint a fishing commissioner to oversee enactment of the convention and to coordinate international efforts to improve the Rhine salmon regime
11. Signatories commit to supporting investigations regarding the "nature and habits" of the varieties of salmon and agree to apprise one another of the results.

These provisions reflected the mature international consensus of an "epistemic community" of salmon scientists. All parties to the convention, upstream and down, had committed themselves to improving the salmon's reproductive potential according to the dictates of "salmon science," both by desisting from their assault on "ripe" salmon and juvenile fish and by augmenting natural reproduction with artificial methods. At the same time, the restrictions placed on Dutch seine fisheries enabled more salmon to make it upriver—to the immediate benefit of German, French and Swiss fisheries and ultimately even of the Dutch.

In France and the German states, parliaments quickly ratified the new convention; the Dutch government, however, had a hard time selling it to its own fishers and to its parliament. In parliament it faced a pragmatic coalition of liberals who opposed any form of state regulation of economic activities and representatives of salmon fishers who viewed salmon fishing on the Rhine as a simple zero-sum game in which salmon forgone by Dutch fishers simply ended up in the nets of German and Swiss fishers. The salmon fishers and their parliamentary representatives categorically denied the science and chose to see the Rhine's salmon as an inexhaustible public good. The government took a dim view of such an atavistic attitude toward what it saw as established scientific fact:

[V]arious members [of the second chamber of the parliament] deem it opportune simply to cast serious doubt on the main points, as do many fishers, instead of becoming acquainted with what has been written on the subject or with the extremely enlightening English parliamentary documents available in the library of the chamber. The government would be disappointed if this unfounded doubt were to lead to rejection of the law—but it can hardly be expected of her that she produce an elaborate ichtyological treatise.[35]

However, the government *was* prepared to explain why Dutch fisheries should be regulated in the context of an international agreement. The basic argument was that Dutch restraint (and hence more fish for upriver fishers) was the price that had to be paid for upriver conservation of salmon stocks. The responsible Dutch ministers argued in parliament that four measures were essential to preserve salmon stocks in the Rhine: regulating fishing on the entire river so that enough salmon could make it upstream to spawn, a prohibition on capturing salmon at spawning grounds, ensuring that the eggs are not disturbed, and provisions to protect juvenile salmon on their long journey to the sea. The government argued that maintaining this "division of restraint" among the various riparian nations demanded that Rhine salmon, like Rhine shipping, become the object of a moral economy, a "community of interest," spanning the entire river:

For the salmon fishery the whole river is a natural unity, of which each part suffers the consequences of what is done or neglected on the rest of the river and of which the lower part, destined by nature for the largest catch, also has the biggest interest in protecting reproduction and young fish in the upper part. It goes without saying that those in the upper part will hardly be inclined to spare the spawning fish and their brood merely to accommodate the fishers of the lower river. There must be a community of interest if it is to be expected that salmon procreation will be fostered everywhere and for this reason too a part of the salmon has to be allowed to migrate upstream unhindered.[36]

This call for what amounted to a salmon moral economy was rooted in a specific "scientific" understanding of the salmon's life cycle, a perspective that was not universally shared in 1870 or even in 1885. It was still easy enough to dismiss the "scientific" methods and conclusions of emerging salmon science as theories invented only to advance the fortunes and interests of self-styled experts and government advisors. Local salmon fishers and fishing entrepreneurs could and did make equal claims for other kinds of practical knowledge based on traditional lore and long experience. There were, in short, many opinions in 1870 and long thereafter about when, where, and how salmon reproduced—and, hence, how fishing practices should be regulated. In 1870, in a contribution to the debate concerning the Mannheim salmon convention, the Dutch salmon entrepreneur Abram Quakernaat van Spijk claimed that, insofar as the vast majority of the fish spawned at sea, intensive fishing would not compromise the Rhine salmon's reproductive potential. He challenged advocates of more stringent regulation to prove that "the greater or lesser reproduction of the salmon would depend on that relatively miniscule number that swims up our rivers and passes through our fisheries on its way upstream to spawn."[37] Quakernaat even questioned whether the many shapes, colors, and sizes of salmon that were observed in the course of the annual cycle of what we now take to be a single species, *Salmo salar*, were in fact manifestations of the same fish.[38] There were clearly different legitimate opinions about salmon breeding habits and these could be selectively mobilized to suit the particular short-term interests of fishers at different locations along the river in maximizing their catches (i.e., redistribution of the common flow)—quite aside from their shared long-term interest in the survival of salmon in the river. If an upriver drift-net fisher felt disgruntled about the number of fish left him by a downriver seine fishery, he merely had to invoke the loss of reproductive potential to transform his particular interest into a common concern. In like manner, of course, a partisan such as Quakernaat could easily expose this noble "concern" as mere self-interested ideology cloaked in the mantle of science. *Salmo salar*'s habits and even its identity were wild cards in the debate on desirable methods and degrees of regulation. The controversy puts one in mind of the present debate on climate change, in which there is more consensus about the observations than about what they might mean and about what forms of action or inaction are appropriate to the situation. As with theories of climate, preferences for one or another theory of the salmon's life cycle often seemed informed more by economic and political interests than by the weighing of evidence.

In the end, the second chamber of the Dutch parliament rejected the convention by three votes. Disillusioned by what they saw as a Dutch refusal to make any concessions, Prussia, Bavaria, Hesse, and France abandoned the convention in dismay. However, by means of a royal decree the Dutch government was able to overrule parliament and enact the terms of the convention into Dutch law on October 10, 1872. As a result, Dutch salmon fishers became subject to the terms of an international treaty that had been rejected by their own parliament and to which the other signatories no longer considered themselves bound. To many people—and certainly to most Dutch salmon fishers—the Dutch government seemed to have buckled under pressure from Prussia.

The government adamantly denied foreign influence, and the minister of justice explained the royal decree as a legal framework for the "equitable bearing of necessary limitations."[39] Without these measures—even if only unilateral—Rhine salmon would ultimately die out. In the minister's words, "The intent is certainly facilitation of salmon breeding, and to this end certain interests and extraordinary advantages have to be sacrificed."[40] The "extraordinary advantages," he explained, were those that had accrued since 1858 to the seine fisheries located just above the point where tides ceased to cause the river's current to turn. By the terms of the royal decree of that year, these fisheries did not have to suspend fishing at incoming tide—something that turned out to be a heavy burden for their tidal downstream competitors. In fact it was precisely during the twice-daily hiatus of their tidal neighbors that the "non-tidal" seine fisheries made their biggest catches. As we saw, the Mannheim rules stipulated that *all* seine fisheries be closed down eight hours out of every 24, thus redressing the balance.

But if the new royal decree reformed the existing Dutch salmon commons on a more equitable basis, it was also seen by its authors as a hopeful step to a transnational salmon commons based on a pragmatic moral economy rather than a strict, inflexible, and possibly non-enforceable treaty. "It is only by limiting our fisheries to a certain extent," the minister of justice explained, "that we can achieve a forceful stimulation of salmon breeding beyond our borders, the advantageous consequences of which we may safely estimate to be greater than the clear and present disadvantages of restrictions. I flatter myself that the measures now taken in the Netherlands will, also abroad, lead to the realization of measures that will enhance salmon breeding."[41] But with the exception of an 1875 agreement among Baden, Elsass-Lothringen, and Swiss Rhine cantons, this proved an idle hope. The Prussian, Hessian, and Bavarian middle section of the river—highly industrialized, but also the site of the biggest German salmon fisheries—remained beyond the pale of what, in spite of the ambiguity of commitments by parliaments and fishers, had still become a cosmopolitan commons, albeit of a patchwork nature.

German Perceptions

In 1870, as a result of the Franco-Prussian war the French provinces of Alsace and Lorraine were incorporated into the newly founded German Empire. This considerably reduced political complexity along the Rhine and opened up new possibilities for regulating salmon fishing. In the same year, the Deutsche Fischerei-verein (German Fisheries Association) saw the light of day, representing the interests of both professional and sport fishers, inland as well as maritime. The Deutsche Fischerei-verein, with headquarters in Berlin, lobbied the imperial government on fisheries issues, published a news bulletin and later a journal, held national and regional meetings, and represented German fisheries in international exhibitions. It kept a close watch on the Rhine salmon fisheries, including the changing and always threatening configuration of seine fisheries in the Netherlands. In this capacity it was instrumental in shaping the fisheries "foreign policy" of the new Reich.

German Rhine salmon fishers and their Verein were interested primarily in catching salmon, but they were convinced that it was necessary to safeguard the breeding of Rhine salmon if they were to continue to catch them. These interests converged in programs of artificial breeding, legal protection of salmon eggs and juvenile salmon, improving accessibility of headwaters for salmon, fighting pollution, and, last but not least, convincing the Dutch to desist from their rapacious fishing techniques.

However, when it came to speculating about what had caused the decline in overall salmon catches since 1850 the Germans did not shrink from acknowledging their own responsibility. This perception may have been encouraged by the fact that trout and other fish native to German streams were also suffering serious declines, which suggested that the increasing scarcity of salmon could be ascribed at least partly to the deteriorating morphologies and water quality of German rivers and lakes. This had been recognized as early the 1840s in connection with the Tulla rectifications. It came to the forefront again in the 1880s.

Industrial and agricultural weirs were thought to be major obstacles for migrating salmon. An 1881 commentary on the Sieg, a Rhine tributary once brimming with salmon, is typical of many others:

In the past, great numbers of salmon entered the Sieg and its tributaries, so that the fish were often quite cheap at Siegen; this has definitively ceased to be the case since the construction of both of the very difficult weirs at Siegburg and Dasberg.[42]

At the same time, Germans were becoming increasingly aware of the insidious pollution of their streams, rivers, and lakes by industries and municipal sewer systems, as evidenced by rapidly declining fish populations. The Sieg's salmon were no exception:

The Sieg suffers from the following disadvantages: . . . Poaching is endemic, including the use of poisons and dynamite; extraction of water by very extensive sewage farms and many manufactures; effluents from tanneries, glue factories, and paperworks at Siegen; the ironworks at Wissen dumps its red-hot slag into the Sieg, to the great detriment of the fisheries. . . . The chemical factory at Eitorf has released very dangerous substances into the Sieg . . . ; the mines at Ziethen and Silistria burden the Sieg with lead-bearing sediments; the Friedrich-Wilhelm ironworks dumps damaging substances in the river. The Sieg's water would be clear and the fish in great measure beneficial, if only it were not polluted.[43]

To be sure, there were efforts to combat the damming and the polluting of rivers. The failed 1869 Mannheim salmon accords had enjoined upstream signatories not only to support programs of artificial breeding and stocking but also to "see to it" that fish ladders be constructed at "appropriate sites" to "facilitate the upstream migration of salmon and trout." Sadly, however, the fish ladders that were constructed proved quite ineffective, though not for lack of ingenuity. The Mannheim treaty also included prohibitions against "throwing, flushing or leaching factory wastes or other substances" that could be harmful to fish into fishing waters (although there were huge loopholes). These articles were incorporated into both of the simulacra of the Mannheim accords: the Dutch royal decree of 1872 and the 1875 treaty between Baden, the North Swiss cantons, and Elsass-Lothringen.[44] On the occasion of the 1882 Fisheries Exhibition in Berlin, the king of Saxony offered a prize for the best work on the pollution of waters and its prevention (with special attention to fish). Alas, as might have been expected, none of the entries were deemed worthy of the prize. Protecting Rhine salmon against the ravages of the "carboniferous" Rhine seemed to be a losing battle.

The Salmon Convention of 1885

The year 1885 proved to be one of the best years on record for salmon catches. Oddly, this coincided with a Prussian initiative to realize a new international salmon convention. In June of 1885, representatives of Prussia (acting for the German Rhine states), the Netherlands, and Switzerland met in Berlin to hammer out a protocol. The signatories had been "guided by the wish to increase the salmon population in the Rhine," and had gathered to "regulate salmon fishing in the Rhine watershed in a uniform manner."

As had been the case with its unfortunate predecessor, the Mannheim salmon convention, the new convention had been accompanied by public controversies, not only within each of the countries, but also between the fisheries associations of, especially, the German Empire and the Netherlands. In the Netherlands, the former government advisor, Francois Pollen, excluded from the rather secretive negotiations at Berlin, welcomed the new international initiative because it would bind German fisheries to the same standards as the Dutch. Nonetheless, he severely criticized the absence of mandatory daily breaks in fishing, such as had been stipulated in the previous arrangement (and in fact enacted in the Netherlands by royal decree), claiming that it was essential to have these all along the Rhine if a significant number of salmon were to reach their breeding grounds regularly.[45] Excellent years such as 1885 could be ascribed to poor fishing conditions (high river stages, ice) in the previous year which had allowed more salmon to pass upstream to breed. But consistently good years required careful "cultivation," and on this point, in Pollen's opinion, the new Berlin salmon convention fell short.

The Berlin convention was indeed less stringent than its predecessor. Gone were the minimum mesh sizes, the prohibition against dumping noxious substances into fishing waters, and, as noted, the daily eight-hour hiatus. The latter (amounting to 56 hours per week) was replaced by a weekly closure of 24 hours, lasting from 6 p.m. Saturday to 6 p.m. Sunday.[46] Within the Netherlands, the new rules privileged seine fisheries operating above the tidal zone, inasmuch as they could once again operate around the clock (except on weekends). In sum, Pollen certainly seems to have been correct that on the point of protecting salmon the Berlin convention was much weaker than its predecessor.

Were these weaknesses the reason that, despite speedy ratification of the Berlin salmon convention by all the signatories in 1886, the total salmon catch on the Rhine declined precipitously that very same year—and, with ups and downs, continued to decline until by the 1930s salmon fishing had ceased to be an economically gainful activity altogether? In retrospect, at least, most commentators give overfishing its due, but emphasize that changes in river morphology, deteriorating water quality, and the blocking of tributaries and even the main river by dams and weirs were the decisive factors.[47]

In retrospect, the 1886 salmon convention revealed itself as a misguided plan for a cosmopolitan salmon commons because it was based on the fantasy that salmon fishing could be divorced from the exploitation of other riverine resources. Salmon 2000, a project established in 1986 by the International Commission for the Protection of the Rhine to reinstate a viable salmon population on the Rhine by the year 2000, is a tacit critique of the blinders that led the old salmon convention down the wrong road.[48] Salmon 2000 is not about rules and regulations for salmon fishing, but about re-engineering the river and its waters to re-accommodate the fish— now at the expense of unbridled valorization of other resources.

Figure 10.3
Though after 1885 salmon catches on the entire Rhine declined steadily, it took a long time for dams and pollution to drive salmon from the Rhine completely. This illustration shows the biggest Dutch salmon market at Kralingseveer in 1905, which continued to operate daily well into the 1920s. Source: http://nl.wikipedia.org.

Phenol

Although discoloration, smell, and "miasmas" may have instigated complaints about Rhine water pollution before the 1860s, it seems that deteriorating water quality gained systematic political attention only after a narrative emerged that connected it to the decline of fish stocks—especially of salmon. Evil-looking and bad-smelling rivers were rarely sufficient cause to curb industrial progress, but the decline of fisheries—also an economic interest—regularly mobilized organized political opposition. Ultimately, of course, the powers in favor of allowing effluents to be discharged nearly untreated into the environment carried the day against those championing salmon fisheries, with the consequence that salmon disappeared from the Rhine despite the fact that fishing itself had been subjected to the constraints of a transnational salmon commons.

But even in a moribund state, Rhine salmon continued to draw public and political attention to Rhine pollution—and ultimately became the centerpiece of a water-quality commons on the Rhine. This only came about when salmon were no longer defined as a commodity on which to make a profit, but rather as a natural species whose life-support systems had fallen prey to human avarice and short-sightedness.

However, the main motivation for fighting Rhine pollution after World War II was not the absent salmon, but the increasing costs of rendering Rhine water suitable for the production of drinking water, for the irrigation of fields, and for fighting saline intrusions from the sea—particularly in the Netherlands. Nonetheless, at a crucial juncture it became plain that the pursuit of these kinds of narrow economic interests simply pitted the Rhine-riparian governments and civil societies against one another as they defended their particular interests as upstream perpetrators and downstream victims of pollution. As we shall see, an ecological disaster and a common resolve to bring back *Salmo salar* provided the framework for rapprochement and the establishment of a vigorous clean-water commons.

The International Commission for the Protection of the Rhine against Pollution

By 1946, as the war-torn Rhine economies slowly chugged into gear again, Dutch waterworks, greenhouse farmers, and water managers were once again facing serious degradation of Rhine water quality—including a pernicious increase in salinity. For lack of a suitable international forum, the Dutch government argued its case to the Central Commission for the Navigation of the Rhine in April of 1946. The commission declared the matter beyond its jurisdiction, but the delegates agreed to bring the matter to the attention of their respective governments in anticipation of an international conference to be organized by the Netherlands.

Fortunately for the Dutch, the Swiss came to their rescue on the international stage in response to an appeal made in a diplomatic note in July of 1946.[49] The main Swiss protagonist here was the biologist Otto Jaag, a water conservationist and a propagandist for sewage treatment, who in 1952 became the first director of the Swiss Federal Agency for Water Supply, Sewage Treatment, and Water Conservation. Jaag, one of the first to address the issue of riverine pollution on a European scale, was sympathetic to the plight of downstream countries like the Netherlands:

Each country is responsible for the purification of polluted waters in its territory, but neither waterways nor subterranean waters stop at borders. . . . Each country is forced to bear the waters that reach them from countries situated upriver. The Rhine, for example, carries to Holland the residues of a great number of European countries. . . .[50]

The Swiss debated the issue of Rhine pollution at a conference of their Rhine cantons in October of 1946. They concluded that the Berlin salmon convention of 1886 was the royal road to getting Rhine pollution on the international agenda. None of the signatories had ever repealed the Berlin convention, despite the fact that the salmon were as good as gone. By the terms of that treaty, the national salmon commissioners were still obliged to hold periodical international meetings. The first postwar meeting was held in Zurich in August of 1948. Rhine pollution was the final item on the conference agenda. The delegates, aware that the issue involved much more than the nearly extinct Rhine salmon, proposed that the Rhine-riparian states create a new international commission on the model of the existing Central Commission, but now for the protection of the Rhine against pollution.

This idea took root. After exchanges of diplomatic notes between the Swiss government and the governments of Germany, France, Luxemburg, and the Netherlands, a first meeting of the International Commission for the Protection of the Rhine against Pollution (ICPRP, currently ICPR) was held in the spring of 1950. Like the Central Commission, the ICPRP was built on the premise of national sovereignty over the Rhine and its resources. It could not propose measures except by unanimous consent of the signatories, nor could it implement any except through existing national agencies.

With postwar industrial reconstruction as a universal European priority, none of the Rhine nations seems to have had plans or means to actually

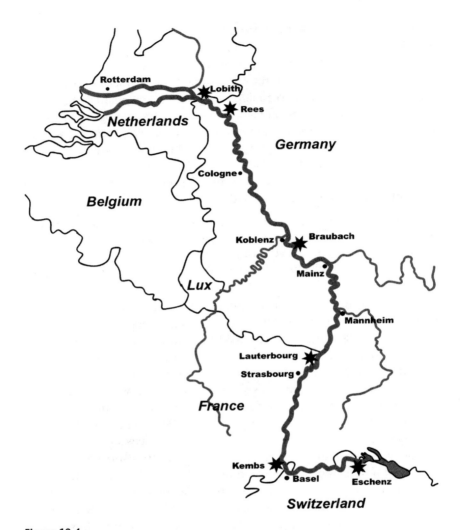

Figure 10.4
By 1950, the International Commission for the Protection of the Rhine against Pollution had established six international measuring stations—indicated on this map by
stars—to sample the Rhine's water on a daily basis according to a standard protocol.
The locations were chosen so as to provide global evidence of the sources of pollution. Map by author.

curtail pollution. For the time being, the new commission dedicated its efforts to painting as complete a picture of the state of the Rhine's waters as was possible. Under Otto Jaag's inspired leadership, an expert subcommittee was quickly able to establish a consensus on international measurement protocols. By mid 1953, nine sampling stations from the Bodensee to Vreeswijk were taking standardized biweekly assays of chemical and biological composition at specified depths and positions across the breadth of the river. In 1956 the results of the first two years were published after ratification by the ICPRP plenary assembly. "The Rhine is so heavily burdened," the report concluded, "that all means must be mobilized in order to improve the situation as quickly as possible."[51]

The Chemicals and Chloride Conventions

In 1963, with the signing of the Treaty of Bern, the ICPRP at last achieved a firm grounding in international law and a clear mandate. In addition to publishing on the nature and extent of Rhine pollution, the commission was now also charged with advising the governments on possible countermeasures. And its mandate was no longer limited strictly to water-quality issues but extended to "protection" in general. However, the new legal status of the commission brought few changes in its practical work, let alone in the alarming degradation of the Rhine's water quality.

The tide turned on the heels of a catastrophe and under the benign influence of an emergent environmentalist consciousness among experts and publics alike. An accidental spill of the insecticide Endosulfan by Hoechst Chemicals near Mainz in June of 1969 killed thousands of fish, forced the closure of waterworks intakes in the Netherlands, and got massive press coverage. The Rhine acquired the sobriquet "the sewer of Europe." With environmental quality becoming an increasingly potent electoral issue, the Rhine governments felt pressed to commit themselves to at least a show of good intentions. In 1972 they organized the first of what would be many Rhine ministers' conferences. Attended by the highest-ranking politicians with responsibility for environment and water quality, these conferences drastically streamlined the circumcuitous policy-making procedures of the ICPR and came to function as a kind of executive board for the commission, outlining goals and setting the agenda for the work of plenary meetings and specialist committees.

Pursuant to a mandate formulated at the First Conference of Rhine Ministers, the ICPR produced two treaty texts in 1976, both of which were summarily signed and ratified: a Chemicals Convention and a Chloride Convention. The Chemicals Convention committed the signatory states to regulating and in some cases terminating discharges of toxic or otherwise harmful substances in their own national territories. The ICPRP had initially stipulated 83 chemicals for the to-be-terminated "black list." Implementation of the convention consisted of recommendations to member states by the ICPRP of effluent limits (measured in terms of concentrations in the river). These first had to be unanimously passed by the ICRP plenary council and subsequently enacted into national law by the individual member states. This turned out to be a tedious and conflictual process, the complexity of which was exacerbated by the requirement of harmonization with European measures. Ten years on, emissions limits had been agreed upon for only twelve substances. Commentators attribute the failure of the Chemicals Convention to a combination of technical, scientific, and economic uncertainty about the effects of various concentrations of toxins, mutual distrust among the signatories, and fears of compromising economic welfare.[52]

Oddly, however, while the international politics of Rhine water pollution was stalemated in the decade after 1976, the Rhine's actual water quality improved markedly. Healthy river water contains about 10 milligrams of oxygen per liter. The Rhine's all-time low of 4.7 mg/l was measured in 1971 at Lobith. By 1976 it was already up to 6.3 mg/l, and 10 years later it had topped 9 mg/l. In the same period, the number of species of aquatic invertebrates increased from 30 to 110.[53] Figures for lead, ammonium, and many other pollutants show similar trends. How was this possible?

We must recall that, in general, international environmental politics merely provides more or less obligatory frameworks for environmental lawmaking by national and sub-national governments. Nothing prevents nation-states or subnational governments from enacting legislation in the absence of international agreements. In fact, nothing prevents even individual citizens or corporations from voluntarily reducing their effluent loads—even in the absence of legal constraints at the local, the national, or the international level. In short, it is important to appreciate the possibility of cosmopolitan commons as loosely-jointed concatenations of institutions and practices at many levels.

The improvement in Rhine water quality despite near-paralysis on the international stage was in fact attributable to massive investments in effluent-management and sewage-treatment plants by Swiss and German governments and industry in the course of the 1970s. This might seem irrational for upstream nations, but in fact most Swiss and Germans themselves suffered from pollution originating in their own countries. Swiss lakes and fish stocks had long been threatened by eutrophication due to agricultural runoff, while the pollution of the Rhine made its water unsuited to irrigation and increasingly costly for industrial processes—even in much of Germany. As early as 1957 both the Swiss and the Germans had passed laws protecting the quality of their surface waters, followed by France in 1964 (and the Netherlands only in 1970).

In the 1960s, some big industrial firms, including BASF at Ludwigshafen, also began to see the light. In 1964 Rhineland-Pfalz passed a law aimed specifically at curtailing BASF's effluent load on the Rhine. Vast quantities of Rhine water (800 million cubic meters in 1960, nearly one-third of the average annual minimum flow of the entire river) passed through this plant, and some of that water was returned heavily polluted by waste chemicals. However, BASF also spent huge sums on treating Rhine water for use in its own production processes. For example, in 1968, treating boiler feedwater, which accounted for only 1.5 percent of all BASF water use in that year, cost no less than 7.5 million DM.[54] These figures explain why the Pfalz government encountered a cooperative BASF interested in new purification technologies rather than a recalcitrant chemical giant. In the late 1960s, BASF completely overhauled the drainage system at its Ludwigshafen works in order to be able to pass all its effluents through a new wastewater treatment facility of its own design. When the plant came on line in 1975, the ICPR almost immediately recorded an amazing 2 mg/l jump in the Rhine's oxygen content at the Dutch border station at Lobith, the largest single increase ever. The original investment of 150 million DM has paid off handsomely; nowadays BASF has a strong "green" identity and is a market leader in industrial water-purification and effluent-treatment technologies.

The point here is that apparently a transnational moral economy aimed at conserving Rhine water quality could emerge despite the failure to codify this in formal international agreements—or rather, despite the failure to implement agreements that had already been made. The political and economic incentives to take measures against water pollution on regional and national levels were apparently strong enough that actors could proceed without a legal guarantee that other riparian actors would cooperate. However, there were limits to such a "voluntarist" commons. Chemically and biologically speaking, there was still a lot of intricate and tough work to be done on the Rhine, and a moral economy based purely on mutual positioning and de-centralized voluntarism tended to avoid the stickiest wickets. It could also easily founder on the inhospitable shores of suspicions of free riding, unfair advantage, and other forms of profiteering. In short, a robust effort to clean up the Rhine still lacked a foundation in a compelling international agreement. But what if the Chemicals and Chloride Conventions were not the way?

Disillusionment and the International Water Tribunal

There was much frustration and anger about the non-implementation of the 1976 conventions, especially in the Netherlands. The unexpected withdrawal of the French from the Chloride Convention dashed hopes of any speedy reduction in the Rhine's salinity. One commentator characterizes the ten years subsequent to 1976 as a "period of unresolved issues."[55] In 1978, a major Dutch newspaper headlined: "Action against pollution minimal. Complacency in parliament concerning Rhine treaties."[56]

The legalistic stalemate inspired the Dutch environmental movement to focus on the ethical and legal dimensions of Rhine pollution. In a daring move, the Reinwater Foundation took the chloride-dumping French potash mines to civil court—a brave undertaking in view of the feeble state of international environmental law at the time. A second initiative was a self-styled International Water Tribunal that sought

to transcend everyday legal wrangling and to have evidence presented in the clear light of human ethical duty towards the environment; and of the moral culpability of actions to the detriment of that environment. The law courts shirk these issues, seeking refuge in the niceties of legal wording and in the complexities of modern chemical science.[57]

The tribunal, held from October 3–8, 1983, called major Rhine polluters to account for their failure to abide by standards of environmental morality. Identifying the big polluters by name also exposed the way national legislation and international treaties provided environmental "cover," enabling big polluters to persevere in their harmful effluent policies. Responses to invitations to defend themselves before the tribunal (an opportunity declined by all) suggest that the polluters considered the degree to which they degraded the quality of the Rhine's water "legal" and "within the law."[58]

The International Water Tribunal articulated a fundamental shift in the conception of river pollution. Previously, water pollution—at least on the international stage—had been seen as an unfortunate "negative externality" of industrial progress which imposed extra costs on (generally downstream) users in the acquisition of clean water. But the International Water Tribunal rejected this narrow economic model, which had been the tacit framework for both the Chemicals and Chloride Conventions, in favor of an "environmentalist" conception that saw the unchecked dumping of wastes and effluents as an assault on fragile ecosystems—and ultimately a threat to the survival of human and other life-forms. This was expressed in the International Water Tribunal's summary report as follows:

A law agreed on internationally has to be transformed into national legislation and can only then be enforced. That national governments are tardy in doing so does not relieve citizens, and certainly not industries, of the responsibility to create aquatic environments with the utmost care. The use of the term "responsibility" rather than the legally more relevant "liability" is an indication that not just damage expressible in money terms is concerned but rather the harm done to the whole of our habitat.[59]

After Sandoz: The Rhine Action Plan and a "Touch of Nature"

It took another major pollution incident to propel the Rhine clean-water commons into a new ecological phase. On the night of November 1, 1986, near Basel, a chemicals storage facility of the pharmaceutical firm Sandoz caught fire. The water used to fight the blaze flushed a deadly cocktail of highly toxic chemicals directly into the Rhine. The public outcry that accompanied the wave of poison as it made its way downriver, killing fish and other wildlife and forcing water intakes to be closed, was so vociferous that it took only twelve days to convene the Rhine ministers at Zurich, even before the last of the toxins had reached the North Sea. That meeting, and another held a month later (December 19) acknowledged that treating the Rhine's waters as an object of rivalrous economic and chemical accountancy had become a dead-end. Squabbles about tolerable levels of chemicals had degenerated into a "deadlocked game" in which the players were trapped in the incommensurability of reified national interests straitjacketed in upstream-downstream asymmetries.[60] In this impasse, the Sandoz disaster provided an electorally opened window of opportunity.

The Dutch, as a perennial plaintiff, played a prominent role in orchestrating the new international order. At the November 12 meeting of Rhine ministers, the Dutch Minister of Public Works and Water Management, Neelie Smit-Kroes, was the first to take the floor. She had been advised against excoriating the very nervous Swiss delegation for the Sandoz tragedy, noting their excellent previous record on Rhine pollution.[61] In a successful bid to create trust rather than foster enmity, she publicly commiserated with the Swiss about the Sandoz incident, calling for new international measures to prevent such accidents and their effects in the future. The ICPRP was asked to work out a proposal for discussion at the next conference of ministers, which was scheduled for December 19. The challenge for the Dutch was to make sure that this proposal would protect their national interest as downstream users of Rhine water.

In the mid 1980s, Dutch water management was going "integral"—that is, seeking to incorporate not only as many stakeholders as possible, but also all the aspects of water, particularly the central function of clean water in sustaining ecosystems.[62] One of the routine tools that had emerged in the course of developing this paradigm was the use of "indicator species" to measure the quality of ecosystems. Such species were chosen on the basis of sensitivity to crucial parameters of the ecosystem: oxygenation of water, salinity, susceptibility to drought, and so on. Their relative presence or absence thus provided insight into basic ecosystem parameters. But they could also be used to define goals for ecosystems, in the sense that ecosystems able to support a particular species had to meet certain critical physical, chemical, and morphological standards.

This mindset brought salmon back to the Rhine, though as a "consummation devoutly to be wished" rather than a present reality. On the flight back from the Zurich ministers' meeting, the biologist Henk Kroes, a member of the Dutch delegation, argued that Dutch policy should aim at "bringing salmon back into the Rhine." Minister of Public Works and Water Management Neelie Smit-Kroes immediately recognized the political appeal of such an absurdly ambitious goal and adopted "Salmon 2000" as the slogan of Dutch Rhine water-quality policy. Bringing back the salmon by the year 2000 shifted the problem of Rhine pollution from a narrowly chemical footing to a broad biological and hydromorphological one. It encouraged a vision of the Rhine's waters as part of a degradable (and indeed already quite degraded) ecosystem—a common-pool resource for which all the states in the Rhine watershed bore stewardship responsibility.

The problem was getting the other delegations in the ICPRP to jump on this bandwagon, despite initial skepticism bordering on disbelief. The Dutch civil engineer Pieter Huisman, executive secretary of the ICPRP from 1976 to 1981 and an advisor to Minister Smit-Kroes in the mid 1980s, recalls the head of the German delegation somberly pronouncing that "the Rhine will never again become a salmon river."[63] Stern measures were called for. At this juncture, Huisman recalled the old Berlin salmon convention of 1886, forgotten but still binding on the signatories. Article VII stipulated that, in the interest of "increasing salmon stocks in the Rhine basin," signatories commit themselves to "opening and making accessible the natural spawning locations in the tributaries for salmon swimming upriver." The disingenuous argument of the Dutch delegation to the ICPRP delegations and the upcoming ministers' conference was that they were in fact legally obliged to "bring salmon back into the Rhine" by, among other things, improving the quality of the river's water.

But inasmuch as this was an attempt to institute a new moral economy, the exchange of costs and benefits had to be equitable for upstream and downstream riparians. While upstream riparians were being burdened with ensuring water purity and building fish ladders, the Dutch also assumed obligations by virtue of the fact that, with respect to salmon, *they* were the upstream party. So where previously the Dutch had been the perennial plaintiffs in regard to effluents, the turn to salmon transformed them into gatekeepers with something to offer—namely access from the sea to the Rhine. Before the Salmon 2000 project, this had been a non-issue simply because there no longer were any Rhine salmon. But had there been salmon, the Dutch would long ago have gotten into big trouble over their systematic closure of the Rhine estuaries after the 1953 flood. The huge dams and one-way sluices would effectively have prevented salmon from swimming into the Rhine estuaries from the North Sea, save for Rotterdam's still-open New Waterway. Now that a Rhine salmon population was to be re-established, the Dutch would also have to make sacrifices by reopening—at least at regular intervals—the sea sluices in the Haringvliet dam and the closure dike in the IJsselmeer.[64]

The Rhine Action Plan was adopted by the Rhine Ministers and the EU in October 1987 as the roadmap for this new moral economy. Pieter Huisman, instrumental in its creation, described it as follows:

With the RAP the Rhine-riparian states have chosen a new approach to combat damaging pollution. Instead of establishing detailed emission norms for individual firms and industries with its associated bureaucratic red tape, the Rhine-riparian states chose a strategic target that could be met in a limited time span: by 1995 every Rhine-riparian state has to reduce pollution by 50% compared to 1985. Every country is free in the choice of means. This strategic decision has no binding force in international law. It is no more nor less than a political commitment based on mutual understanding and trust.[65]

The Rhine Action Plan was successful precisely because it harnessed existing idiosyncratic national dynamics to a common goal rather than trying to force them into the Procrustean bed of an international treaty. It succeeded in maintaining and intensifying national and regional projects to reduce a broad range of harmful effluents. Levels of pollution from point sources (mostly industrial firms and sewage treatment plants) continued to decline from the mid 1980s onward, and gradually the Rhine was restored to a "clean" river with stretches that nowadays exhibit a respectable riverine ecology.[66] Obstacles to salmon migration have been mitigated and some spawning grounds restored. Atlantic salmon haven't yet entered the river in droves, but with human assistance a native population seems to be taking hold. There is considerable optimism about reaching the goals of what now has become Salmon 2020.

As a consequence of its practical success, the Rhine Action Plan has been extolled by commentators as a new kind of international regime that has succeeded in reducing waste emissions even in a situation of asymmetric "externalities." In an article titled "The River Rhine: From Equal Apportionment to Ecosystem Protection," André Nollkaemper speaks of a "radical break from the pre-existing history of overdevelopment of the river basin towards a more sustainable use of the Rhine."[67] Tun Myint notes the shift from the hegemony of the nation-state to the inclusion of sub-state and transnational actors: "In the Rhine's history, RAP was the first of its regime type that began to enhance multilayer governance by incorporating participation of non-state actors into what used to be an *inter*-national state-centric regime."[68] From our own perspective in this volume, it can be ventured that the RAP—with its focus on the restoration of a common biological heritage, its inclusion of a wide range of practically committed non-state actors, and its emphasis on a moral economy based on trust and reciprocity—has provided the institutional framework for a truly sustainable transnational commons for water quality on the Rhine. The European Water Framework Directive of 2000, inspired by the RAP, but into which the RAP's aims have now been subsumed, can be seen as an attempt to improve on the RAP by extending the commons to actors in the entire river basin while also diffusing its lessons to the management of water bodies throughout the European Union.

Conclusion

The Rhine as Avenue of Modernity

Cosmopolitan commons of both national and transnational scope have emerged in the Rhine basin over the past two centuries, and they currently frame the way humans interact with almost all the flows on the river. This is a surprising outcome in view of the persistence of state sovereignty over the river, deep divisions between upstream and downstream actors, and a chronic dearth of perspicacity. But the historical outcome makes more sense when we recognize that the Rhine has been one of Europe's premier avenues of modernization, a transport corridor along which European industrial modernity acquired the first inkling of its transnational, continental potential. The Rhine was a stage on which local constellations of actors gradually came to see themselves as interdependent elements in a complex system of flows distributed across time and space, and ultimately as beneficiaries of cosmopolitan arrangements to manage these flows in common.

Because rivers intrinsically "dislocate space from place," they can be seen as beachheads of modernity.[69] The theorist Anthony Giddens conceives of the transition to modernity as centered on "time-space distantiation," i.e., the increasing interpenetration of the local here-and-now by other places and times. In the context of Western Europe, places on the Rhine have been at the vanguard of this development. All the resources embedded in the river's flows—potable water, fish, waste removal, tourist revenues, cheap bulk transport, hydropower, gold flakes, and so on—were and are valorized locally, i.e., transformed into use values or rents at some specific location. The same was true for the risks implicit in some of the flows, including flooding and pollution; these too were inevitably discounted locally. Nonetheless, the nature of the flows at any location and how the associated valorizations and risks played out depended on what went on at other locations on the river—that is, where the flows came from and where they went. With respect to any particular locality, the flows were cosmopolitan phenomena that connected upstream and downstream locations in ways that increasingly challenged traditional absolute senses of place and time.

But even though the riverine flows of modernity increasingly connected riparian locations in complex material and institutional webs of time-space interdependencies, the perspicacity, the knowledge, and the political spans of control that would eventually enable local actors to create cosmopolitan commons spanning the entire river took time to develop and often lagged behind more fundamental economic and demographic developments. However, these reflexive "regulatory" impulses were just as native to modernity as capitalism, commerce, industrialization, and urbanization. Paradoxically, they owed much to the emergence of the modern state, with its centralizing tendencies and its mission to "rationalize" economy and society—often by regulating or limiting the operation of market forces. In view of nineteenth century German governments' early complicity in the re-engineering and international management of the Rhine, it is no coincidence that many authors have singled them out (especially Prussia) as precursors of the modern interventionist state.[70]

In this chapter I have examined how Rhine-riparian actors gradually learned to know (and in some cases manage) the novel ways in which local fates became entangled with cosmopolitan structures as the Rhine basin rapidly industrialized and urbanized after the middle of the nineteenth century. I have demonstrated that, despite formidable obstacles, actors along the Rhine by and large succeeded in breaking out of their tragedy-bound "deadlock games" and "prisoner's dilemmas" to create sustainable transnational cosmopolitan commons. They were abetted by the possibility of

tradeoffs among different flows, issue linkages, new measurement and surveillance technologies, new forms of scientific perspicacity, Germany's and
to a lesser degree France's status as both upstream and downstream riparian,
the proliferation of European institutions and infrastructures, the wake-up
calls of "normal accidents," and the general turn to "ecological modernization." However, there is clearly no single story for all the Rhine's flows. The
accounts of efforts to manage ships, salmon, and pollutants in common
have shown that transnational commons making stands or falls with a deep
appreciation of the material and historical specificity of the resource (or
risk). Though ships, fish, clean water, floods and hydropower are all flows
in and on the Rhine, each of them has its own story to tell—though, as we
have also seen, the stories tend to get inextricably intertwined.

Space vs. Resource-Spaces

The historical specificity of resources and their management brings into
sharp relief how the mundane and apparently uniform geographic space of
the Rhine basin became many different spaces depending on the resource
in question. Håkon With Andersen comes to the same conclusion in his
chapter on the North Sea. The space of Rhine shipping included chiefly the
main river channel and a few of the larger tributaries. Critical locations and
foci of cosmopolitan concern in this space were those where the river was
chronically shallow, narrow, or unusually turbulent, such as at the infamous Binger Loch or the Lorelei. The space of fish entailed the entire river
system, including the smallest tributaries, with the downriver seine-fishing
sites and the upriver spawning grounds the most contentious locations.
The space of clean water included not only the entire river system but all
the fields, forests, and human settlements in the watershed too. Foci of collective concern here were the big industrial and mining complexes and the
conurbations lining the river and its major tributaries.

Knowledge

For most of the cases described in this book, including the navigational
commons described in this chapter, the creation of cosmopolitan commons seems to have been chiefly a political and institutional challenge,
with right knowledge relegated to a back seat. For the case of salmon and
pollutants in the Rhine, the relationship is clearly inverted. This also applies
to medium-range transboundary air pollution, discussed in the chapter by
Arne Kaijser. Knowledge certainly plays a role in other cosmopolitan commons, though more often only in the actual production of value (as in
producing weather forecasts or navigating airplanes in bad weather) rather

than in shaping the cosmopolitan commons itself (i.e., helping to specify the moral and technological constraints the commoners should live by). By contrast, I have shown how different conceptions of the salmon life cycle pointed to different rules for the commons and how chemical knowledge was essential in determining whether and how much action should be taken against polluters. In fact, every single pollutant—feces, mercury, nitrates, and so on—challenged chemists, physicists, biologists, and limnologists to develop specific life-cycle theories describing the substance's origins, its dispersion, its chemical biography in river water, and its effects on ecosystems and economies.

Hence, as in Kaijser's chapter, "epistemic communities" acquired a forceful and central role in assessing what risks were being run and in suggesting possible collective countermeasures. Particularly significant from the perspective of modernization is the growing credence attributed by governments to "scientific" theories of (in this case) salmon and chemicals in preference to "lay" theories like those professed by fishermen and captains of industry and their representatives. The experts who developed these theories on the basis of observation and controlled experiment were themselves peripatetic cosmopolitans maintaining extended international networks through exchange of publications, scientific congresses, and personal acquaintance. They sometimes acted like a cognitive fifth column, often utilizing their direct access to policy-making circles in their native countries to foster policies serving long-range cosmopolitan interests rather than short-term national ones. In time they came to inhabit hybrid locations, such as that of Salmon Commissioner or that of a member of an expert committee of the Central Rhine Commission or the ICPRP, which enabled them to influence commons policy quite directly on the basis of their cosmopolitan scientific insights. Such epistemic communities of "cosmopolitan technocrats" have become increasingly decisive in shaping the politics of the European Union—a development that began on the Rhine.

Integrating Technology and Commons into History

Because this chapter covers a long time span and addresses the common management of three different resources (as does With Andersen's chapter), it offers insights into how the approach of cosmopolitan commons can be a way to bridge the gap between history of technology and more conventional political, economic, and cultural history. The crux is that I show how the three cosmopolitan commons (ships, salmon, and water) are not simply three parallel systems of transnational resource-governance that happen to coexist in the same conventional space, but rather that their histories are

intertwined—and not only with each other, but also with the contemporary histories of markets, technologies, politics, and moral orders. Though
the history of modern commons is still being written, it strongly suggests
that cosmopolitan commons and the technologies that bring them into
being and sustain them are not deviant legal-economic islands distinct
from "normal" history but, quite the contrary, are thoroughly embedded
in the conventional institutional and moral order in myriad ways—are, in
fact, an essential though still underrated element of the history of Europe
and, ultimately, of the whole world. As our encroachments on the ecosystems that support our existence on this planet become more grave and
irreversible, we shall certainly have occasion to ponder the depth of Shakespeare's adage that "One touch of nature makes the whole world kin."[71]

Acknowledgments

I thank Eda Kranakis for her critical support in the course of writing this
chapter. The two anonymous referees pointed out inconsistencies and
shortcomings in an earlier version, as did my fellow authors in the course
of the several meetings that accompanied the making of this volume. I also
thank the participants in a seminar held at the Institute for Social Sciences
at the University of Lisbon on May 7, 2009, at which an early version of a
portion of this chapter was presented.

Notes

1. The pioneering effort to analyze a river as an energetic entity is Richard White,
The Organic Machine: The Remaking of the Columbia River (Hill and Wang, 1995).

2. This point is a conceptual centerpiece in Lucien Febvre's transnational history of
the Rhine. Febvre's historical account occupies the first half of a joint work. See
Albert Demangeon and Lucien Febvre, *Le Rhin, Problèmes d'histoire et d'économie*
(Armand Colin, 1935), especially chapter 3.

3. Ibid. Febvre saw the Rhine basin as a "hyphen," a link between peoples rather
than a frontier. He also emphasized that it was a "valley of cities."

4. Mark Cioc, *The Rhine: An Eco-Biography, 1815–2000* (University of Washington
Press, 2002).

5. Hugo Grotius, *The Rights of War and Peace*, volume 2 (Lawbook Exchange, 2004;
facsimile of 1738 London translation), p. 151.

6. Ibid., pp. 161–162. Grotius does not consider this ownership a legitimate basis for
levying usurious tolls simply to fill the ruler's coffers. He argues on pp. 154–155 that
tolls should be levied only to defray the costs of facilitating transport.

7. See also chapter 7 in this volume.

8. Friedrich Pfeiffer, *Rheinische Transitzölle im Mittelalter* (Akademie-Verlag, 1997).

9. Ibid., p. 332.

10. Roy Gardner, Noel Gaston, and Robert T. Masson, Tolling the Rhine in 1254: Complementary Monopoly Revisited (unpublished paper, 2002) (available at http://scholar.google.com).

11. Edwin J. Clapp, *The Navigable Rhine* (Houghton Mifflin, 1911), p. 6.

12. Robert Mark Spaulding, "Anarchy, Hegemony, Cooperation: International Control of the Rhine River, 1789–1848" (n.d.) (available at http://www.ccr-ikzr.org).

13. Ibid.

14. Ibid.; W. J. M van Eysinga, *La Commission Centrale pour la Navigation du Rhin* (A.W. Sijthoff, 1935). The website of the Central Commission (http://www.ccr-zkr .org) also includes a valuable collection of downloadable articles documenting its early history.

15. Cioc, *The Rhine*, chapter 4.

16. David Blackbourn, *The Conquest of Nature: Water, Landscape and the Making of Modern Germany* (Norton, 2006).

17. Thomas Ashworth, *Essay on the Practical Cultivation of a Salmon Fishery, Addressed to the President and Council of the International Congress, to Promote the Cultivation of Fisheries, held at Arcachon, 1866* (Judd and Glass, 1866).

18. Andrew Young, *The Natural History and Habits of the Salmon: With Reasons for the Decline of the Fisheries, and also How They can be Improved, and Again Made Productive: Also an Account of the Artificial Incubation of the Salmon* (Longman, Brown, Green and Longmans, 1854).

19. Petition J. Sepers and 12 others regarding destructive methods of salmon fishing, 2nd Chamber Dutch Parliament, 1856–7, 21st session, November 22, 1856, p. 182. Regarding the reference to concessions, it should be noted that the fishermen did not own the waters in which they fished and hence had no *a priori* claim to the fish they caught. Fishing rights belonged to wealthy noble or patrician families and after the establishment of the Kingdom of the Netherlands in 1815, increasingly to the Dutch state. Traditionally, a percentage of the catch was demanded (usually 20 percent, but running up to 50 percent in extreme cases). See T. Bakker, "De Ammerstolse zalmvisserij 'De Snackert'," in *Historische Encyclopedie Krimpenerwaard* (Stichting Krimpenerwaard, 1976). Under state ownership, fishing rights, and even complete fisheries were leased at auction for periods of five or more years. The prices that were fetched depended on the expectations investors had of the sizes and value of their catches, so concession prices can be taken as a rough measure of salmon-fishing prosperity.

20. A. Quakernaat van Spijk, *Eenige beschouwingen over onze zalmvisscherijen. Naar aanleiding der gesloten overeenkomst te Mannheim dd. 27 november, 1869* (H. J. van de Garde, 1870), p. 7.

21. Francois P. L. Pollen, *Eene wettelijke regeling voor Nederland van de zalmvisscherijen in den Rijn, beschouwd in verband met de Conventie van Mannheim (1869) en die van Berlijn (1885). Memorie opgedragen aan de beide kamers der Staten-Generaal* (Scheveningen, 1886). The price comparison has to be corrected for 5 percent inflation between 1840 and 1868. (See http://www.iisg.nl/hpw/calculate2-nl.php.) Some idea of the value of a salmon can be gained from figures from 1930, when salmon were fetching 3 guilders per kilogram. At an average of 10 kilograms, a fish would bring in 30 guilders. A farmhand was reported to earn 3 guilders a week, so catching a salmon was a major windfall for a fisherman. Piet Hartman, *Zalm Vernomen! De Zalmvisserij Ten Tijde Van Weleer* (Vèrse-Hoeven, 1995).

22. Jan Blom, *De Zalmvisserij van Ammerstol* (Europese Bibliotheek, 1989). Though rest periods coincided with the incoming tide in the rivers' tidal reaches, the effects of the tide itself were not felt at Ammerstol. The enforced rest was legally required by virtue of the salmon convention then in force.

23. H. A. Visser, "De zalmvisserij, verleden tijd," *Kwartaalblad van de Historische Vereniging "West-Alblasserwaard"* 11, no. 4 (1992): 14–15.

24. "Communications from Provincial Estates, municipalities and chambers of commerce and petitions from fishermen concerning 1. the decline of the salmon fishery 2. its causes 3. the means of recovery," appendix to *Memorandum of Reply to the Provisional Report of the Committee of Investigation regarding adoption of the Act of Mannheim of Nov. 27, 1869*, Second Chamber of Parliament proceedings 1869–1870, p. 1755.

25. As part of a major overhaul of the Dutch river system, many of the kills in the Biesbosch were dammed up in the late 1860s, then, in 1870, replaced by a new section of tidal river called the Nieuwe Merwede. This sounded the death knell for the *steek* fisheries, but provided new opportunities for even more deadly seine fisheries. It probably was this threat—which in the event did not materialize—that moved the government of Baden to call for an international salmon convention in 1869.

26. In 1878 the entire Prussian salmon catch on the Rhine—from the Dutch border to Bingen—was estimated to amount to about 1/20 of the Dutch catch. See Max von dem Borne, *Die Fischerei-Verhältnisse des Deutschen Reiches, Oesterreich-Ungarns, der Schweiz und Luxemburgs* (Deutsche Fischerei-verein, ca. 1880).

27. Werner Böcking, *Nachen und Netze. Die Rheinfischerei zwischen Emmerich und Honnef* (Rheinland Verlag, 1982).

28. Otto Beck, *Beschreibung des Regierungsbezirkes Trier. Zur Erinnerung an die 50jährige Jubelfeier der Königlichen Regierung zu Trier am 22 April 1866*, volume 1 (Trier, 1868).

29. Ibid., p. 548. The pointed reference to "jusqu'a la mer" refers to the long-standing conflict between Prussia and the Netherlands about the geographical limits of the Rhine navigation commons. The Congress of Vienna stipulated that the Rhine should be freed of tolls and duties. The Dutch wanted to retain their remunerative harbor taxes in Rotterdam and Amsterdam and so interpreted the phrase "jusqu'a la mer," the downstream limit of the navigational commons, to mean "up to the sea," i.e., only up to the point where tides began to exert an effect, thus placing the tidal North Sea harbors beyond the pale of the international agreements. This remained a serious bone of contention and paralyzed the navigational commons until Dutch resistance folded before the signing of the Treaty of Mainz in 1831. See Eysinga, *La Commission Centrale*. It is also interesting that Beck uses the word "rauberisch" (thievish), the same word that was used to refer to the "robber barons" who illegally exacted exorbitant tolls along the medieval Rhine. The parallel was clearly intended to emphasize how Dutch fishing practices were criminally undermining the viability of salmon populations (and fisheries) on the Rhine—just as the "robber barons" had threatened profitable shipping and the finances of the Holy Roman Empire in the Middle Ages.

30. Grossherzogliche Oberdirection des Wasser- und Strassenbaues, Baden, *Correction des Rheins von Basel bis zur Grossherzoglichen Hessischen Grenze* (Karlsruhe, 1863); Blackbourn, *The Conquest of Nature*; Cioc, *The Rhine*; Christophe Bernhardt, "Die Rheinkorrektion. Die Umgestaltung einer Kulturlandschaft im Übergang zum Industriezeitalter," *Der Buerger im Staat, Zeitschrift der Landeszentrale für Politische Bildung Baden-Württemberg* 2 (2000): 76–81. One trenchant critic of the project, the publicist Fritz André, argued in 1828 that the corrections were harmful for the salmon population because they would increase the current. He was right about the harm, but, as it turned out, for the wrong reasons—he believed that the fish could not breed in rapidly flowing rivers. See Fritz André, *Bemerkungen über die Rectification des Oberrheins und Schilderung der furchtbaren Folgen, welche dieses Unternehmen für die Bewohner des Mittel- und Unterrheins nach Sich ziehen wird* (C. J. Elder, 1828).

31. *Internationale Ausstellung, Mailand 1906. Abteilung: Deutsche Binnenfischerei* (Deutsche Fischerei-verein, 1906), p. 14.

32. Petition J. Sepers and 12 others. op. cit. Note the apparent belief that salmon also spawned on the lower Dutch reaches of rivers which of course also served as a tacit repudiation of claims by German and Swiss fishers to a share of the catch—or at least of their claims that salmon had to be let through by the Dutch in order to reproduce at all.

33. In the Dutch system a "royal decree," unlike a law, does not have to be passed by parliament. It is, however, subject to scrutiny by the courts once enacted and may be recalled for legal shortcomings.

34. Allgemeine Landesarchiv Karlsruhe 233/30830, cited on p. 58 of Götz Kuhn, *Die Fischerei am Oberrhein. Geschichtliche Entwicklung und gegenwärtiger Stand* (Eugen Ulmer, 1976).

35. Proceedings of the Second Chamber, Session 1869–70. Approval of Articles of the Treaty signed at Mannheim, *Memorandum of Reply*, p. 1756.

36. Ibid., 1756.

37. Quakernaat van Spijk, *Eenige beschouwingen*, 4.

38. Ibid., p. 8.

39. Proceedings of the Second Chamber, Dutch Parliament, 1871–72. Session November 21, 1871. *Debate on the Budget 1872*, 301.

40. Ibid.

41. Ibid.

42. Max von dem Borne, *Die Fischerei-Verhältnisse des Deutschen Reiches*, p. 142.

43. Ibid.

44. The Berlin salmon convention of 1885 dropped all references to pollution, probably in an effort to minimize chances of parliamentary rejection in industrializing states like Prussia and the Netherlands—even though this was hardly realistic from the point of view of improving salmon stocks. In the Netherlands the new treaty simply replaced the royal decree of 1872 with the result that probitions against polluting fishing waters were struck from the books. On the Upper and High Rhine, however, the governments of Baden, Switzerland, and Elsass-Lothringen opted to retain and even fortify the anti-pollution article in the new tripartite treaty they ratified in 1887.

45. François Pollen, *Eene Wettelijke Regeling voor Nederland*.

46. Pollen argued that the weekly closure provided the salmon not only with less total "offtime," but also gave them insufficient time to run the gantlet from the most seaward *steek* fisheries to upstream spawning locations. Salmon setting out at 6 p.m. on Saturday ran a good chance of being caught Sunday night or Monday before they made it to their destination. However, fishers of souls up and down the river embraced the weekend closure since it encouraged salmon fishers to spend the Lord's day in church.

47. Deutsche Fischerei-verein, *Internationale Ausstellung, Mailand 1906*, pp. 10–12; Kuhn, *Fischerei am Oberrhein*; Horst Johannes Tümmers, *Der Rhein—Ein Europäischer Fluß Und Seine Geschichte* (C. H. Becksche Verlagbuchhandlung, 1994); Carel Dieperink, "From Open Sewer to Salmon Run: Lessons from the Rhine Water Quality Regime," *Water Policy* 1, no. 5 (1998): 471–485. Early Prussian concerns about industrial water pollution culminated in the Emscher Association of 1899. On the latter, see Cioc, *The Rhine*, pp. 77–108.

48. Cioc, *The Rhine*, pp. 185–193.

49. Memorandum Alfred Matthey-Doret, chief of the Swiss Agency for Water Protection, Dec. 3, 1959," personal archive, Pieter Huisman.

50. Otto Jaag, "La Crise des Lacs et Rivières de l'Europe Centrale," *Bulletin Français de Pisciculture* 27, no. 177 (1955): 129–140.

51. Cited in N. L. Wibaut Isebree-Moens, "Internationaal Onderzoek naar de Mate van Vervuiling van het Water van de Rijn," *Water, Bodem, Lucht* 46, no. 3 (1956): 52–57.

52. Dieperink, "From Open Sewer to Salmon Run"; Pieter Huisman, "Rijnoeverstaten en Europese Unie Ondernemen Acties na Vergiftigings- en Overstromingsrampen," *Tijdschrift voor Waterstaatsgeschiedenis* 16, no. 1 (2007): 46–55; André Nollkaemper, "The River Rhine: From Equal Apportionment to Ecosystem Protection," *Review of European Community and International Law* 5, no. 2 (1996): 152–160; Pieter Huisman, Joost de Jong, and Koos Wieriks, "Transboundary Cooperation in Shared River Basins: Experiences from the Rhine, Meuse and North Sea," *Water Policy* 2, no. 1–2 (2000): 83–97; Tun Myint, Strength of "Weak" Forces in Multilayer Environmental Governance: Cases from the Mekong and Rhine, PhD dissertation, Indiana University, 2005.

53. *Upstream: Outcome of the Rhine Action Programme* (ICPR, 2003).

54. Hans-Jürgen Gessner, *Wasserversorgung und Umweltschutz in der chemischen Industrie—Dargestellt am Beispiel der Badischen Anilin- und Sodafabrik AG (BASF), Ludwigshafen am Rhein* (Akademie für Raumforschung und Landesplanung, 1973).

55. Dieperink, "From Open Sewer to Salmon Run," p. 475.

56. *Volkskrant*, April 19, 1978.

57. *Casebook of the International Water Tribunal*, chapter 6, p. 1 (cited on p. 141 of Myint, Strength of "Weak" Forces).

58. Although the International Water Tribunal formally stood outside the state system, it was financed in part by grants from five Dutch ministries and the city of Rotterdam. Apparently the IWT was also seen as pursuing a Dutch national interest in addition to its "universal" aim of saving the environment. See Myint, Strength of "Weak" Forces.

59. *Brochure International Water Tribunal* (Drukkerij Rob, n.d.) (cited on p. 150 of Myint, Strength of "Weak" Forces).

60. Thomas Bernauer, "Explaining Success and Failure in International River Management," *Aquatic Science* 64 (2002): 1–19.

61. Huisman, "Rijnoeverstaten en Europese Unie Ondernemen Acties."

62. Cornelis Disco, "The Ecological Turn in Dutch Water Management" *Science, Technology and Human Values* 27, no. 2 (2002): 206–235; Alex van Heezik, *Strijd om*

de Rivieren. 200 jaar rivierenbeleid in Nederland (Rijkswaterstaat, 2006); *Omgaan met Water. Naar een Integraal Waterbeleid* (Ministerie van Verkeer en Waterstaat, 1985).

63. Interview of Pieter Huisman, October 20, 2011.

64. The "sacrifices" consist of transforming the large fresh-water basins behind the tidal sluices—former estuaries—into brackish and tidal waters again. This will require very costly adaptations to harbors and fresh-water intakes. In the past two years, a vigorous political debate on whether the Netherlands will honor these obligations pitted the Liberal–Christian Democratic coalition government against the "epistemic community" of civil engineers and ecologists with their professional stake in the Rhine clean-water and ecological commons.

65. Huisman, "Rijnoeverstaten en Europese Unie Ondernemen Acties," p. 52.

66. Ine D. Frijters and Jan Leentvaar, *Rhine Case Study* (UNESCO, 2003).

67. Nollkaemper, "The River Rhine," p. 157.

68. Myint, Strength of "Weak" Forces, p. 157.

69. Anthony Giddens, *The Consequences of Modernity* (Stanford University Press, 1990).

70. See, for example, G. W. F. Hegel, *Elements of a Philosophy of Right* (Cambridge University Press, 1991); James C. Scott, *Seeing Like a State: How Certain Schemes to Improve the Human Condition Have Failed* (Yale University Press, 1998); Blackbourn, *The Conquest of Nature*.

71. This line is often quoted out of context. The usual understanding is that Shakespeare is arguing that nature, and particularly our human nature, can move us to feel our common humanity and brotherhood. In fact, Ulysses is consoling the out-of-fashion Achilles by emphasizing how our common human nature seduces us to the slavish worship of novelty, however tawdry, and makes us disparage boring old virtues like those professed by the neglected Achilles. I conflate the two interpretations by arguing that while our "touch of nature" certainly makes us grand despoilers of our planet's natural bounty as we ceaselessly pursue riches and novelty it also opens up the possibility of enacting our "natural kinship" by establishing the cosmopolitan commons necessary to create a sustainable future.

11   Conclusions

Nil Disco and Eda Kranakis

Cosmopolitanism seeks to explain the reorientation of national and international affairs to accommodate proliferating interdependencies, yet the task of adding empirical depth and weight to this theoretical framework has only begun[1]—particularly in regard to the roles of nature and technology in cosmopolitan dynamics. The present book attempts not only to enlarge cosmopolitan theory along these lines, but also to provide it with empirical foundations by showing how cosmopolitan commons enhance and regulate the shared use of large natural structures and human-built infrastructures. The case studies reveal the recurring dynamics of cosmopolitan commons—the continually evolving and multiplying international treaties, the networked moral economies that build trust and set negotiated limits to autonomous action, the massing of interests to overcome risks or enhance action and resource valorization, and the layering of transnational actor networks in relation to large resource-spaces (the Rhine, the Vuoksi, the North Sea, European airspace, the spaces of crop diversity) or in relation to infrastructure-based activities (broadcasting, weather forecasting).

To the extent that cosmopolitan theory has considered technology, it has seen it largely as a way to connect points A, B, and C (across borders) more rapidly and with greater frequency than before, or as a source of risk and environmental degradation.[2] Our studies go beyond this perspective by also exploring what it takes to enable technologies to connect spaces on a continuing basis, what it takes to manage and expand the ongoing resource extractions that allow these technologies even to exist and function in the first place, and what it takes to control the routine, quotidian risks associated with economic activities such as resource extraction and the production and utilization of goods and infrastructures at industrial scales.

Surveying cosmopolitan commons in Europe—understanding their tendency, as evidenced in this book, toward continual upscaling to encompass larger spaces, more organizations, more and larger actor networks, and new domains of regulation—ultimately provides a point of entry into the study of globalization. More immediately, however, it offers a new perspective on the history of European integration. There have been growing criticisms, implicit and explicit, of the limited scope of many accounts of European integration, with their focus on the elite political bargains that defined the European Community as a political and economic entity.[3] These grand bargains—the treaties that created the Common Market and eventually the European Union—are fundamental, but where did they come from? The Treaty of Rome (1957), which founded the Common Market, is a significant case in point. It represented an enormous integrative leap, even though initially for only six countries. Yet can this leap be adequately understood, as it is in classic textbook and scholarly accounts,[4] as an offshoot of the European Coal and Steel Community? As the outcome of encouragement by the United States, following precepts embodied in the Marshall Plan? As predominantly a tool for the reconstruction of nation-states? As a response to the failure of European military integration (expressed in the failure of the proposed treaty for a European Defense Community)? As a response to the decline of European empires? Certainly these interpretations all have explanatory weight, but the fact remains that the Treaty of Rome was most immediately about organizing a common market, with the aim of freeing the movement of goods, people, capital, and services among the member states, making production more efficient, and, according to the original vision,[5] enabling European transport systems to work efficiently at larger scales. These immediate, stated aims are particularly congruent with frameworks that had already been established by European cosmopolitan commons.

It is not just that cosmopolitan commons made production and free movement technically possible and meaningful at ever-larger scales. To function well, markets require foundations of law and trust. The studies in this volume show that these foundations have important roots in the cosmopolitan commons that have been built up painstakingly, across a plurality of domains, since the nineteenth century. Leaps of integration such as the Treaty of Rome make more sense, and indeed seem more necessary, when viewed against the backdrop of the dense, multifarious histories of Europe's cosmopolitan commons. It makes sense that the proliferation and upscaling of these commons not only helped make a Common Market possible, but also made the establishment of a larger, more formal, overtly political framework of European cooperation increasingly necessary, in order to manage and govern Europe's growing range of transnational commons, their interactions, and their associated realms of commerce.

One could take this argument even further: the original Common Market can be seen as displaying, itself, many features characteristic of cosmopolitan commons—for example, a moral-economy foundation linked to an international treaty structure, attendant epistemic communities, and a web of interdependent actor networks. Two things that differentiate the Common Market and make it unique are its much broader political aim and the fact that, even at the outset, its scope encompassed multiple resource-spaces, infrastructures, and social purposes. Further, its formal institutions, modeled on the institutions of nation-state governance, were much broader in their scope and range of powers than those of cosmopolitan commons. Moreover, as the Common Market has evolved into the European Union, its institutions—the European Court of Justice, the European Parliament, the Council, the European Commission—have come to play governance roles in all of Europe's commons.

Commons and Modernity

Cosmopolitan commons, like cosmopolitanism itself, are creatures of modernity. "Modernity" is hardly a settled concept, but there seems to be agreement on at least the idea that under modernity the temporal and spatial scales of interactions have increased. Cosmopolitan commons are at the heart of this process. Social theorists have denoted this phenomenon—or constituent elements of it—in several distinct ways: Ferdinand Tönnies famously distinguished between *Gemeinschaft* and *Gesellschaft*, and Emile Durkheim between "mechanical" and "organic" solidarity; Norbert Elias referred to "longer chains of interdependence"; Anthony Giddens speaks of "time-space distantiation." The idea underpinning these diverse models is that local settings are increasingly embedded in dense networks that span ever-greater times and distances. What happens at time/space location A now has repercussions at B and C, and vice versa.

Looking at cosmopolitan commons helps us get a handle on this phenomenon, to fathom the dynamics of increasing scale from local to regional, national, transnational, and ultimately global infrastructures, networks, and governance regimes. Cosmopolitan commons presuppose accretion; they typically expand in space, in membership, and in the variety and complexity of their undertakings. Indeed, the scope and the scale of modern production and commerce could hardly be maintained in the absence of correspondingly extensive cosmopolitan commons. And as "longer figurations"—extended chains of interdependencies—continue to be cobbled together, uncontrolled mutual effects among nominally independent entities chronically proliferate, encouraging further upscaling of commons regimes. At the same time, entities outside the commons find it increasingly advantageous to join in (or disadvantageous to be left out), resulting in yet further expansion of the commons. These dynamics are visible in all of the case studies. From this perspective, the building of cosmopolitan commons can be understood as the fundamental precondition for the expansion of human activity across space and time in the era of modernity.

Yet the basic political constituents of cosmopolitan commons have been the nation-states. That is why this book is so full of bilateral, trilateral, and multilateral treaties. This has been the basic form of legally binding agreement among states and, by extension, their inhabitants. In many of our accounts, nation-states therefore assume the role of "commoners," acting more or less at the behest and putatively in the long-term interests of their subjects. Several of the authors, however, address the limits of this statist model of transnational governance, arguing that sustainable cosmopolitan commons are transnational, rather than merely international—i.e., that sub-state and super-state actors must also pragmatically commit themselves to a moral economy of which treaties and conventions are only the outward legal and political manifestation. Kranakis' institutionally exuberant story of European airspace makes it clear that many kinds of organizations and "epistemic communities" were needed to shape and maintain a European airspace commons based on the conditional abeyance of national sovereignties. Disco's account of Rhine pollution shows that some of the remaining "stubborn" forms of Rhine pollution began to decrease significantly only after the statist model (in the form of the stringent Chemical and Chloride Conventions) was abandoned and replaced by the much more inclusive, voluntary, and ecologically legitimized Rhine Action Plan. Similar points are made in all the other chapters.

All of this suggests that cosmopolitan commons are at the vanguard of an alternative modernity that has been characterized as "neo-medieval."[6] This approach argues that European integration is proceeding not only through the integration of nation-states but also at levels above and below that of the nation-state. As a result, the state grows "porous" as its absolute hegemony— even over its own territory—is challenged by a complex of supra-state institutions and infra-state non-governmental organizations and "epistemic communities." This is yet another way in which the case studies in this book contribute to a new perspective on European integration. Today's Europe, as the chapters in this volume make clear, is much more than an alliance of nation-states. It is also a dense transnational and supranational network of agencies, professional networks, lobby groups, citizens' initiatives, and so on—the "usual suspects," from a cosmopolitan-commons perspective.

Technology and Cosmopolitan Commons

To gain a comprehensive theoretical appreciation of cosmopolitan commons, it is important not only to look upward—to perceive their relationship to Europeanization, globalization, and the structuration of modernity—but also to look downward—that is, to analyze their constitution and workings. Technologies are crucial in this regard. The chapters in this book make it clear that technologies play several roles in the constitution of cosmopolitan commons. Their first and rather traditional role is making nature amenable to human needs. Technologies are the ruses and artifices that allow us to beguile our natural environment into doing our bidding—more or less. In this most basic, Marxian, sense, technologies are "forces of production" that enable us to valorize natural resources and turn them into social utilities. This modality is prominent in all of the case studies.

But we are all sorcerers' apprentices, and our technologies more often than not get the better of us. What economists call "negative externalities" (and the rest of us simply "risks" or "unwanted and unintended side effects") are integral and seemingly inescapable parts of technologies as "forces of production." However, we need not hang our heads in despair. Commons, and by implication cosmopolitan commons, can also emerge from a shared refusal to let the chips fall where they may—that is, from a shared determination to combat unintended risks by concerted and cooperative action. This can assume many forms. A number of studies in this volume feature "risk commons" arising to protect both humans and the natural environment from the ravages of industrial production, and to control risks or unwanted side effects linked to natural phenomena or to resource exploitation—risks like calamitous weather, genetic erosion, water and air pollution, or degradation of services due to insufficient coordination or unregulated competition. Each of the studies in this book deals with risk and technological side effects in one way or another.

Aside from mediating (and continually revolutionizing) our relationship with existing "natural" resources, technologies also have the nearly magical property of creating new resources. This penchant for renewal reveals the extent to which even prosaic "natural" resources aren't so natural after all. Several of the chapters describe cosmopolitan commons based on valorization of a "natural" resource that entered human history thanks to new knowledge and associated technological ruses and artifices. Wormbs' chapter on the European radio-frequency spectrum, Kranakis' on airspace, and Saraiva's on gene banks as a by-product of a new conceptualization of "genetic diversity" are telling examples, as is Edwards' analysis of the role of computers and climate models in the creation of a meteorological commons.

A final but extremely important role for technologies is to provide the means to monitor the behavior of commoners and to enforce behavior that sustains rather than undermines the commons. Obedience to shared rules comes neither cheaply nor spontaneously. Moral economies feed on mutual trust, and nothing fosters trust better than clear rules, limits, and symmetrical "panoptical" transparency. Commoners must know their rights, their responsibilities, and the limits of their legitimate demands on the commons. Above all, they should be able to assess at all times (at least in principle) whether the other commoners are abiding by the rules. The piracy and free riding that otherwise tend to flourish corrode and demoralize commons. Of course, achieving mutual panoptical transparency is not only a matter of material technology. Administrative techniques (e.g., licensing, registration, and formal identification systems) also are important, though even these increasingly depend on material technologies like sensors and elaborate information storage and retrieval systems.

Technologically sophisticated monitoring and enforcement become more and more indispensable for sustaining a cosmopolitan commons as production processes become more complex, resource valorization intensifies, and "negative externalities" proliferate in unpredictable ways. Examples discussed in this book include inspection of fisheries, monitoring of air and water quality, monitoring of aircraft, standardization of meteorological observations, calibration of measurement devices, storage protocols for seed banks, and monitoring of water levels and of the frequency stability of transmitters. These "second-order" technologies shore up cosmopolitan commons, keep them operating, and protect them against erosion. They are innovations in their own right, and they are as crucial to the survival of cosmopolitan commons as are legal and institutional provisions.

Epistemic Communities and Transnational Governance

Symmetric panoptical monitoring is a necessary but not a sufficient condition for maintaining commons, a conclusion that can be drawn from Garret Hardin's original morality tale of the tragic pastoral commons. The drama of Hardin's village pasture depended tacitly on all the villagers' knowing (or assuming that they knew) what was going on in the pasture. Who was grazing how many cattle, and how often? Who contributed to maintenance of the pasture, and to what extent? In Hardin's original "unregulated" version of the village commons, this knowledge, in combination with short-sighted self-interest, in fact brought about the collapse of the commons through overgrazing. Everyone felt compelled to valorize as large a share of the pasture's grazing potential as he could, perceiving that the neighbors were doing the same. But in Hardin's later work, and certainly in the writings of his major critics, knowledge can, under certain conditions, also assume the more perspicacious aspect of "knowledge of the whole"—that is, knowledge of the mechanisms that produce "tragic" outcomes, and hence knowledge that provides a means of reflexively avoiding such outcomes, either proactively or retroactively.

The accounts in this book attest to the fact that knowledge plays a similar role in the more extended, modernist "figurations" of cosmopolitan commons. We might say that the knowledge incorporated in "blind" technologies of production, communication, and transportation is unreflexive knowledge of the kind that easily gives rise to tragedies of the commons, or at least to underproductive commons. Consider the mechanized over-fishing of salmon on the Rhine, the casual emission of sulfur dioxide into the European atmosphere from high smokestacks, the technical capacity to harness the Vuoksi's hydropower, the erosion of crop diversity thanks to "Green Revolution" commercial hybrids, new technologies of sub-sea oil exploration and extraction, and the increased range and power of radio transmitters. In every case, what has mitigated or prevented tragedies of the commons, and has fostered something like optimal valorization of resources, has been the emergence of more perspicacious knowledge, knowledge of the ways in which single-minded pursuit of short-term self-interest ultimately leads to less for everyone.

Knowledge is not a disembodied *Geist*. As Karl Mannheim argued long ago, knowledge is a collective "property" of the class of the knowledge-able. Under modernity, the preeminent bearers of knowledge have been scientists, engineers, professional workers, and experts of many stripes. Such groups are "introverted" in the sense that they base their identity on abstruse, esoteric knowledge central to their expertise, and on the maintenance of partly autonomous institutions for its reproduction and expansion. At the same time, they are "extroverted" in the sense that they intervene (or are called on to intervene) in the creation of new technologies, infrastructures, and institutional arrangements. Peter Haas gave the name "epistemic communities" to groups of experts that "settled into" particular sociotechnical niches as resident specialists (innovators, designers, analysts, and advisors).[7]

A typical form of this "settling in" is for epistemic communities to function as "hired guns" for powerful and wealthy parties, including especially large firms and governments, to help them pursue their specific interests. We can assume that neither the complex technical systems and infrastructures described in this book nor the regulations, treaties, and institutions called into being to control and manage them could exist without the participation of such "embedded" epistemic communities.

Despite their organic roles as justifiers of the status quo or as conceptual enablers of local organized interests, epistemic communities always retain a critical reflexive potential by dint of the fact that they maintain their own cosmopolitan structures of education, communication, and circulation— quite apart from the corporate and state liaisons their members enter into. From the viewpoint of employers, this double loyalty makes them potentially treacherous accomplices. The quiescent professional paradigms that serve to advance the interests of powerful patrons such as firms and states may, at any moment, succumb to the corrosive discourses of Cassandras and other critics. This volume is replete with examples. To name just three: Edwards argues that the very same meteorological expertise that produces weather forecasts, and that daily saves lives and millions of dollars, has become transmogrified into the climate-change discourse, a reflexive critique of the carbon addiction of modern industrial societies; Kaijser shows how the productive paradigms of soil science and atmospheric chemistry became a *j'accuse* against British coal-fired power plants; and Kranakis details how the heterogeneous knowledge needed to "furnish" a commercially robust airspace morphed into a critique of aggressive forms of nationalism. In these examples, epistemic communities appear to cross the line from being the eyes of the state to being its conscience, a transition that implies a vision of a greater whole whose collective welfare would automatically revert to the benefit of its parts.

The emancipatory potential of epistemic communities hardly makes them a consistent force for democracy, however. They typically earn their salt as corporate advisors or as protagonists of state power. And thanks to their monopolies on knowledge, epistemic communities have been able to insinuate themselves into positions of great influence on the international stage, mediating negotiations among states and defining the technological and institutional frameworks of new transnational orders of governance. In doing so, they have often gone against the grain of popular sentiment, and have even undermined existing democratic process as they facilitated the visions of international elites, yet in some instances, we should add, happily so. Just in case we had forgotten, Tony Judt has forcefully reminded us how bitterly vengeful most Europeans were after World War II.[8] Everywhere there were wounds to be healed and scores to be settled. International cooperation could scarcely find poorer soil in which to flourish. But, as Korjonen-Kuusipuro's account of the Vuoksi shows, international epistemic communities were instrumental in forging new links despite the deep acrimony among the local populations. Korjonen-Kuusipuro shows how the hydroelectric community on the Vuoksi—and indeed the Finnish elite as a whole—sought cooperation and rapprochement with their former enemy on a very practical basis. She also emphasizes how a small and elite epistemic community of engineers and politicians developed relations of mutual trust and respect, and in some cases even friendship, despite the nationalistic hatred in which their dealings were immersed. This is of a piece with similar postwar rapprochements on the Rhine and the North Sea and later accords on transboundary air pollution; indeed, these were clearly the precursors of formal European integration. But there is also a darker side to this kind of epistemic moral vanguardism, evident, for example, in Saraiva's description of how the epistemic community of plant breeders succeeded in eliminating the farmer's plot as an authentic source of genetic variety and a much-needed grass-roots (literally) alternative to "sanitized" seed banks such as the one at Svalbard.

The Resilience and Fragility of Cosmopolitan Commons

Cosmopolitan commons are agents of sustainable resource governance, yet they cannot work if organized in a way that inhibits their own functioning. The chapters by With Andersen, Disco, Kranakis, Wormbs, and Saraiva show that malfunctioning and breakdowns are not infrequent, and that these are often related to decisions about who is to count as a commoner. Wormbs shows that the failure to include Germany as an etherspace commoner after World War II negatively affected the performance of this commons. Similarly, in the case of airspace, a European commons became viable only when Germany—at first formally excluded from the Paris Treaty—was brought in through the back door. Even more tellingly, Saraiva posits that the seed-bank commons will have little chance of sustaining biodiversity as long as it excludes farmers.

In cases of malfunction, the builders of cosmopolitan commons must keep on revising the commons, and must find new ideas and methods to make them work. Our case studies corroborate Elinor Ostrom's insistence that continual revision is the norm for commons management, and that this is the only way to maintain resilience. Laws and treaties can be revised. Obstacles in treaties can be circumvented and lacunae can be filled by drawing up additional bilateral or multilateral treaties. Also, new organizations can be founded or new technologies developed to improve perspicacity and surveillance. Together these options lead to the elaboration of more "neo-medieval" governance webs. Thus, cosmopolitan commons are inherently works in progress, and that is one of their strengths.

When formal elements of cosmopolitan commons prove inadequate or deleterious and yet cannot be formally changed (often for political reasons), solutions are sometimes found through flexible interpretation of rules, through unilateral national legal changes, through bilateral or multilateral side agreements, or even through selective non-observance of treaty provisions (as in the case of frequency allocations for post-World War II Germany, discussed by Wormbs). Yet sneakier approaches of selective nonconformance with commons rules are a slippery slope. Rule evasion remains an important source of the fragility of commons, as the continued problems of illegal fishing (described by With Andersen), broadcasting interference (analyzed by Wormbs), and the Germans' use of civil aviation to prepare for war (described by Kranakis) illustrate. In short, although cosmopolitan commons need room to evolve, and require some flexibility, surreptitious nonconformance can spell danger.

Accretion of actors and resources likewise has an ambiguous effect on the welfare of a cosmopolitan commons. On the one hand, accretion makes commons stronger by expanding adherence to their rules (e.g., by adding new signatories to treaties), by adding new organizations or actor networks that support the commons, or by enlarging their domains of action and accountability. All of the case studies demonstrate positive effects of accretion. On the other hand, accretion can be unsettling and divisive. Cosmopolitan commons embody efforts to moderate asymmetries (e.g., upstream-downstream asymmetries or power asymmetries among states or other entities) in ways that work to the net advantage of the commons community and its sustainability projects. Accretion (whether in the form of geographic upscaling, utilization of new technologies, or new ways of using an existing resource-space) can put commons at risk by fostering a shift to lower-common-denominator solutions or by weakening levels of commitment to rule abidance. This is yet another reason why cosmopolitan commons cannot remain static.

The resilience of cosmopolitan commons can also be fostered by elaborating positive rallying points—persuasive concepts or idealistic visions that encourage people to come together and think about interests and concerns larger than their own. Perspicacious knowledge is crucial here, but finding viable rallying points may also require something less tangible. Disco's study shows that the Rhine clean-water commons languished until it was reconceived around an idealistic vision of what bringing back the salmon would require. The combination of perspicacious knowledge and commitment to larger ideals and conceptual frameworks bolsters resilience by creating spheres of common, transnational dialog and debate, which in turn lay foundations for the emergence of larger "public consciousness."

Finally, our understanding of the resilience and fragility of cosmopolitan commons is deepened by seeing them as social arrangements within larger human society rather than bracketing them as exotic social phenomena. These particular arrangements have certain specific elements, as we have argued: international treaties, moral economies, multiform technologies, interdependent actor networks, and epistemic communities, all structured around particular resources or resource-spaces, including infrastructural resource-spaces, embodying specific social purposes (e.g., weather forecasting). Yet viewing cosmopolitan commons as simply elements of *society* encourages us to recognize that they display all of the characteristics and contradictions, the strengths and weaknesses, of human social arrangements in general. Cosmopolitan commons embody rationality, but they don't always function rationally. They embody formal rules, but they rely equally on informal social networking. Cosmopolitan commons depend on moral economies too, but they do not always produce morally defensible outcomes, nor can they entirely eliminate intransigence, hypocrisy, under-handedness, or the numerous moral dilemmas that plague human society. Cosmopolitan commons are a foundation for sustainable human interaction with the Earth in the present era of globalization and expansive networks, but they are not self-regulating and never will be. As the case studies of this book show, they require continual effort, commitment, monitoring, reassessment, and the cultivation of a pragmatic and non-dogmatic approach to the implementation of common interests and ideals.

Notes

1. Most of the literature on cosmopolitanism is predominantly theoretical—see, e.g., Ulrich Beck, *The Cosmopolitan Vision* (Polity, 2006); Ulrich Beck and Edgar Grande, *Cosmopolitan Europe* (Polity, 2007). Approaches with a more empirical focus tend to explore cultural and (large-scale) political aspects of cosmopolitanism. Admittedly, there are plenty of empirical studies on everything from environmental problems to global finance that can be interpreted through a cosmopolitan lens, but there are very few that consciously or reflexively make use of cosmopolitan theory.

2. See, e.g., David Held, *Cosmopolitanism: Ideals and Realities* (Polity, 2010); Garrett Wallace Brown and David Held, eds., *The Cosmopolitanism Reader* (Polity, 2010).

3. Examples: Thomas Christiansen, Knud E. Jorgensen, and Antje Wiener, *The Social Construction of Europe* (Sage, 2001); Sabine Saurugger, "Sociological Approaches in

EU Studies," *Journal of European Public Policy* 16, no. 6 (2009): 935–949; Ulrich Beck, "Reinventing Europe: A Cosmopolitan Vision," in *Cosmopolitanism and Europe*, ed. C. Rumford (Liverpool University Press, 2007); Gerard Delanty, "The Idea of a Cosmopolitan Europe: On the Cultural Significance of Europeanization," *International Review of Sociology* 15, no. 3 (2005): 405–421.

4. E.g., Alan Milward, *The European Rescue of the Nation-State* (University of California Press, 1992); Pierre Gerbet, *La construction de l'Europe* (Imprimerie nationale Éditions, 1994); M. R. Stirk and David Weigall, eds., *The Origins and Development of European Integration: A Reader and Commentary* (Pinter, 1999); Desmond Dinan, *Ever Closer Union: An Introduction to European Integration*, third edition (Lynne Rienner, 2005).

5. Intergovernmental Committee on European Integration, The Brussels Report on the General Common Market (an abridged English translation of the document commonly called the Spaak Report), 1956 (available at http://aei.pitt.edu). See also Andrew Moravcsik, *The Choice for Europe: Social Purpose and State Power from Messina to Maastricht* (Cornell University Press, 1998).

6. Jörg Friedrichs, "The Meaning of New Medievalism" *European Journal of International Relations* 7, no. 4 (2001): 475–502. See also Jan Zielonka, *Europe as Empire* (Oxford University Press, 2006); James Anderson, "The Shifting Stage of Politics: New Medieval and Postmodern Territorialities?" *Environment and Planning D* 14, no. 2 (1996): 133–154.

7. Peter M. Haas, "Introduction: Epistemic Communities and International Policy Coordination," *International Organization* 46, no. 1 (1992): 1–35.

8. Tony Judt, *Postwar: A History of Europe since 1945* (Pimlico, 2007).

Index